ALIPHATIC ALKYLATION

ALIPHATIC ALKYLATION
In Petroleum Refining

WEIZHEN SUN
WEIZHONG ZHENG
LING ZHAO
JINZHU LI
HUANXIN GAO

ELSEVIER

Elsevier
Radarweg 29, PO Box 211, 1000 AE Amsterdam, Netherlands
125 London Wall, London EC2Y 5AS, United Kingdom
50 Hampshire Street, 5th Floor, Cambridge, MA 02139, United States

ISBN: 978-0-443-21984-9

For information on all Elsevier publications
visit our website at https://www.elsevier.com/books-and-journals

Publisher: Candice Janco
Acquisitions Editor: Rafael Guilherme Trombaco
Editorial Project Manager: Soumya Yadav
Production Project Manager: Sruthi Satheesh
Cover Designer: Vicky Pearson Esser

Typeset by STRAIVE, India

Contents

5. Reactors, catalyst deactivation, and regeneration

6. Alkylation process intensification

7. Alkylation process safety

About the authors

Weizhen Sun is a professor at the East China University of Science and Technology (ECUST), Shanghai, China. He graduated from East China University of Science and Technology (ECUST), majoring in chemical engineering, and obtained his master's and doctorate degrees in chemical engineering from ECUST in 2004 and 2009, respectively. He was engaged in molecular simulation for one year in the group of Professor Berend Smit at the University of California, Berkeley. His main research interests include chemical reaction engineering, multiscale simulation, and applications of nanomaterials in energy and medicine. He excels at using chemical engineering theory and methods to study the reaction kinetics of complex systems, and multiscale simulation, optimization, and design of large-scale industrial reactors, especially focusing on the surface and interface properties and reaction-transfer coupling relationships of heterogeneous reaction processes using multiscale methods. He has published more than 80 publications in international chemical journals such as AIChE J., Chem. Eng. Sci., Chem. Eng. J., Ind. Eng. Chem. Res., and so on.

Weizhong Zheng is a distinguished researcher at the East China University of Science and Technology (ECUST), Shanghai, China. He graduated from ECUST, majoring in chemical engineering, and obtained his bachelor's and doctorate degrees in 2014 and 2020, respectively. He was engaged in ultrafast dynamics for one year in the group of Professor Michael David Fayer, Stanford University. He was mainly engaged in the research on the design of green solvents and the process intensification of heterogeneous systems. He excels at combining multiscale molecular simulation, in situ characterization, machine learning, and experimental methods to study interfacial microstructure and mass transfer characteristics, guiding the rational design of green solvent and green process development. He has published more than 40 papers in journals including J. Am. Chem. Soc., AIChE J., Chem. Eng. Sci., Chem. Eng. J., and Ind. Eng. Chem. Res. He was awarded the China Petroleum and Chemical Industry Federation CPCIF-Clariant Sustainable Development Youth Innovation Award.

Ling Zhao is a professor at the East China University of Science and Technology (ECUST), Shanghai, China. She graduated from ECUST, majoring in chemical engineering, and obtained her master's and doctorate degrees in chemical engineering from ECUST in 1993 and 1996, respectively. After graduated from school in 1993, she has been working at the United Institute of Chemical Reaction Engineering in ECUST. From 2002 to 2003, she was awarded the KC Wong Fellowship by the Royal Society of England and worked as a postdoctoral fellow at the Department of Chemical Engineering, University of Birmingham. Her main research interests includes supercritical fluid-assisted polymerization and polymer processing, and heterogeneous reaction and transfer process intensification, such as polyester, nylon, polycarbonate, acrylonitrile polymerization, aromatic hydrocarbons, liquid-phase oxidation, and sulfuric acid alkylation. She has participated in the National Science and Technology Progress Prize and received the Provincial and Ministerial Science and Technology Progress Prize. She has published more than 400 papers in international chemical journals such as AIChE J., Chem. Eng. Sci., and Chem. Eng. J.

Jinzhu Li is a senior engineer at SINOPEC Zhenhai Refining & Chemical Company, Ningbo, China. He graduated from Fushun Petroleum Institute with a bachelor's degree and from East China University of Science and Technology with a master's degree in chemical engineering. He was engaged in the operation and management of petroleum refining plant for 24 years, and responsible for the construction and operation management of a 500,000 tons/year liquefied gas separation plant, and the expansion of the MTBE plant from 20,000 tons/year to the 40,000 tons/year, then to 100,000 tons/year, and finally to 150,000 tons/year. He was also responsible for the operation and technical management of the gasoline-diesel hydrogenation unit and the wax oil hydrocracking unit, as well as for the technical comparison, construction, start-up, shutdown, and operation management of the 300,000 tons/year sulfuric acid alkylation and the supporting acid regeneration unit. He has rich experience in oil refinery operation safety and technical management.

Huanxin Gao is a senior engineer at SINOPEC Shanghai Research Institute of Petrochemical Technology Co., Ltd., Shanghai, China. He graduated from Lanzhou Institute of Chemical Physics, Chinese Academy of Sciences, majoring in physical chemistry. His main research interests include zeolite acid catalysts and selective oxidation catalysts, and their applications in the petrochemical industry, especially focusing on the alkylation of benzene with alkene and epoxidation of propylene. His group has developed and commercialized several catalysts and licensed several process technologies. He has published more than 30 publications in international journals.

Acknowledgment

I am deeply thankful to the exceptional individuals whose dedicated contributions made the completion of this book possible. Their unwavering commitment and professional guidance were essential in ensuring the smooth and timely delivery of this project.

First and foremost, I want to thank Mr. Zhihong Ma for his outstanding work in editing and summarizing the entire content of the book. His meticulousness and patience have played a crucial role in enhancing the quality of this book. I also want to extend special thanks to Ms. Bihong Li for her editorial corrections and improvements to the article content. Her expertise and attention to detail have added many highlights to the accuracy and fluency of the publication.

Furthermore, I am grateful to Mr. Ming Wen for his contribution to organizing the references. His rigor has provided solid support for the academic integrity of this book. Lastly, I would like to thank Mr. Longmeng Luo, Mr. Yong Yang, Mr. Xinpeng Zhou, and Ms. Xuemei Han for their contributions to reference organizing and image sourcing in this book. Their enthusiastic participation and professional standards have added infinite highlights to the visual presentation of this book.

I extend my deepest gratitude and respect to each of them for their hard work and selfless support.

CHAPTER

1

Introduction of aliphatic alkylation

1.1 Definition of aliphatic alkylation

Aliphatic alkylation is the chemical process related to the introduction of an alkyl group onto an aliphatic molecule with an open chain, which involves the replacement of a hydrogen atom in an aliphatic chain with an alkyl group to form a new C–C bond. Specifically, a typical aliphatic alkylation process usually refers to the alkylation reaction between isobutane and low-molecular-weight C3-C5 olefins to produce branched chain paraffins, and alkylate gasoline. Most of the isobutane and C3-C5 olefins mainly come from the fluid catalytic cracking (FCC) process [1], or other catalytic and thermal conversion processes in refineries. The product, referred to as alkylate, is considered as a high-quality blending material for the gasoline pool due to its clean burning and exceptional antiknock properties with high octane number and low Reid vapor pressure (RVP).

The alkylation process was developed in the 1930s and was initially used to produce high-octane aviation gasoline [2]. Later, the alkylate became more important, as spark-ignited engines became more powerful with higher compression ratios and required fuels with higher octane numbers. Recently, the environmental regulations limited the content of benzene and total aromatics in the gasoline pool. Alkylation is becoming increasingly popular as an octane booster than catalytic reforming, as alkylate does not contain any olefins or aromatics, which is an ideal blending component of gasoline.

Initially, the alkylation process was carried out at high temperature (400–500°C) and high pressure (17–30 MPa) without the presence of the catalyst, known as "thermal alkylation." This process was replaced by "catalytic alkylation" with a strong acid catalyst due to the high level of side reactions such as cracking and the poor quality of products. Liquid and solid acid catalysts were used for the alkylation process, including

Aliphatic Alkylation
https://doi.org/10.1016/B978-0-443-21984-9.00002-7

sulfuric acid (H_2SO_4), hydrofluoric acid (HF), phosphoric acid, aluminum silicate zeolite, boron fluoride, anhydrous aluminum chloride, alumina-platinum oxide, and sulfonic acid resins.

The widespread use of liquid acid catalysts such as H_2SO_4 and HF in the alkylation process has reduced production costs and brought significant benefits to petroleum refining, but it has also brought notable environmental risks, equipment corrosion, or high toxicity. Nowadays, H_2SO_4 or HF-catalyzed aliphatic alkylation technologies are still the main processes for producing alkylate. Acidic ionic liquid catalysts have shown good catalytic performance in C4 alkylation reactions due to their high catalytic activity and low corrosiveness. However, the high synthesis costs and the waste residues generated during the alkylation reaction make the industrialization of acidic ionic liquid-catalyzed alkylation difficult. To reduce the risks associated with liquid acid-catalyzed alkylation, environmentally friendly solid acids with adjustable acidity are applied to produce aliphatic alkylate compounds. Several solid catalyst processes, including the UOP Alkylene, CB&I AlkyClean, and KBR K-SAAT processes, are commercially available but have not been industrialized on a large-scale due to the rapid deactivation and difficult regeneration of solid acids.

1.2 History of aliphatic alkylation

Alkylate obtained from the C4 alkylation was discovered in the early 1930s [3,4]. Before that, alkanes (e.g., isobutane) were considered chemically inert organic compounds. Herman Pines et al. hypothesized that alkanes might be active in acids with the finding anomalies in the olefin assay based on H_2SO_4 extraction, which was proved by the full conversion of ethylene into saturated hydrocarbons with $AlCl_3$ [5]. Over the next few years, the combinations of the strong acids such as $AlCl_3$ or BF_3 with the proton source such as HCl, HF, H_2SO_4, or water [3] were developed for the C4 alkylation. The alkylate was initially found to possess exceptional aviation gasoline properties, with the highest octane rating among known fuel components, excellent motor octane characteristics, and the strong response to lead. The demand for high-octane aviation gasoline during World War II stimulated the fast development of the aliphatic alkylation process to produce high-octane alkanes. In order to meet the demand for aviation gasoline at that time, H_2SO_4 and HF became the dominant catalysts for commercial alkylation [6–8]. The first commercial H_2SO_4 alkylation unit was built by Humber Oil Company in 1938 at Baytown, Texas [4]. In the following year, six oil companies jointly developed alkylation technology and successfully developed the H_2SO_4 and HF process for high-octane aviation fuel [4]. On this basis, the first commercial hydrofluoric acid alkylation unit was started in 1942 at the Phillips Petroleum

Company refinery in Boger, Texas [9]. The unit expands the range of olefin feedstocks to include propylene, butene, and pentene. The traditional H_2SO_4 and HF-catalyzed alkylation process is still used in the petroleum refining industry today. The HF process typically operates at a reaction temperature of 30–40°C, and the heat generated can be removed using circulating water without the need for refrigeration. It boasts a fast reaction rate, requiring only about 10 s of residence time, resulting in the smaller reactor sizes and cost savings. The use of distillation for regenerating waste acid leads to lower operating costs. Additionally, the HF catalyst has low viscosity, making it cost-effective to achieve a large and dispersed interface. These advantages have driven the rapid development of HF-catalyzed alkylation. After 10 years into industrialization, it accounted for 25% of the total alkylation capacity, and by the early part of this century, this proportion remained at around 50%, establishing its dominant position during this period. In 1942, the Soviet Union built the first industrial alkylation unit at the Neftegaz plant of Grozny [10]. From 1938 to 1945, the number of alkylation units (mainly H_2SO_4 and HF) put into operation in the United States increased from 1 to 42 [10]. By the end of World War II, there were 30 hydrofluoric acid alkylation units in operation in US plants, producing approximately 17,500 tons of alkylate per day [10]. In addition, from 1938 to the 1940s, the H_2SO_4 alkylation process experienced rapid and widespread industrial development due to its advantages over other alkylation methods.

In the middle of the 1950s, the automobile industry developed rapidly, and alkylate became part of automobile fuels. However, the demand for alkylate has been strongly constrained by a high costs compared to other components during the 1950s–1960s [10]. The emergence of higher-performance automotive engines created a demand for improved gasoline production and quality within the refining industry. The introduction of catalytic reforming technologies such as the UOP Platforming process provides refiners with a critical tool for producing high quality gasoline [4]. In addition, lead reduction programs adopted by the United States Environmental Protection Agency in the 1970s and 1980s resulted in an increasing demand for alkylate as blending components of automotive fuels [10]. As a result, many refineries expanded their alkylation operations and incorporated propylene (and sometimes even pentene) into the alkyl unit feedstock along with butenes to meet the growing demand.

With the gradual phasing out of leaded gasoline and the emergence of environmentally friendly gasoline, the importance of alkylates and their production has increased, since alkylates feature high unleaded octane numbers, low volatility, low sulfur, and almost zero olefins and aromatics, making them crucial for producing high quality, environmentally friendly gasoline. Alkylates typically make up around 15% of the gasoline blend pool but can reach up to 60% for low-sulfur formulations (https://www.nextwaveenergy.com/news/next-wave-energy-partners-announces-commercial-operations-innovative-ethylene-alkylate). Methyl tert-butyl ether (MTBE) in gasoline was phased out gradually in 2004. According to the Energy Policy Act of 2005 and the Energy Independence and Security Act of 2007, ethanol is blended into the gasoline pool at an average of about 10%. The low Reid vapor pressure (RVP) of alkylates is essential to offset the high vapor pressure of added ethanol while maintaining a high-octane number [4].

In addition to the quality of alkylate, the safety and economy of the catalyst have also been included in the scope of research. Anhydrous HF has toxicity, high vapor pressure, and the tendency to form aerosols when released, making the alkylation process a great safety hazard [11]. Considering that HF is a good and cheap catalyst for C4 alkylation, attempts have been made to improve the safety of HF alkylation. One of the approaches proposed in the 1990s was the use of additives in HF acid catalysts to reduce the vapor pressure and minimize the amount of HF released into the atmosphere in the event of an accident [3]. Texaco and UOP announced a launch of their Vapor Pressure Reduction Assist Technology (Alkad) technology, which uses "onium" compounds, such as HF salts of amines, as additives [12]. Likewise, Phillips and Mobil have developed an additive technology called Reduced Vapor Pressure (ReVAP). The names of the additives used by Phillips and Mobil have not yet been released [13,14]. However, Phillips has considerable patent coverage involving HF-catalyzed alkylation in the presence of sulfones, particularly sulfolane, and the recovery of these materials from the acid-soluble oil produced in HF-catalyzed alkylation [15–25]. Following the acquisition of ConocoPhillips' HF alkylation technology in 2007, UOP LLC is now the sole licensor for automotive fuel HF alkylation [4]. The acquired technologies include the Hydrisom pretreatment process, HF gravity feed process, and HF alkylation-enhanced processes such as Staged Olefin Feed Technology (SOFT), ReVAP, and Acid Management Process (IMP). These technologies were combined and optimized with all the existing technologies of UOP, including Staged Feed Sequencing Reactor Technology (SFSR) and Alkad, to create AlkyPlus Technology [26]. Therefore, UOP can provide a variety of alkylation pretreatment technologies, HF alkylation technology packages, and various improvement solutions. Compared with traditional methods, ReVAP/Alkad technology can reduce the hazards of HF acid leakage and volatilization by 60%–90%, increase the product octane number by 0.8 units, and reduce the dry point. IMP technology uses a gravity acid precipitation tank, which can safely and quickly transfer acid to an HF acid tank within 90 s, reducing catastrophic acid leaks by approximately 50% and reducing hazards by 90%. SOFT/SFSR introduces multiple olefin feed points into the reactor, allowing the reaction to occur at a higher alkane/olefin ratio, increasing the octane number of the product and increasing unit production capacity by 20%–70%. These improvements further enhance the safety and affordability of AlkyPlus technology. However, the volatility, corrosivity, and toxicity of HF catalysts remain major concerns, and their use is strictly regulated by environmental agencies such as the United States. As a result, new alkylation units have largely moved away from HF over the past 30 years [27].

Compared to HF-catalyzed alkylation, H_2SO_4-based alkylation technology is generally considered relatively safe. When H_2SO_4 is used as a catalyst, C8 compounds, especially trimethylpentane (TMPs), exhibit high selectivity as target products, and there is no need for isomerization of 1-butene in the feedstock [26]. Due to optimized reaction mechanisms, isobutane consumption is relatively low. However, technologies using H_2SO_4 as a catalyst generate a significant amount of waste acid that must be regenerated. The transportation of waste acid and fresh acid to and from H_2SO_4 regeneration units has raised some concerns and increased the pressure on refineries to establish H_2SO_4 regeneration units near their alkylation facilities. The maturity and integrated optimization of waste acid regeneration processes have reduced the disadvantages of catalyst regeneration using H_2SO_4. The strict limitations on the use of HF have provided significant development opportunities for the increasingly mature and refined H_2SO_4 alkylation, and in recent years, the most newly constructed alkylation units have adopted the H_2SO_4 method. There are primarily four H_2SO_4 alkylation technologies in use, including DuPont's STRATCO effluent indirect refrigeration, ExxonMobil's series stirred tank self-cooling process, CB&I Lummus's CDAlky low-temperature alkylation process, and the SINOALKY H_2SO_4 alkylation process by Sinopec [26]. Due to the low-temperature reaction conditions required for H_2SO_4 alkylation technology, the high viscosity of concentrated H_2SO_4, which serves as the catalyst, can hinder effective contact with the feedstock. Therefore, much of the focus in the development of H_2SO_4 alkylation processes has been on the development of reactors and related internal components, including the improvement of good mixing and contact of materials within the reactor at low temperatures, and the enhancement on mass and heat transfer within the reactor.

As shown in Fig. 1.1, the global alkylate production capacity reached 115.6 million tons per year (Mt/a) in 2019 [26]. Based on the regional distribution of capacity and the technologies employed, approximately 44% of the alkylate production capacity is in North America, predominantly utilizing liquid acid alkylation technology, with an equal split between HF and H_2SO_4 methods. The alkylation production capacity of Asia and Europe accounts for 36.5% and 8.5% of the global capacity, respectively. In Asia, the H_2SO_4 method is predominantly used, with some hydrofluoric acid and other alkylating units. In Europe, the HF method is the main technology, while H_2SO_4 units account for less than 30% of the

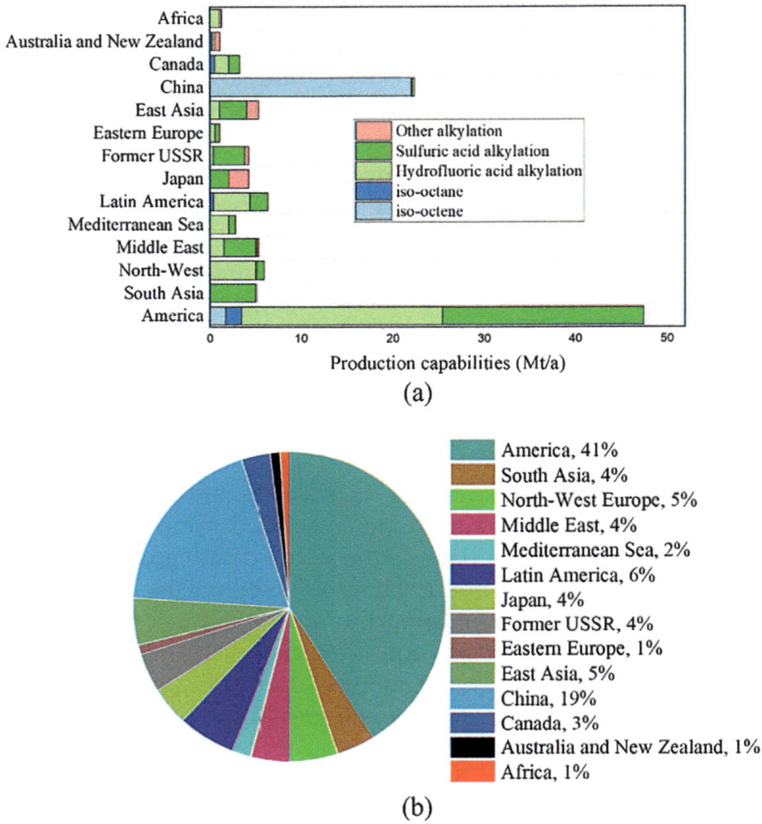

FIG. 1.1 The global alkylate production capacity in 2019. (A) Alkylated oil production capacity in various regions of the world and (B) Percentage of world alkylated oil production capacity distribution.

capacity. In 2019, the alkylate production capacity in the United States reached 47.4 Mt/a, representing 41% of the global capacity, while China reached a capacity of 22.36 Mt/a, accounting for 19.3% of the global capacity.

Despite H_2SO_4 and HF alkylation technologies having been improved and optimized over the past two decades, they still exhibit notable disadvantages such as high process costs, high acid volume-specific consumption, corrosion, toxicity, and acid residue recovery, which make these technologies less attractive.

To address the challenges posed by the aforementioned technologies, Haldor Topsøe has devised a solution by combining both liquid and solid phases within a single catalyst for the alkylation process [28]. Specifically, fixed-bed and moving-bed alkylation technologies incorporate the liquid acid phase, such as trifluoromethanesulfonic acid or other sulfonic acids, into a solid porous matrix. While these approaches were developed and even implemented on a commercial scale in the past, they have since fallen out of use. UOP, on the other hand, has successfully ventured into the integration of liquid and solid phases in their catalyst. In this context, Honeywell UOP and Chevron collaborated to create the ISOALKY process, which operates at temperatures below 100°C and transforms typical streams from FCC into high-octane blend components. This innovative process employs nonvolatile ionic liquids with aluminum chloride anions for online regeneration, continually introducing halogens to mitigate the environmental impact of automotive gasoline. Notably, Sinochem Hongrun in China and Chevron have embraced the ISOALKY process to produce 5000 barrels of cleaner automotive fuel daily.

It is also reported that the ISOALKY process is replacing the previously implemented technology in the subalkylation process of UOP [28]. This process is based on the use of chloroaluminate catalysts containing platinum loaded onto α- or γ-aluminum oxide. Catalyst reactivation and thermal regeneration are critical steps in the subalkylation process, similar to those used in continuous catalyst regeneration in catalytic reforming. In addition, UOP has recently developed a slurry catalyst alkylation process that conducts alkylation reactions through a distillation tower configuration [28]. In this distillation tower setup, finely dispersed solid catalysts are introduced into the reactor and pass through the trays of the distillation tower. However, there is no data available regarding pilot plant testing or commercialization of this technology.

In 2005, China National Petroleum Corporation (CNPC) demonstrated the ionic liquid alkylation process, developed by China University of Petroleum and licensed by Well Resources Corporation, using isopropyl CILA for the first time during the modification of their existing H_2SO_4 alkylation unit; then, in 2013, the independent refinery, Deyang Chemical Industry Co., built the first 100,000-ton/year ionic liquid unit [28]. In 2018, CNPC, Sinopec Limited, and China University of Petroleum jointly introduced a new composite IL (CIL) for the Harbin Petrochemical alkylation unit [28]. This project employs IL alkylation technology to produce alkylate fluid from C4 liquefied gas. Compared to other alkylation processes, it is said that the CIL process has the advantage of significantly lower waste generation and less equipment corrosion, allowing the use of carbon steel as the construction material for process equipment. The Harbin Petrochemical 150,000-ton/a alkylation unit produces high-quality isobutene [29].

In the 1990s, the development of solid acid catalysts, such as sulfonic acid-functionalized resins and acidic ion exchange resins, allowed for even more efficient and selective alkylation reactions. These catalysts were less corrosive and easier to handle than traditional liquid acid catalysts and allowed to produce a wide range of alkylate compounds with high selectivity. In 1996, ABB Lummus Global and Albemarle Corporation collaborated to develop a solid acid catalyst and the corresponding alkylation process. Subsequently, Neste Oil Corporation aimed to promote this technology globally, optimized the process, and established an industrial demonstration unit. This technology was later named AlkyClean, with joint ownership of the technology held by Albemarle, Lummus, and Neste. In the 2000s, researchers began to focus on developing more sustainable and environmentally friendly alkylation processes [27]. One area of interest was the use of heterogeneous catalysts, which are more environmentally benign than traditional homogeneous catalysts. The world's first solid acid alkylation unit using AlkyClean technology successfully started operations in 2015 at Huifeng Petrochemical in Zibo, Shandong Province of China, with a capacity of 100 kt/a. The research octane number (RON) of alkylate is around 95.0, and the olefin conversion rate is 100%. In 2008, UOP commissioned a 220 kt/a Alkylene industrial unit at the Baku Heydar Aliyev Refinery in Azerbaijan. It is reported that the HF alkylation unit could potentially be retrofitted into an Alkylene unit, but there have been no further reports on this matter. In May 2015, KBR Inc. entered into a strategic partnership agreement with Exelus Inc., securing exclusive licensing rights for Exelus's solid acid catalyst (ExSart) for KBR's K-SAAT solid acid alkylation technology. In early 2018, a 100 kt/a solid acid alkylation unit based on the K-SAAT process was completed at the Hike Ruilin Chemical facility in Dongying, Shandong Province of China. There have been limited publicly available reports on the operational status of these two units, and it is said that these facilities have seen limited operations in recent times. Heterogeneous catalysts are also easier to recycle and can be reused multiple times, making them more economically viable.

1.3 Role of the aliphatic alkylation unit in the refinery

The alkylation process is one of the first processes introduced in the refinery to produce high-quality aviation gasoline, and it is widely used nowadays as a process that improves the quality of motor gasoline. It is used to convert isobutane and low molecular weight olefins (mainly a mixture of propylene and butylene) into alkylate, a component of high-quality gasoline [30]. This process occurs in the presence of a strong acid such as H_2SO_4 or HF as a catalyst. The alkylation combines two shorter hydrocarbon molecules by mixing isobutane with light olefins, such as propylene or butene, from the Fluid Catalytic Cracking Unit (FCCU) into a longer chain of gasoline-range molecules, thereby producing high-quality gasoline blends [31].

Since crude oil typically contains only 10%–40% of the hydrocarbon content in the gasoline range, refineries typically use FCCUs to convert high molecular weight hydrocarbons into smaller, more volatile compounds, which are then converted into liquid gasoline-sized carbon hydrogen compounds. Other low molecular weight olefin and isoparaffin molecules are also produced as by-products of the FCC process, which are undesirable. Alkylation converts these by-products into larger isoparaffin molecules with a high-octane number While FCCUs are a very common unit in modern refineries, it is uncommon for a refinery to have an alkylation unit [31].

The product alkylate consists of a mixture of high-octane number, branched-chain alkanes (primarily isoheptane and isooctane). The octane number of the alkylate depends primarily on the type of olefin used and the operating conditions. For example, isooctane is made from butene mixed with isobutane and has an octane rating of 100 by definition. However, there are other products present in the alkylate effluent, so the octane number changes accordingly.

One of the key advantages of the alkylation process is its flexibility in handling various feedstocks. The aliphatic alkylation unit can process a wide range of feedstocks, including light olefins, isobutane, and other hydrocarbons. This flexibility allows refineries to adjust their production to meet changing market demands and to take advantage of different feedstock prices. The alkylation process can also improve the quality of gasoline blending components by reducing their sulfur content. Sulfur is a harmful pollutant that contributes to air pollution and can cause health problems in humans. The aliphatic alkylation unit can help refineries reduce the sulfur content of their gasoline blending components, making them more environmentally friendly and safer for consumers [32]. Furthermore, the aliphatic alkylation unit can play a critical role in ensuring the safety and reliability of the refinery. The alkylation process can help to prevent the formation of harmful chemicals such as acids, which can corrode equipment and cause safety hazards. By using alkylation to produce high-quality gasoline blending components, refineries can reduce the risk of equipment failure, reduce maintenance costs, and ensure the safe and reliable operation of the facility [33]. Finally, the aliphatic alkylation unit can help to reduce greenhouse gas emissions by producing high-quality gasoline blending components that improve fuel efficiency. As governments around the world seek to reduce carbon emissions and combat climate change, the alkylation process offers a way to produce cleaner, more efficient gasoline that can help to meet these environmental goals.

In summary, the aliphatic alkylation unit is a critical process unit in a refinery that plays a vital role in producing high quality, excellent performance gasoline while meeting changing regulatory and consumer demands. The alkylation process offers several advantages, including flexibility in handling various feedstocks, reducing the sulfur content of gasoline, improving the safety and reliability of the refinery, and reducing greenhouse gas emissions.

1.4 Other alkylation processes

Alkylation reaction refers to the reaction of introducing an alkyl group (-R) into carbon, nitrogen, oxygen, and other atoms in an organic molecule. Alkylation reactions in the petroleum refinery can be categorized into several types based on the reactants and the nature of the reaction. Besides the aliphatic alkylation reaction, the main types of alkylation reactions in the petroleum refinery include aromatic alkylation, phenol alkylation, oxygenate alkylation, and heteroatom alkylation. These various types of alkylation reactions allow the petroleum refinery to produce a wide range of products with improved properties, such as higher-octane gasoline components or specialty chemicals. In aromatic alkylation, alkyl groups are added to aromatic compounds, such as benzene or toluene. Aromatic alkylation is important for producing alkyl aromatic compounds, such as ethylbenzene, which is mainly used to produce styrene, then used to produce polystyrene, acrylonitrile-butadiene-styrene resin, styrene-acrylonitrile resin, and other products. In addition, ethylbenzene can also be converted into acetophenone, cellulose acetate, diethyl benzene (DEB), ethyl anthraquinone, ethylbenzene sulfonic acid, propylene oxide, and methyl benzyl alcohol, which has significant economic value in the chemical industry [34].

Aromatic alkylation mainly includes traditional acid-catalyzed processes using $AlCl_3$, HF, or H_2SO_4 as catalysts, solid acid-catalyzed processes using molecular sieves, clays, and supported heteropolyacids, as well as ionic liquid catalysis processes [35,36]. Traditional acid-catalyzed processes have the advantages of high production efficiency and high product quality. However, the catalysts used in these processes can cause severe equipment corrosion, and the production process generates a large amount of waste acid, leading to high posttreatment costs. In the current industrial production of linear alkyl benzene, HF is still the mainstream catalyst for benzene alkylation due to technological limitations and existing equipment conditions, occupying a significant market share [37]. On the other hand, there are relatively more industrial examples using solid acids to catalyze the alkylation of aromatic hydrocarbons with short-chain olefins. However, due to the tendency of solid acids to coke and deactivate, there are very few industrial examples of using solid acids for the alkylation of aromatic hydrocarbons with long-chain olefins [38]. As for ionic liquid catalysis processes, ionic liquids have the advantages of a high density of active sites, high catalytic activity at low temperatures, good flow ability, low volatility, and environmental friendliness [39–41]. However, poor stability and difficulties in recycling and reuse are the main factors hindering their industrial applications [42].

Research and commercial development of benzene alkylation to produce ethylbenzene have a history of over 90 years, starting with the Friedel-Crafts catalyst ($AlCl_3$-HCl) in the 1930s [43]. The $AlCl_3$-HCl process caused corrosion issues, leading UOP and Mobil Badger to develop supported catalysts and zeolite-based catalysts in the 1960s and 1970s [44,45]. The third-generation Mobil Badger process commercialized the use of ZSM-5 as a catalyst and introduced a second reactor for transalkylation in the 1990s [37]. Over the past two decades, research institutions and companies in mainland China have undertaken extensive efforts in developing their own benzene alkylation technologies. China Petrochemical Corporation (Sinopec) has invested significantly in benzene alkylation research and development during recent decades. Some zeolites, such as MFI, FAU, BEA, and MWW, are used as catalysts for benzene alkylation. Benzene alkylation can generate in the vapor phase and liquid phase, respectively [37].

The Mobil-Badger ethylbenzene process was the first commercial process [37]. Specifically, benzene is alkylated with pure ethylene in a gas-phase fixed bed reactor to produce ethylbenzene, where the product ethylbenzene is separated as a product and the EBs are recycled back to the reactor to undergo transalkylation. The Mobil-Badger ethylbenzene process was improved by adding a transalkylation reactor in the 1990s. The service life of the catalyst increased from about 3 months to 1 year [46]. SRIPT initiated a research project on the pure ethylene process in 1994, which mainly addressed two issues: catalyst deactivation caused by coke deposition during ethylene polymerization and the removal of xylene by-products from ethylbenzene [47]. SRIPT studied the gas-phase benzene alkylation reaction using ZSM-5 catalysts with different SiO_2/Al_2O_3 molar ratios and crystal sizes and found that increasing the SiO_2/Al_2O_3 molar ratio inhibited ethylene polymerization and improved ethylbenzene selectivity and ZSM-5 stability.

FCC exhaust gas contains 10%–30% ethylene and can be used as a source of ethylene for ethylbenzene production, benefiting refineries and reducing carbon emissions [37]. However, when FCC tail gas is directly fed into the alkylation reactor as a source of ethylene, impurities like propylene can cause side reactions and catalyst deactivation. Separating ethylene from FCC tail gas is advantageous for alkylating benzene, but it will increase the overall investment and operational costs. To address this, Mobil-Badger has improved its pure ethylene process to use FCC waste gas as a feedstock [37]. The dilute ethylene process involves a complex raw material pretreatment system. After removing water, sulfur and separating propylene from the FCC waste gas feed, the stream is sent to the alkylation reactor to produce ethylbenzene. This process was first commercialized at the Shell refinery in Stanlow, United Kingdom, with an annual ethylbenzene production capacity of 160 tons in 1991. Dow Chemical built a larger industrial facility in the Netherlands with a capacity of 450 kt/a [48]. There are many other benzene alkylation processes, which will not be discussed here.

References

[1] X. Meng, Z. Wang, R. Zhang, C. Xu, Z. Liu, Y. Wang, Q. Guo, Catalytic conversion of C4 fraction for the production of light olefins and aromatics, Fuel Process. Technol. 116 (2013) 217–221.
[2] X.P. Zhao, Q.H. Zhuang, X.D. Wu, X.B. Wang, Y. Zhang, H.F. Qin, A Process for Recovering Light Hydrocarbons from the Tail Gas of a Polyolefin Plant CN101585746 (2009).
[3] S.I. Hommeltoft, Isobutane alkylation: recent developments and future perspectives, Appl. Catal. A Gen. 221 (2001) 421–428.
[4] D.A. Nafis, K.A. Detrick, R.L. Mehlberg, Alkylation in petroleum processing, in: S.A. Treese, P.R. Pujadó, D.S.J. Jones (Eds.), Handbook of Petroleum Processing, Springer International Publishing, Cham, 2015, pp. 1–17.
[5] H. Pines, Saga of a discovery-alkylation, Chemtech 12 (1982) 151–154.
[6] L. Schmerling, in: B.T. Brooks, C.E. Boord, S.S.K. Kurtz Jr., L. Schmerling (Eds.), Alkylation of Saturated Hydrocarbons, Rheingold Publishing Corp, New York, 1955, pp. 363–408.
[7] R.M. Kenedy, Catalytic alkylation of paraffins with olefins, in: P.H. Emmett (Ed.), Catalysis, Alkylaton, Isomerization, Polymerisation, Cracking and Hydroreforming, Rheingold Publishing Corp, New York, 1958, pp. 1–41.
[8] W.L. Lafferty Jr., R.W. Stokeld, Alkylation and Isomerization, Advances in Chemistry Series, 1971, p. 130.
[9] C. Phillips Petroleum, E.O. Champlin Refining Company, Hydrofluoric Acid Alkylation, Phillips Petroleum Company, Bartlesville, OK, 1946.
[10] K.K. Akhmadova, M.K. Magomadova, A.M. Syrkin, N.L. Egutkin, History, current state, and prospects for development of isobutane alkylation with olefins, Theor. Found. Chem. Eng. 53 (2019) 643–655.
[11] R. Diener, Mitigation of HF releases, J. Loss Prev. Process Ind. 4 (1) (1991) 35–43.
[12] G.A. Olah, Environmentally Safe Catalytic Alkyation Using Liquid Onium Poly (Hydrogen Fluorides), US Patent 5,073,674, 1991.
[13] L. Shoemaker, Revap: Phillips reduced volatility alkylation technology, in: AICHE Spring Meeting Houton, TX, 1997.
[14] L. Shoemaker, K. Hovis, B. Randolph, M. Pfile, AM-97-44 NPRA Meeting San Antonio, TX, 1997.
[15] A.D. Eastman, B.B. Randolph, R.G. Abbott, R.B. Eldridge, Alkylation Catalyst Regeneration, US Patent 5,264,647, 1993.
[16] A.D. Eastman, R.G. Abbott, R.B. Eldridge, F.-M. Lee, D.P. Mann, R.J. Mitchell, Alkylation Catalyst Regeneration, US Patent 5,264,649, 1993.
[17] M.A. Better, J.E. Child, T.R. Melli, S. Yurchak, Method for Regenerating HF/Sulfolane Alkylation Catalyst, US Patent 5,264,651, 1993.
[18] M.A. Better, J.E. Child, K.J. Del Rossi, A.A. Gupte, T.R. Melli, S. Yurchak, Sulfolane Purification Method, US Patent 5,276,243, 1994.
[19] A.D. Eastman, R.G. Abbott, Alkylation Catalyst Regeneration, US Patent 5,347,067, 1994.
[20] B.B. Randolph, R.G. Abbott, Method for Suppressing the Evaporation of Hydrogen Fluoride from a Mixing of Hydrogen Flouride and Sulfone, US Patent 5,498,818, 1996.

[21] R.G. Abbott, B.B. Randolph, Isoparaffin-Olefin Alkylation, US Patent 5,569,807, 1996.
[22] R.G. Abbott, B.B. Randolph, Isoparaffin-Olefin Alkylation, US Patent 5,639,932, 1997.
[23] K.W. Hovis, R.L. Anderson, Method for Separating Sulfone From a Hydrocarbon Stream Having a Concentration of Sulfone Producing a Dry Sulfone Product, US Patent 5,648,587, 1997.
[24] R.L. Anderson, B.B. Randolph, Method for Removing Sulfone and Hydrogen Fluoride From a Hydrocarbon Stream and Producing a Sulfone Product Free of Hydrogen Fluoride, US Patent 5,648,588, 1997.
[25] R.G. Abbott, R.P. Williams, M.M. Johnson, J.W. Vanderveen, Isoparaffin-olefin alkylation, US Patent 5,120,897, 1997.
[26] D.X. Cao, X.N. Yang, G.Q. Cao, C4 alkylation technology development trend, Pet. Petrochem. Today 28 (2020) 29–37.
[27] S.J. Luo, A polypropylene production recovery device, CN215969602U.
[28] H. Diaz Velazquez, N. Likhanova, N. Aljammal, F. Verpoort, R. Martinez-Palou, New insights into the progress on the isobutane/butene alkylation reaction and related processes for high-quality fuel production. A critical review, Energ. Fuel 34 (2020) 15525–15556.
[29] A.S. Amarasekara, Acidic ionic liquids, Chem. Rev. 116 (10) (2016) 6133–6183.
[30] L.F. Albright, Alkylation of isobutane with C3—C5 olefins: feedstock consumption, acid usage, and alkylate quality for different processes, Ind. Eng. Chem. Res. 41 (2002) 5627–5631.
[31] A. Brillis, Catalytic Cracking of C8 Aliphatic Hydrocarbons Over Ultrastable Y Zeolite and Its Deactivation, University of London,University College London, United Kingdom, 2002.
[32] A. Szklo, R. Schaeffer, Fuel specification, energy consumption and CO_2 emission in oil refineries, Energy 32 (2007) 1075–1092.
[33] C.S. Hsu, P.R. Robinson, Gasoline production and blending, in: Springer Handbook of Petroleum Technology, 2017, pp. 551–587.
[34] A.N. Emana, S. Chand, Alkylation of benzene with ethanol over modified HZSM-5 zeolite catalysts, Appl. Petrochem. Res. 5 (2015) 121–134.
[35] H. Faghihian, M.H. Mohammadi, A novel catalyst for alkylation of benzene, C.R. Chim. 15 (2012) 962–968.
[36] C.M. Gordon, New developments in catalysis using ionic liquids, Appl. Catal. A Gen. 222 (2001) 101–117.
[37] W. Yang, Z. Wang, H. Sun, B. Zhang, Advances in development and industrial applications of ethylbenzene processes, Chin. J. Catal. 37 (2016) 16–26.
[38] J.H. Clark, Solid acids for green chemistry, Acc. Chem. Res. 35 (2002) 791–797.
[39] G.A. Baker, S.N. Baker, S. Pandey, F.V. Bright, An analytical view of ionic liquids, Analyst 130 (2005) 800–808.
[40] Y. Cui, J. Yin, C. Li, S. Li, A. Wang, G. Yang, Y. Jia, Experimental and theoretical studies on compositions, structures, and IR and NMR spectra of functionalized protic ionic liquids, Phys. Chem. Chem. Phys. 18 (2016) 19731–19737.
[41] A.J. Greer, J. Jacquemin, C. Hardacre, Industrial applications of ionic liquids, Molecules (2020).
[42] R. Sheldon, Catalytic reactions in ionic liquids, Chem. Commun. (2001) 2399–2407.
[43] C. Perego, P. Ingallina, Recent advances in the industrial alkylation of aromatics: new catalysts and new processes, Catal. Today 73 (2002) 3–22.
[44] K. Tanabe, W.F. Hölderich, Industrial application of solid acid–base catalysts, Appl. Catal. A Gen. 181 (1999) 399–434.
[45] A.A. O'Kelly, J. Kellett, J. Plucker, Monoalkyibenzenes by vapor-phase alkylation with silica-alumina catalyst, Ind. Eng. Chem. 39 (1947) 154–158.

[46] N.Y. Chen, W.E. Garwood, Industrial application of shape-selective catalysis, Catal. Rev. 28 (1986) 185–264.
[47] W.M. Yang, H.M. Sun, Q.L. Chen, Development of zeolite catalysts for the gas-phase alkylation of benzene and ethylene to ethylbenzene, Adv. Fine Petrochem. (2002) 12–14.
[48] Y. Wang, Y. Li, X. Hu, Formation and oxidation resistance of silicide coatings for Mo and Mo-based alloys, J. Mater. Sci. Technol. 17 (2001) 479–481.

CHAPTER

2

Alkylation technology

Modern society needs to consume a large amount of energy, and the current energy depends mainly on the burning of fossil fuels in the refining process. From the perspective of sustainable development and environmental protection, there is currently a lot of interests in replacing primary energy fossil fuels with alternative renewable energy sources. However, given the maturity and efficiency of oil use and the difficulty of developing alternative sources of energy on a large scale, the prevalence of refining and petrochemicals is likely to continue in the coming decades, especially given that modern extraction techniques can extract oil from discovered oil fields, tar sands, and more complete recovery of hydrocarbons from unconventional sources.

Over the past decades, aliphatic alkylation processes, especially C4 alkylation, have been extensively employed in petroleum refineries to produce high-quality gasoline blending components. With the official implementation of the world's Standard for Gasoline and the gradual promotion of ethanol gasoline, the C4 alkylation process has received increasing attention from refineries to produce the desired alkylate products primarily composed of C_5-C_{16} iso-paraffins. Furthermore, it is an excellent choice to increase the added value of refinery by-products containing C_3-C_5 components. This chapter introduces the C4 alkylation process with HF, H_2SO_4, ionic liquids, and solid acids as catalysts. The process characteristics, operating parameters, and improvements of C4 alkylation technologies are summarized, and the development trends of C4 alkylation units are discussed.

The classification of alkylation technologies in the world is summarized, as shown in Fig. 2.1.

The general process of the C4 alkylation technology is displayed in Fig. 2.2 with process description as follows:

(1) Pretreatment of olefin and isobutane feeds takes place in a coalescer prior to entering the reaction section to remove water, sulfur, and other contaminants.

https://doi.org/10.1016/B978-0-443-21984-9.00008-8

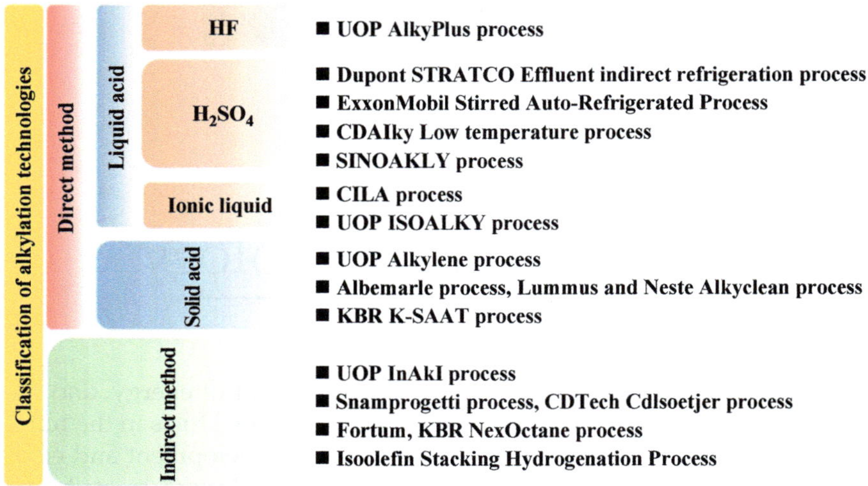

FIG. 2.1 The classification of alkylation technologies. *From D.X. Cao, X.N. Yang, G.Q. Cao, C4 alkylation technology development trend, Pet. Petrochem. Today 28 (2020) 29–37.*

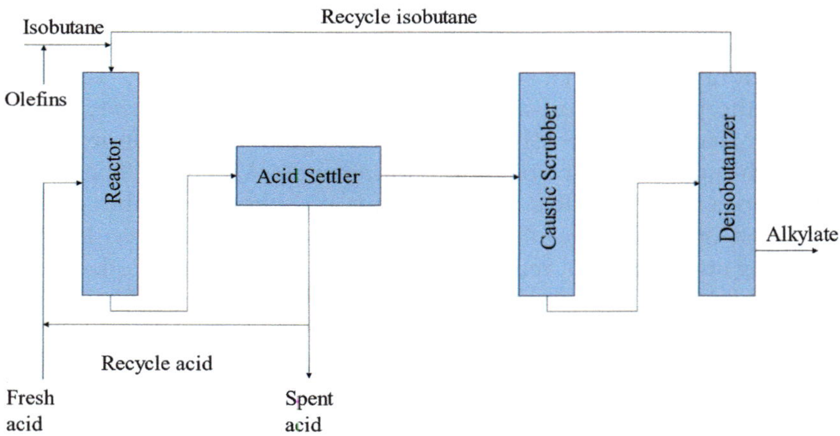

FIG. 2.2 Alkylation process block diagram. *From J.G. Speight, Handbook of Industrial Hydrocarbon Processes, Gulf Professional Publishing 2019.*

(2) The mixture of light olefins (C_3, C_4, C_5, or a combination) and isobutane is introduced into the reactor containing a series of reaction chambers. These chambers are filled with acid catalyst, either H_2SO_4 or HF. The presence of the catalyst facilitates the reaction between isobutane and olefins, resulting in the alkylate products, a naphtha-range product with C_5-C_{16} iso-paraffins.

(3) The effluent from the reactor directly flows into the acid settler, where hydrocarbons are separated from the acid. The acid is then isolated and returned to the reactor, while the hydrocarbons undergo a washing process with aqueous alkali before being sent to the fractionator.

(4) The fractionation section consists of the depropanizer, deisobutanizer, and debutanizer. The isobutane generated from deisobutanizer is returned to mix with raw feedstocks to maintain the high ratio of paraffin to olefin (I/O). Alkylate obtained from the deisobutanizer can be used for motor-fuel blending directly, or it can undergo further processing to produce aviation-grade gasoline.

2.1 H_2SO_4 catalyst

The first H_2SO_4 alkylation unit began operation in 1938 at the Beaumont refinery in the United States. The H_2SO_4 catalyst exhibits high selectivity for C_8 components, particularly for compounds like trimethylpentane (TMP). Furthermore, there is no need for the isomerization of 1-butene in the feedstock due to the fast isomerization between C4 olefins. Despite the high acid assumption of the H_2SO_4 alkylation process, the increasing optimization of waste H_2SO_4 acid regeneration processes has reduced the disadvantages accordingly. However, the use of HF is strictly restricted due to the potential release danger, which offers significant development opportunities for the application of H_2SO_4 alkylation. Most newly constructed alkylation units have adapted H_2SO_4 technology in recent years.

There are mainly four H_2SO_4 alkylation technologies in use, including the Stratco effluent refrigerated process, ExxonMobil series of stirred tank self-cooling process, CB&I Lummus CDAlky low-temperature process, and Sinopec SINOALKY process. Compared to the HF alkylation process, the H_2SO_4 alkylation process exhibits many advantages. For instance, the reduced products of dimethylhexane (DMHs) often result in higher quality alkylate. Savings, contingent on the olefins used, can be as high as 5%–10%. H_2SO_4 technologies often operate with lower I/O ratios of 7–9, compared to that of 12–15 for HF, resulting in considerably lower distillation costs for recovering and recycling unreacted isobutane in H_2SO_4 technologies.

However, due to the low temperature reaction conditions required by H_2SO_4 alkylation technology, the high viscosity of the concentrated H_2SO_4 catalyst significantly hinders the thorough contact between the catalyst and feedstock. As a result, many patent holders have primarily focused on the development of reactors and related internal components to enhance the mixing and contact of materials within the reactor at low temperatures, thus improving mass and heat transfer within the reactor [1].

2.1.1 Stratco effluent refrigerated process

The Stratco effluent refrigerated H_2SO_4 alkylation process is employed worldwide, constituting approximately 80% of the market share. The research octane number (RON) of alkylate products typically falls within the range of 94–97. The characteristic of this process is that the reaction effluent is used as a refrigerant to reduce investment and operation costs. The whole process consists of three sections, i.e., reaction section, refrigeration section, and distillation section.

In the reaction section, the raw feedstocks, mainly isobutane and butene, are mixed with the recycled isobutane, which is then cooled, dehydrated, and mixed with the recycled refrigerant to be charged into the typical alkylation reactor called the Stratco reactor (Fig. 2.3), which is a meticulously engineered unit that serves both as a reactor and a heat exchanger. Within this apparatus, many heat exchange tubes are incorporated to regulate the reaction temperature and dissipate excess heat generated during the reaction. Typically, a series of 2–4 contactor reactors is employed to enhance the productivity of alkylate and minimize acid consumption. The olefin space velocity generally falls within the range of 0.06–0.19 h^{-1}, and the reaction temperature is carefully controlled within the range of 5–18°C [2].

It should be noted that the Stratco horizontal eccentric reactor is the core device of the process. The horizontal pressure vessel is equipped with a high-power agitator, an inner circulation sleeve, and a heat treatment tube bundle. The acid phase and the hydrocarbon phase are rapidly emulsified by stirring the impeller and the emulsion circulates at high speed when the alkylation reaction occurs. Part of the emulsion is introduced into the acid

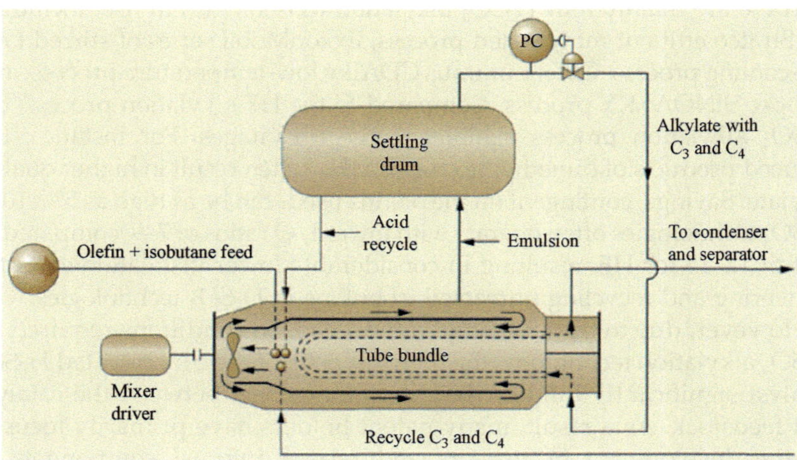

FIG. 2.3 Stratco effluent refrigeration process. *Acquired from C.S. Hsu, P.R. Robinson, Petroleum Science and Technology, Springer, 2019 (for Figure 10.16 in Chapter 10).*

settler through the riser to separate the acid phase and hydrocarbon phase. The dense acid phase sinks to the bottom of the acid settler and return to the reactor to participate in the reaction again. Based on the density difference between the acid phase and hydrocarbon phase, the H_2SO_4 circulates naturally between the reactor and the acid settler. The hydrocarbon phase, containing 70–85 wt% of propane and butane (isobutane and *n*-butane), is separated from the acid phase and flows out from the top of the acid settler. After valve decompression and flash evaporation, the cooled hydrocarbon stream flows through the heat pipe bundle of the reactor to absorb the released reaction heat.

In the refrigeration section, the reaction effluent leaves the Stratco reactor and enters the buffer flash tank for gas/liquid separation. The buffer flash tank is a horizontal vessel with intermediate partitions, one side of which is used for the gas-liquid separation of the reaction effluent, and the other side for the circulating refrigerant. After the gas-liquid separation, the liquid phase of the reaction effluent exchanges heat with the feedstocks and flows into the product-refining section for further treatment. The gas phase of the reaction effluent is compressed and condensed, and a small part is extracted from the device to avoid the accumulation of propane in the process. The other part enters the flash tank, and then the obtained propane-rich gas returns to the compressor. The liquid phase enters the buffer flash tank for another flash to further reduce the refrigerant temperature and then enters the reactor through the refrigerant circulation pump.

The reaction effluent can entrain trace acid and sulfates formed by side reactions, which will decompose to generate SO_2 in the later distillation section and cause equipment corrosion. To deal with this problem, after heat exchange with the feedstocks, the reaction effluent first enters the acid wash settler to remove the sulfates and then enters the caustic wash settler to neutralize the trace acid. Then, the obtained hydrocarbon material exchanges heat with the alkylate products from the deisobutanizer bottom and enters the deisobutanizer for fractionation. The top distillate is condensed by the full condenser. Part of the top distillate is returned to the column top as reflux and the other is cooled and returned to the reactor as recycled isobutane to confirm the appropriate I/O ratio in the reactor feed. The bottom distillate is delivered out of the device as alkylate product after heat exchange, and the *n*-butane stream is extracted from the sideline of the column as the by-product after cooling.

2.1.2 CDAlky low-temperature process

CDAlky technology has been developed and industrialized with numerous improvements relative to the process and equipment. The core of the CDAlky process is the innovative contactor/reactor design, a

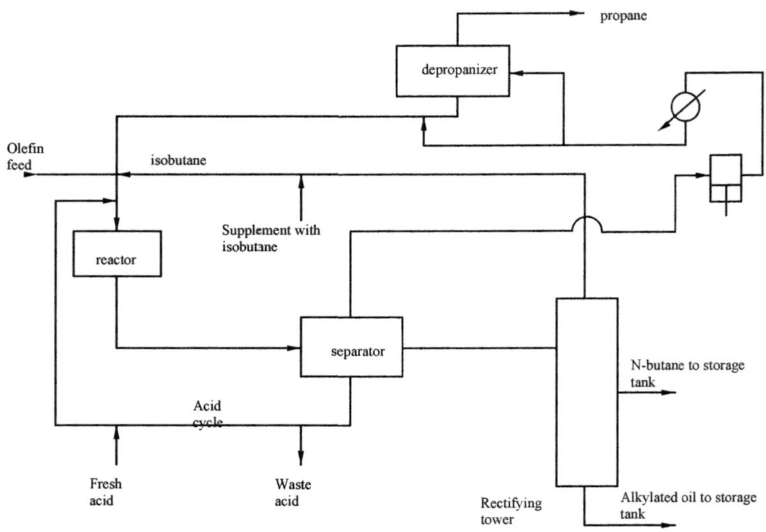

FIG. 2.4 Lummus CDAlky low-temperature process. *Modified from H.X. Ma, F. Zhou, K. Kai, Advances in liquid acid alkylation technology, Chem. Ind. Eng. Prog. 33 (2014) 32–40.*

pivotal section that substantially enhances mass transfer efficiency compared to a conventional reactor. Notably, this design operates at much lower temperatures than standard alkylation processes, thereby pushing the boundaries of energy efficiency and process optimization in the field of alkylation technology, as shown in Fig. 2.4.

The CDAlky catalytic alkylation technology is an advanced, low-temperature H_2SO_4 alkylation process. The reaction temperature is low (typically −3°C, significantly lower than DuPont technology at 7°C), effectively suppressing side reactions and achieving high selectivity for C_8 products. The use of a vertical reactor facilitates the contact between the acid and hydrocarbons on proprietary structured packing, which can eliminate the rotating stirrers, reduce the possibility of operational failure, and minimize acid and energy consumption. The utilization of efficient poly-condensers eliminates the acid washing, water washing, and alkali washing steps for the whole process, thereby reducing plant investment and equipment maintenance costs.

The world's first 200 kt/a alkylation unit using CDAlky technology commenced operation in May 2013 at Shandong Dongying Shenchi Chemical. This technology generates an alkylate with a RON of 97.3. Currently, six alkylation units utilizing the CDAlky process have been commissioned in China. Isobutene tends to undergo exothermic polymerization reactions in alkylation. In traditional alkylation processes, these feedstocks with a high content of isobutene usually lead to the increase in temperature in the reactor due to the increased isobutene oligomerization, which reduces

TABLE 2.1 The comparison of alkylate obtained from the CDAlkyPlus process and the Stratco process.

Parameter	CDAlkyPlus	Traditional alkylate
RVP (kPa)	18.6–31.0	27.6–37.9
RON	93–94	92–94
[(R+M)/2]		
T50 (°C)	107–111	105–107
T90 (°C)	177–178	121–152
End point (°C)	<215	188–216

Data from H.X. Ma, F. Zhou, K. Kai, Advances in liquid acid alkylation technology, Chem. Ind. Eng. Prog. 33 (2014) 32–40.

the RON and yield of the alkylate products. To address this issue, CDTech has integrated the industrialized processes of CDOPT and CDAlky to form the CDAlkyPlus process, capable of processing pure isobutene. The CDOPT process, a pretreatment of C4 feed, process serves as an isobutene pretreatment process with many advantages, including the suppression of oligomerization reactions, elimination of the effects of exothermic side reactions on acid consumption, and reduction in the energy consumption of the refrigeration process. The CDAlkyPlus process optimizes olefin space velocity and I/O ratio to enhance the quality of alkylate, reduces plant construction investment, and shortens the investment payback period. Table 2.1 compares the difference of alkylate obtained from the CDAlkyPlus process and the Stratco process [1,3].

2.1.3 ExxonMobil stirred auto-refrigerated process

The ExxonMobil stirred reactor auto-refrigerated process has a history spanning over 50 years. This process involves mixing and cooling feed olefins and recycled isobutane before entering the reactor in separate and parallel sections, with recycled H_2SO_4 and refrigerant introduced in the first section. Each individual reaction section utilizes a vigorously stirred reactor and is separated and interconnected with vertical baffle plates. The reaction mixture flows from one mixing zone to the next, and the acid-hydrocarbon emulsion in the final reactor section enters a settler for separation. The whole process flowsheet is displayed in Fig. 2.5 [3].

The stirred auto-refrigerated process by ExxonMobil employs a horizontal and a parallel reactor, as shown in Fig. 2.6. Each section of the reactor is stirred. The olefin feed is premixed with propane and subsequently mixed with the recycled isobutane before being fed parallely into the reactor. The acid only enters the first zone of the reactor and then cascades to

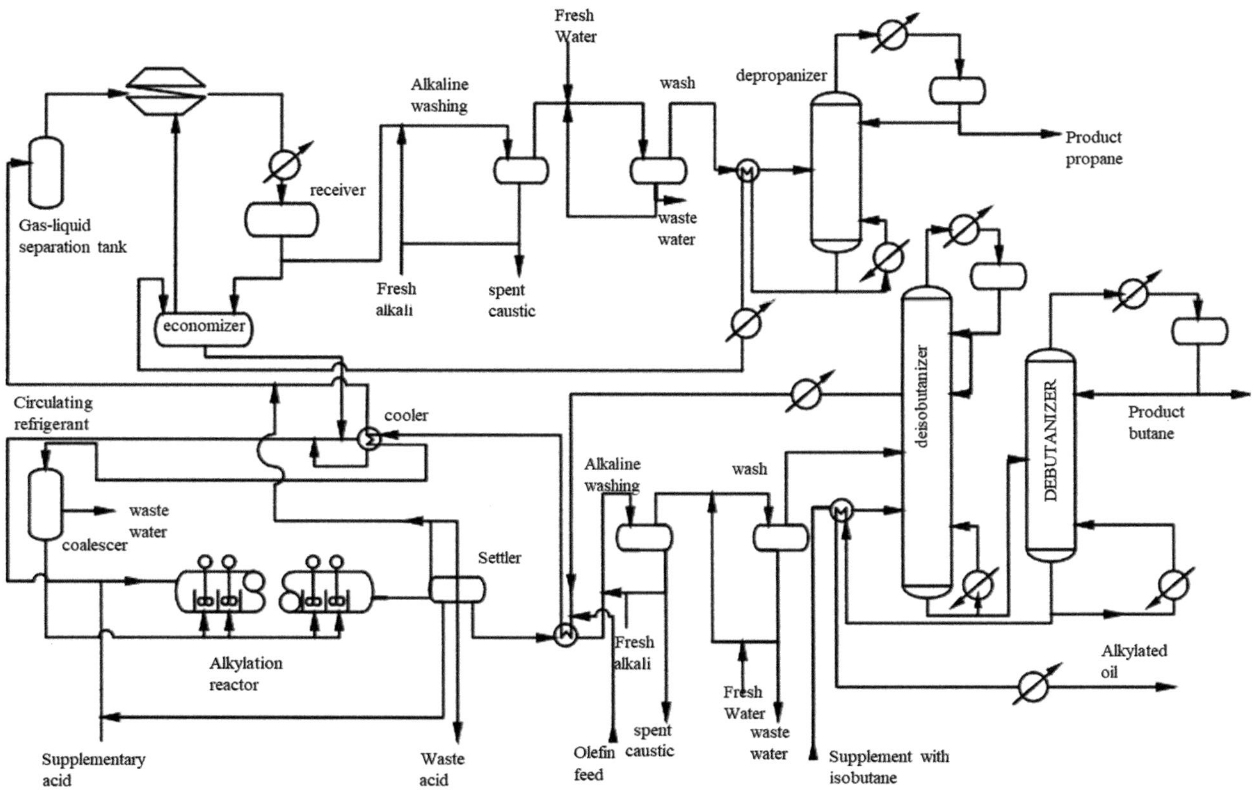

FIG. 2.5 ExxonMobil stirred reactor auto-refrigerated process. *Modified from H.X. Ma, F. Zhou, K. Kai, Advances in liquid acid alkylation technology, Chem. Ind. Eng. Prog. 33 (2014) 32–40.*

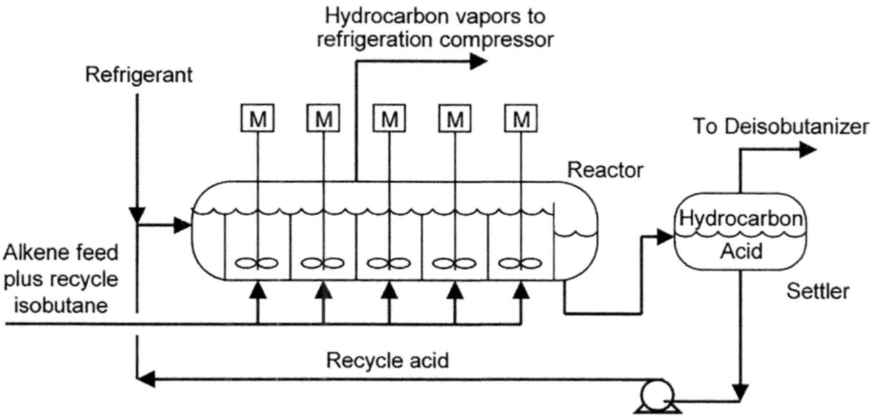

FIG. 2.6 The reactor applied in the ExxonMobil stirred reactor auto-refrigerated process.

other zones [3]. The temperature is regulated through the evaporation of propane and the introduction of isobutane. To remove temperature elevation within the reactor, it operates under a pressure differential. A portion of propane is diverted to prevent accumulation in the system, and the settler is employed for liquid product separation. The operational temperature is typically maintained at 5°C to minimize side reactions. The olefin space velocity stands at approximately $0.1\,h^{-1}$, and the acid concentration is usually held at 90–92 wt% [2].

The refrigeration temperature in this system is higher compared to the indirect refrigeration systems, which allows for the use of small refrigeration compressors, contributing to reduced energy consumption. Additionally, the segmented design of the reactor enables operation at lower temperatures (4.4°C) and an olefin space velocity of $0.1\,h^{-1}$ compared to the DuPont process. The advantages of this process include the substantial capacity of individual reactors, dependable process operation, and lower investment and operating costs.

Although this technology is not currently employed in worldwide facilities, the ExxonMobil H_2SO_4 process is used in approximately 16 installations in other countries. India's Reliance Industries' Jamnagar refinery employs this technology, with a production capacity of 3.57 Mt/a, making it the largest alkylation installation in the world [3].

2.1.4 Sinopec SINOALKY process

The Sinopec SINOALKY process primarily encompasses an N-type multistage, multisection static mixing reactor, self-vaporizing acid-hydrocarbon separator, efficient vaporization packing, and special separation materials for agglomeration, as shown in Fig. 2.7. The new technology is advantageous in reducing side reactions, achieving rapid separation of the reactor effluents into the gas phase, hydrocarbon phase,

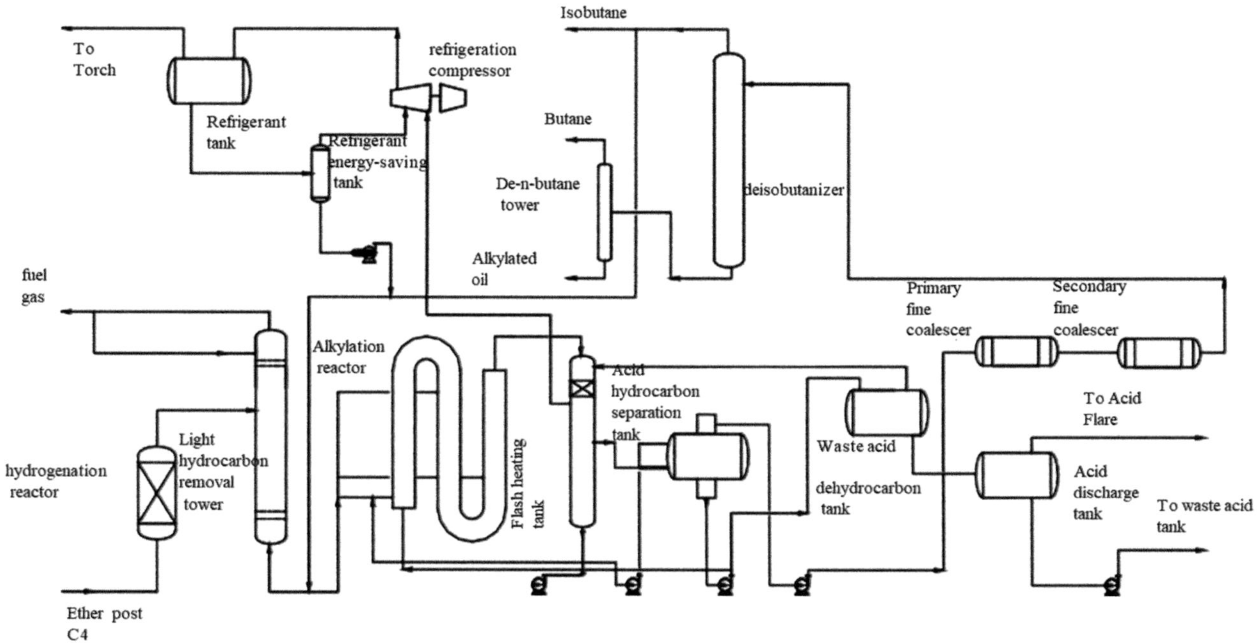

FIG. 2.7 Sinopec SINOALKY sulfuric acid alkylation process.

and acid phase, and quickly extracting reaction heat. It exhibits advantageous characteristics of low temperature, low acid consumption, and excellent product distribution [1,4].

In June 2018, the first industrial unit employing this technology with a capacity of 200 kt/a was commercialized in Shijiazhuang, Hebei Province, China. The key process operating parameters and the quality of alkylate in the Sinopec SINOALKY process are shown in Table 2.2 [4]. Under the

TABLE 2.2 Key process operating parameters and the quality of alkylate in SINOPEC SINOALKY process.

Project	Condition 1[a]	Condition 2[b]	Design value
Raw material feed volume (t/h)	28.2	33.5	28.03
Prehydrogenation reactor inlet temperature (°C)	51.4	51.2	45.0
Pressure of light hydrocarbon removal tower (MPa)	1.77	1.77	1.72
Bottom temperature of light hydrocarbon removal tower (°C)	103.0	104.0	103.0
Alkylation reactor inlet temperature (°C)	0	0.4	0
Compressor inlet pressure (MPa)	0.016	0.018	0.02
Deisobutane tower bottom temperature (°C)	134.0	133.0	126.0
Bottom temperature of the debutanizer tower (°C)	130.0	131.5	163.0
Alkylate properties			
Distillation range (°C)			
Initial Boiling Point	44.5	43.5	≥32.0
50%	104.5	106.5	110.0
End point	188.5	198.0	205.0
Density (20°C) (kg/m^3)	696.7	695.2	690
Vapor pressure (37.8°C) (kPa)	46.25	42.50	35.0~50.0
Water-soluble acid-base	–	–	–
Copper strip corrosion (grade)	1a	1a	1a
RON	97.0	96.5	96.0
ω(Sulfur) (μg/g)	<1.0	<1.0	<5.0
Acid consumption (kg/t)	58.9	58.9	60.0

[a] *The device is fully loaded with feed.*
[b] *The device is at full capacity.*
Data from M.H. Dong, B.N. Zong, SINOALKY Sulfuric acid alkylation process technology and its industrial application, Pet. Process. Petrochem. 50 (2019) 29–32.

stable full-load operation, the RON of the alkylate maintains in the range of 96.5–97.0 with the final boiling point of 189–199°C, vapor pressure of 35.0–45.5 kPa, copper corrosion rating of 1 μg/g, sulfur content below 1.0 μg/g, acid consumption of 58.9 kg/t, and energy consumption of 98.5 kgEO/t. Over a year of industrial operation, the alkylation reactor was confirmed to be safe and efficient. The self-vaporizing, acid-hydrocarbon separator performs well in heat extraction and acid-hydrocarbon separation. It has eliminated the traditional processes of acid washing, alkali washing, and water washing, significantly reducing high-salinity wastewater discharge and alkali consumption and thus achieving clean products [1,4].

2.1.5 Operating parameters

As is well-known, C4 alkylation is a typical complex and exothermic reaction. Several simultaneous reactions take place in the alkylation process, including alkylation, polymerization, cracking, disproportionation, and self-alkylation. This results in the presence of over three dozen iso-paraffins, along with the corresponding carbonium cations. Changes in operating parameters, such as olefin type, temperature, reaction time, and acid strength, have a substantial impact on the quality of the alkylate.

2.1.5.1 Effect of type of olefin

The quality of alkylate varies with different olefins as feedstocks. It can be seen from Table 2.3 that the alkylate produced by butene as the olefin feedstock is the best with the lowest acid consumption compared to the propene and pentene.

From previous studies, the distribution of alkylate produced by 1-butene and 2-butene is very similar with the similar octane number

TABLE 2.3 Effect of different olefins on the alkylation.

Feed	Propene	Butene	Pentene
Olefin consumption	1.27–1.32	1.10–1.16	0.96–1.14
Reaction heat (kJ/kg)	1.96	1.43	1.16
Yield (g/(g olefins))	1.76–1.78	1.70–1.74	1.55–1.60
Acid consumption (kg/(t alkylate))	137–171	51–102	102–137
RON	89–90	94–97	92–93
MON	88–90	92–94	92–93

Data from B.L. Gao, High Octane Gasoline Component Production Technology, China Petrochemical Press, Beijing, 2006.

TABLE 2.4 Effect of type of butene on the octane number of alkylate.

Types of butenes	RON	MON
1-Butene	97–98	93–94
2-Butene	97–98	93–94
Isobutene	90–91	88–89

Data from B.L. Gao, High Octane Gasoline Component Production Technology, China Petrochemical Press, Beijing, 2006.

(Table 2.4), which indicates that 1-butene and 2-butene have a similar reaction pathway in C4 alkylation. However, the octane number of the alkylation with isobutene is lower. It is observed that 1-butene and 2-butene can quickly approach the equilibrium of the isomerization reaction during the H_2SO_4 alkylation by the following Eq. (2.1):

$$CH_2CHCH_2CH_3 + HX \rightarrow CH_3CHXCH_2CH_3 \rightarrow CH_3CHCHCH_3 + HX \tag{2.1}$$

The equilibrium constant of the isomerization reaction is 37.3 at 283.15K, which means that most of 1-butene can be isomerized into 2-butene under industrial conditions. To gain a deeper understanding of the isomerization reactions involving butenes, Cao et al. [5] conducted energy and geometry calculations for the reactants, transition states, and products during the isomerization of 1-butene, 2-butene, and isobutylene using the density functional theory (DFT) calculation, as illustrated in Fig. 2.8. The transition states are calculated where 1-butene transforms into 2-butene, 2-butene isomerizes into isobutylene, and 1-butene changes into isobutylene. Subsequently, the activation energies were calculated as 63.65, 63.43, and 64.42kcal/mol for the isomerization of 1-butene to 2-butene, 2-butene to isobutylene, and 1-butene to isobutylene, respectively. It is evident that the activation energies for the three butenes are similar, which highlights the ability of C4 olefins to isomerize into one another [5].

2.1.5.2 *The effect of impurities in the feedstock*

The impurities in the feedstocks, such as ethylene, butadiene, sulfides, H_2O, dimethyl ether, and methanol, can reduce the catalytic activity of H_2SO_4. Butadiene as one of impurities with the strongest side effects in alkylation does not participate in the alkylation, but butadiene can react with H_2SO_4 to form acid soluble oil (ASO). ASO is a type of viscous heavy

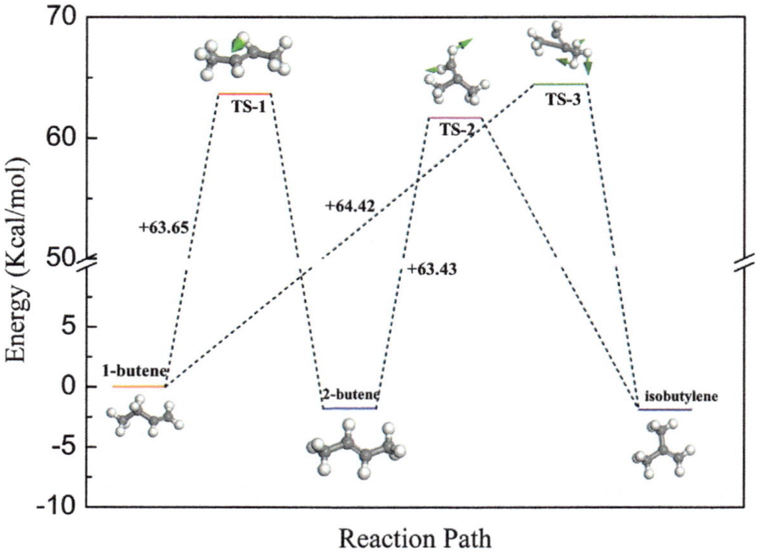

FIG. 2.8 Optimized geometries and potential-energy profile of the reactants, transition states, and products for the isomerization of butenes. The energy of 1-butene is set to zero [5].

oil with relatively high molecular weight, whose existence in H_2SO_4 increases the dry point of the product and decreases the RON of the alkylate. The C4 fraction produced by catalytic cracking usually contains about 0.5 wt% butadiene. If the reaction temperature in the catalytic cracking reaction is relatively high, the content of butadiene can reach up to 1.0 wt%.

For the C4 alkylation catalyzed by H_2SO_4, impurities in the raw materials tend to combine with the acid, thereby causing contamination of H_2SO_4 and subsequent diminishment of catalytic activity. It is conventionally necessary to replace the spent acid when the acid concentration decreases to 90 wt% using fresh H_2SO_4 with 98.5 wt% as the acid source. The detrimental impurities within the H_2SO_4 alkylation process and the requisite quantity of fresh acid to be added per ton of impurities are outlined in Table 2.5.

(1) Ethylene

In the event that a significant quantity of liquefied gas is introduced into the catalytic cracking dry gas, coupled with inadequate removal of the C2 components by using the gas fractionation unit, there exists a distinct possibility of ethylene ingress into the alkylation unit. This tendency is further increased during the alkylation of mixed olefins containing propylene and butene.

TABLE 2.5 The acid consumption depending on different impurities.

Impurities	Acid consumption/(mass ratio of sulfuric acid/impurities)
Ethylene	20.9
Butadiene	13.4
Sulfides	15–60
H_2O	11.8
Dimethyl ether	11.1
Methanol	26.8

Data from B.L. Gao, High Octane Gasoline Component Production Technology, China Petrochemical Press, Beijing, 2006.

When ethylene is introduced into the alkylation reactor, it reacts with H_2SO_4, yielding ethyl hydrogen sulfates, which is weakly acidic compound. This contrasts with the anticipated alkylation reaction between ethylene and isobutane. Ethyl hydrogen sulfate dissolves in the acid phase, consequently exerting a dilution effect on the H_2SO_4. It is worth noting that the influence of ethylene impurities exhibits an accumulating nature. Even trace quantities of ethylene within the feedstock can translate into daily influxes of ethylene into the acid phase on the order of several hundred kilograms, thereby culminating in the production of several tons or potentially exceeding ten tons of spent acid. The sudden introduction of a substantial volume of ethylene into the alkylation reactor not only can impede alkylation reactions but also encourage secondary reactions. Consequently, it is imperative to intensify operational oversight and analytical scrutiny of upstream facilities.

(2) Butadiene

The C4 fraction produced by catalytic cracking typically contains approximately 0.5% butadiene. However, if the feedback of the catalytic cracking unit contains a substantial amount of residual oil or operates at elevated temperatures, the butadiene content can reach 1%. During the alkylation process, butadiene does not undergo alkylation with isobutane; instead, it reacts with H_2SO_4 to form ASO, which leads to an increase in the dry point of the alkylate and a reduction in octane number and yield. Butadiene has a boiling point very close to that of other C4 components, signifying that distillation is an ineffective separation method.

(3) Sulfides

Prior to the introduction of feedstock into the alkylation unit, a standard practice involves subjecting the alkylation feed to liquefied gas desulfurization and thiol removal treatments, aiming to reduce the sulfo compound content to levels below 10 μg/g. The effect of sulfo compounds on H_2SO_4 is noteworthy, as exemplified by the data presented in Table 2.5, indicating that each metric ton of sulfide compounds can result in the scrapping of 15–60 metric tons of H_2SO_4. In cases where methyl mercaptan serves as the sulfo compound, following equimolar molecular reactions, each metric ton of mercaptan sulfides can dilute approximately 53.7 metric tons of 98.5 wt% H_2SO_4 down to 90 wt%, with practical consumption approximating 45 metric tons. The influence of H_2S on sulfuric acid consumption remains relatively marginal.

Beyond their role in expediting acid disposal, sulfo compounds also tend to incline the catalytic function of sulfuric acid toward polymerization reactions and other undesired side reactions, thus deviating from the intended pathways. Consequently, the operational efficacy of the liquefied gas desulfurization assumes paramount importance, significantly impacting the performance of the alkylation unit.

(4) H_2O

The consequences of the presence of water in the feedstock, leading to the dilution of H_2SO_4, are self-evident. Furthermore, feedstocks with elevated water content pose a potential risk of equipment corrosion. When water exists in a saturated state within liquefied gas, it typically registers at approximately 500 μg/g. However, it is crucial to highlight the presence of supersaturated free water carried by the C4 fraction, which demands more attention. In cases of suboptimal operational practices in upstream units, the quantity of free water carried by the C4 fraction can surpass that of dissolved water by several folds, thus leading to a notably rapid acid dilution rate.

The removal of free water is facilitated through a preentry process, in which the feedstock traverses a packed container known as a coalescing dehydrator. This procedure enables the coagulation of minute water droplets, subsequently separated from the lower section of the coalescing dehydrator. Enhanced efficiency can be achieved if the feedstock is cooled using a reactor effluent before entering the coalescing dehydrator. In circumstances where the cost of H_2SO_4 is comparatively elevated, molecular sieves or activated alumina can be deployed to eliminate dissolved water from the alkylation feedstock, thereby reducing the water content to levels below 10–20 μg/g. Furthermore, it is noteworthy that isobutane recycled from the fractionation segment back into the reaction section can contain a substantial moisture load. Dry contact between the recycled isobutane and the spent acid proves both economically viable and efficacious.

(5) Oxygen-containing compounds

A significant portion of the feedstock for alkylation in most refineries originates from methyl tert-butyl ether (MTBE) units. The remaining C fractions from MTBE synthesis typically contain approximately 500 to 2000 μg/g of dimethyl ether and 50 to 100 μg/g of methanol. Dimethyl ether and methanol are prominent impurities in the alkylation process, contributing to acid consumption, and subsequently reducing both the yield and octane number of alkylate.

The relative vapor pressures of dimethyl ether and C4 components exhibit discernible differences, while methanol can form azeotropic mixtures with C4 fractions. Consequently, conventional distillation techniques cannot be applied for the separation of dimethyl ether and methanol from C4 fractions. Additionally, it is noteworthy that water can also form azeotropic mixtures with C4 fractions. Therefore, during the distillation-based removal of dimethyl ether and methanol, there exists the potential for the concurrent elimination of trace amounts of water present in the feedstock.

2.1.5.3 *Effect of temperature*

Alkylation, as a typical exothermic reaction, usually requires a relatively low reaction temperature, which is usually maintained in the range of 2–18°C, with a designed reaction temperature of 10°C. The reduction in reaction temperature effectively inhibits polymerization reactions and other undesirable side reactions, leading to the improvement in the quality and yield of alkylate and the higher RON. At the same mixing efficiency, the increase in the reaction temperature from 2°C to 10°C causes a decrease in the RON by 1–3 units. With a further increase in reaction temperature to 15–20°C, the olefins tend to polymerize into a gel-like substance, which leads to a decrease in the yield of alkylate. However, an excessively low reaction temperature, such as below 2°C, causes an increase in the viscosity of H_2SO_4, which results in the difficult emulsion of the alkane and the increase in the consumption of stirring power. Therefore, the optimal operating temperature of the reactor lies between 4–10°C, with the maximum of 15°C and the minimum of 2°C, based on the aforementioned reasons [6].

Albright et al. examined the effects of temperature on C4 alkylation in a continuous-flow reactor with vigorous agitation. Temperature influences the rates and types of chemical reactions that determine the precise composition of the alkylate. The alkylate with the same RON exhibits similar compositions, regardless of the specific types of C4 olefins, although these compositions do exhibit some variation with temperature, as shown in Fig. 2.9. The elevated temperatures have a more notable effect on the quality of the alkylate produced using either 1-butene or 2-butene compared to those produced using isobutylene [7].

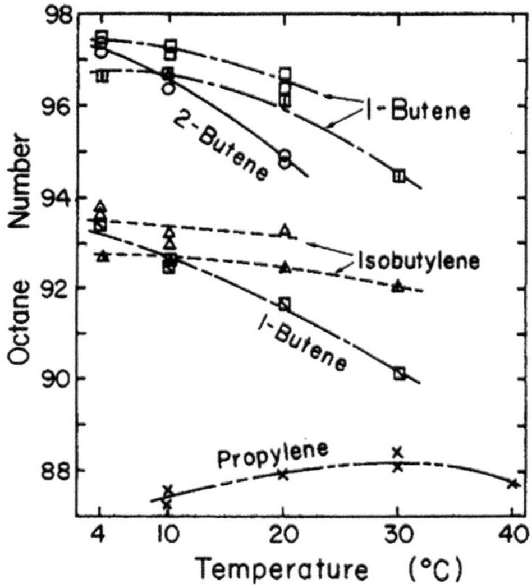

FIG. 2.9 Effect of temperature on the RON of alkylate. *From K. Li, R.E. Eckert, L.F. Albright, Alkylation of isobutane with light olefins using sulfuric acid. Operating variables affecting both chemical and physical phenomena, Ind. Eng. Chem. Process. Des. Dev. 9 (1970) 441–446.*

2.1.5.4 Effect of reaction time

The influence of the contact time between H_2SO_4 and hydrocarbons on the yield and quality of alkylate is also significant. To ensure optimal yield and quality of the alkylate, the reaction time should be maintained at a level above the time required for the complete emulsification of H_2SO_4 and hydrocarbons. A shorter reaction time has adverse effects on both the yield and quality of the alkylate, while a longer reaction time not only can significantly impede the production capabilities of the plant, but also lead to the occurrence of degradation and other undesirable side reactions that compromise product quality. Industrial operations typically control the feed velocity of olefins to regulate reaction time. Usually, the olefin space velocity generally falls within the range of 0.06–0.19 h^{-1} [8].

2.1.5.5 Effect of acid strength

The acid strength of H_2SO_4 is the key factor that determines the quality of the alkylate. In 1970, Albright et al. [7] firstly systemically investigated the effect of acid strength on C4 alkylation catalyzed by H_2SO_4, as shown in Fig. 2.10 and Table 2.6. For each olefin, the optimum acid strength of H_2SO_4 is different. The optimum strength of H_2SO_4 required for 2-butene is apparently slightly lower compared to 1-butene. Approximately 96–97 wt% acid strength of H_2SO_4 is preferred for making high-

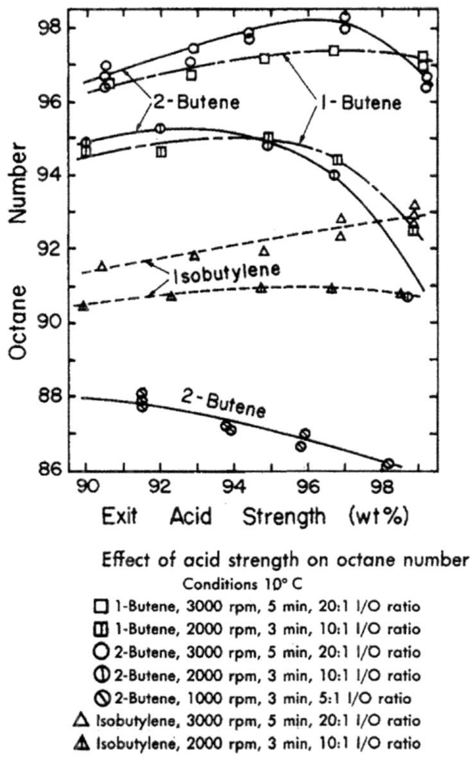

FIG. 2.10 Effect of acid strength of H_2SO_4 on the quality of alkylate. *From K. Li, R.E. Eckert, L.F. Albright, Alkylation of isobutane with light olefins using sulfuric acid. Operating variables affecting both chemical and physical phenomena, Ind. Eng. Chem. Process. Des. Dev. 9 (1970) 441–446.*

quality alkylate from either 1-butene or 2-butene. Fresh 99 wt% acid strength of H_2SO_4 produced the best quality of alkylate for isobutylene.

Furthermore, acid strength affects the chemical reaction steps to a considerable extent. Xin et al. further confirmed the impact of acid strength of H_2SO_4 on the alkylation through recycle experiments [9]. The initial reactions involving isobutylene led to the rapid generation of large carbonium ions (C_9^+) through oligomerization. Excessive acidity in the catalyst results in undesirable cracking reactions, such as β-dissociation of C_9^+, and leads to the formation of LEs. As the acidity decreases, the extent of β-dissociation of C_9^+ and the selectivity of LEs also decreases. However, during the gradual deactivation stage, the accumulation of impurities has an adverse effect on the acidity and catalytic activity of H_2SO_4. This, in turn, reduces the desired TMP selectivity and facilitates side reactions like oligomerization. The quality of the alkylate significantly deteriorated when the acid strength of H_2SO_4 fell below 93 wt%, as depicted in Fig. 2.11.

TABLE 2.6 Effect of acid strength of H_2SO_4 on the quality of alkylate with different types of butene.

		Feed acid strength, wt%					
	Octane	**99**	**95**	**91**	**99**	**95**	**1**
	No.	**Total octanes, wt%**			**Heavy ends, wt%**		
1-Butene and 2-butene	98	88	88	...	4.5	4.5	...
	97	81	81.5	82	9.0	9.0	9.0
	96	74	75	76.5	11.5	11.5	11.5
	95	67.5	69	71.5	15	15	15
	94	61	63	65.5	19	19	19
	92	48.5	50	...	27	28	...
	90	36.5	38.5	...	36.5	38.5	...
2-Butene	88	27.5	29.5	...	48	50.5	...
	87	23	25	...	54	56.5	...
Isobutylene	92.5	54.5	58.5	64	21.5	19.5	17.5
	91.5	48	52	56	26	23.5	21
	90.5	42	45	48	30	27.5	25

Data from K. Li, R.E. Eckert, L.F. Albright, Alkylation of isobutane with light olefins using sulfuric acid. Operating variables affecting both chemical and physical phenomena, Ind. Eng. Chem. Process. Des. Dev. 9 (1970) 441–446.

2.1.6 Additives

Over the past few decades, numerous approaches have been explored to improve the quality of alkylate and reduce acid consumption, such as the reconfiguration of the alkylation reactor, refinement of feed-stream purification, optimization of operating conditions tailored to specific olefins, regulation of the iso-paraffin/olefin ratio, and the lowering of reaction temperatures [10–14]. These methods are mainly aimed to enhance the dispersion of liquid-liquid phases and decrease reaction temperatures in order to yield higher quality alkylate. However, it is important to note that the redesign of the alkylation reactor can be cost-prohibitive and not be readily applicable within existing operational processes. In pursuit of improved catalytic performance in C4 alkylation, there has been a notable focus on novel additives. The introduction of these additives has emerged as a pivotal technique for enhancing interfacial properties during C4 alkylation [15–19]. This approach offers several advantages, including minimal additive usage, reduced by-product formation, and the feasibility of direct co-feeding with H_2SO_4.

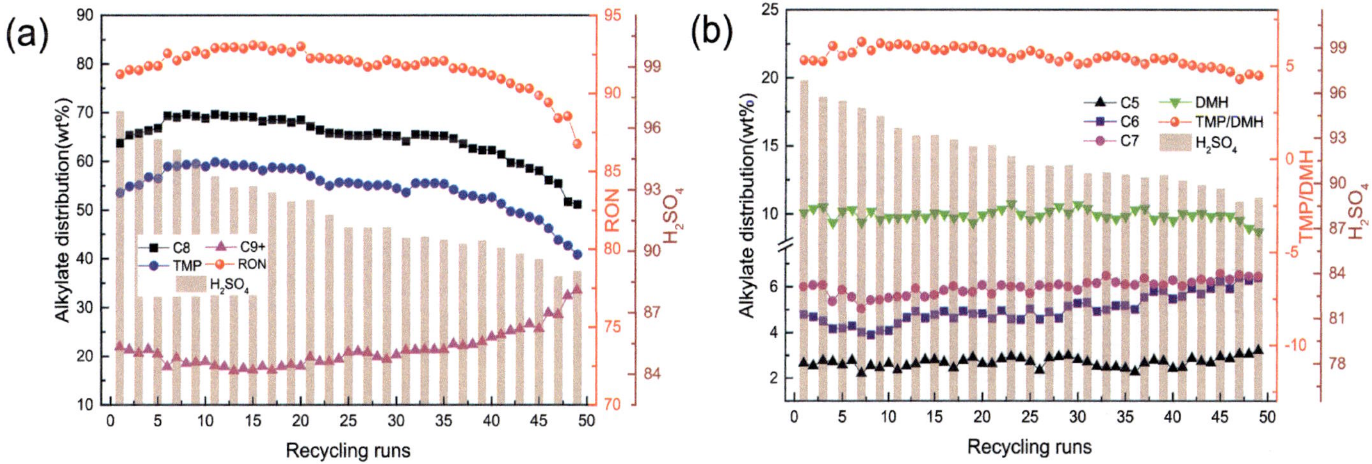

FIG. 2.11 Effects of a recycling run of H_2SO_4 on the quality of the alkylate. *From Y. Xin, Y. Hu, M. Li, K. Chi, S. Zhang, F. Gao, S. Jiang, Y. Wang, C. Ren, G. Li, Isobutane alkylation catalyzed by* H_2SO_4*: effect of* H_2SO_4 *acid impurities on alkylate distribution, Energ. Fuel, 35 (2021) 1664–1676.*

2.1.6.1 Ionic liquid additives

The effects of IL additives on the H_2SO_4-catalyzed C4 alkylation were investigated in detail by Tang et al. [16], followed by Zhang et al. [17,20], Liu et al. [21], and Sun et al. [18,22–24]. IL additives can significantly intensify the interfacial properties and catalytic performance of H_2SO_4-catalyzed C4 alkylation.

Tang et al. proposed a new process for 1-butene/isobutane alkylation catalyzed by sulfuric or trifluoromethanesulfonic acid by the introduction of acidic imidazolium ILs. The IL/acid mixtures outperform the pure acids, achieving equivalent or higher conversion (>95%), greater selectivity for C_8 components (>70%), and higher selectivity for TMP/DMH (TMP/DMH > 7). Six types of 3-methyl-imidazolium ILs were examined. The acidic groups of each IL are introduced through the cation with sulfonic acid groups, the anion (hydrogen sulfate), or a combination of both [16], as shown in Fig. 2.12.

Trace amounts of [Bmim][SbF_6] ILs into H_2SO_4 show a substantial positive impact on the catalytic performance of H_2SO_4 with enhancement in the selectivity of TMP and C_8 alongside augmenting the RON. Furthermore, it can act as a buffer to effectively mitigate fluctuations in acidity

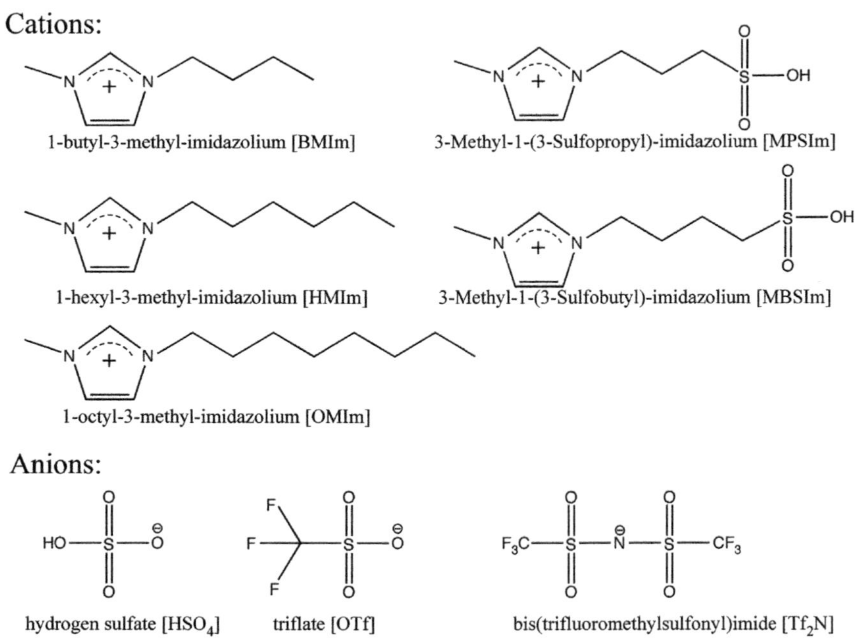

FIG. 2.12 Structures of the cations and anions of the ILs used in Tang's work. *From S. Tang, A.M. Scurto, B. Subramaniam, Improved 1-butene/isobutane alkylation with acidic ionic liquids and tunable acid/ionic liquid mixtures, J. Catal. 268 (2009) 243–250.*

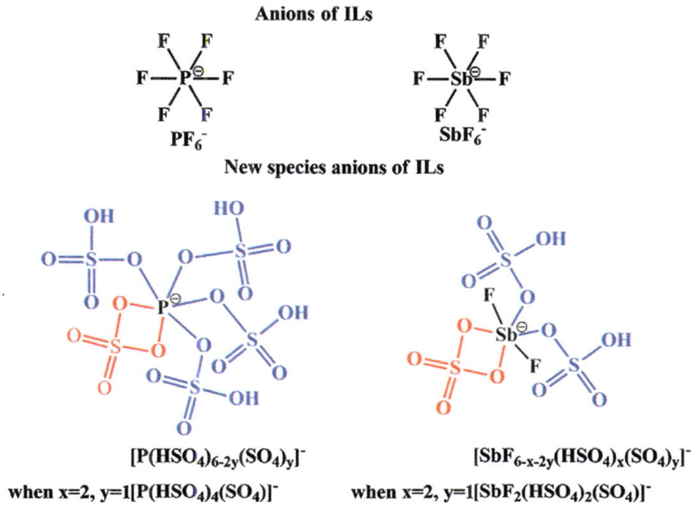

FIG. 2.13 Anions of ILs in the catalytic mixed systems. *From H. Zhang, R. Liu, Z. Yang, F. Huo, R. Zhang, Z. Li, S. Zhang, Y. Wang, Alkylation of isobutane/butene promoted by fluoride-containing ionic liquids, Fuel 211 (2018) 233–240.*

and retard the formation of ASO, thereby significantly prolonging the operational lifetime of H_2SO_4 up to twice its original duration. Consequently, the integration of [Bmim][SbF_6] results in enhanced catalyst utilization and subsequently reduces acid consumption. The H_2SO_4/[Bmim][SbF_6] composite catalyst holds great promise for C4 alkylation with notable environmental and economic benefits [17]. Furthermore, the most interesting feature of these fluoride-containing ILs is their capability to undergo acidolysis in H_2SO_4. The acidolysis products were proved to consist of stable anionic species of the types $[P(HSO_4)_{6-x-2y}\ (SO_4)_y]^-$ and $[SbF_{6\text{-}x\text{-}2y}(HSO_4)_x(SO_4)_y]^-$; When $x=2$ and $y=1$, they were transformed into $[P(HSO_4)_4(SO_4)]^-$ and $[SbF_2(HSO_4)(SO_4)]^-$, respectively (Fig. 2.13). An additional feature that improves the TMPs selectivity is the formation of HF from the acidolysis of the ILs, which makes more protons available for the formation of carbocations. [BMIm][PF_6] is the best-performing cocatalyst in terms of TMPs selectivity and RON due to the larger extent of acidolysis and thus higher concentration of HF formed during the alkylation process [20].

A series of adamantane-based ILs with $[MF_x]^-$ anions, typical quaternary ammonium ILs, amino acid ILs, and imidazolium ILs were synthesized as cocatalysts for C4 alkylation. For the adamantane-based ILs, the optimized RON of alkylate is up to 98. The interfacial properties of the reaction system have been improved to a certain extent, and the service time of the catalyst was prolonged by at least up to two times [25].

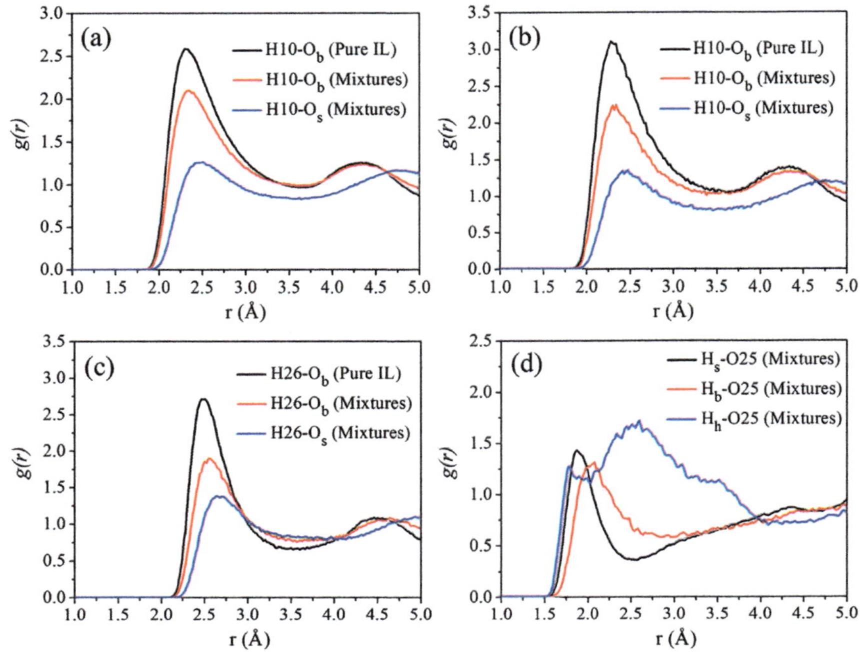

FIG. 2.14 Site-site RDFs between the proton (H10), as well as the sulfonic acid proton (H26), and the oxygen atom on HSO_4^- (O_b) and H_2SO_4 (O_s), and the proton of HSO_4^- (O_b), H_2SO_4 (O_s), and H_3O^+ (H_h) with the oxygen atom of the sulfonic acid groups (O25) for the pure IL and IL/sulfuric acid mixture systems with an IL mole fraction of 0.3: (A) the [PMim][HSO_4] systems and (B–D) the [PSMim][HSO_4] systems. *From W. Zheng, C. Huang, W. Sun, L. Zhao, Microstructures of the sulfonic acid-functionalized ionic liquid/sulfuric acid and their interactions: a perspective from the isobutane alkylation, J. Phys. Chem. B 122 (2018) 1460–1470.*

The SO_3H-functionalized ILs (1-methyl-3-(propyl-3-sulfonate) imidazolium bisulfate ([PSMim][HSO_4]) were further investigated as cocatalysts to enhance the H_2SO_4-catalyzed C4 alkylation. Compared to non-SFIL (1-methyl-3-propyl imidazolium bisulfate ([PMim][HSO_4])), a novel, strong interaction site involving the sulfonic acid group of SFIL and an H_2SO_4 molecule was identified, characterized by a strong hydrogen-bonding interaction, as depicted in Fig. 2.14.

To further probe the property relationship of SFILs together with H_2SO_4, Sun et al. investigated the catalytic performance of SFILs with different alkyl chains ([C_nPSIm][HSO_4], $n=1, 2, 4, 6, 8$) [26]. The SFILs with longer alkyl chains exhibit enhanced catalytic performance, achieving C_8 selectivity up to 75.73% and RON up to 95.66, respectively. This improved catalytic performance is ascribed to the better dispersion of isobutane in the SFIL/H_2SO_4 system facilitated by the longer alkyl chains of SFILs due to the strong interaction confirmed by MD simulation (see Fig. 2.15).

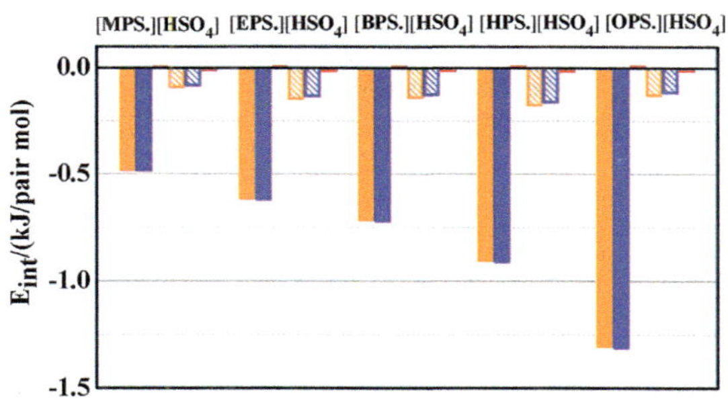

FIG. 2.15 Interaction energy between isobutane and SFILs in different simulation systems. *Yellow* represents total potential energy, and *blue* stands for van der Waals energy. *Red* represents electrostatic energy. The filled bars represent the (isobutane + cation) interaction energy, and the patterned bars represent the (isobutane + anion) interaction energy. *From W. Sun, W. Zheng, W. Xie, L. Zhao, Understanding structure–property relationship of SO_3H-functionalized ionic liquids together with sulfuric acid in catalyzing isobutane alkylation with C4 olefin, Ind. Eng. Chem. Res. 57 (2018) 15310–15318.*

With the deepening understanding of the interfacial microenvironments between catalysts and C4 hydrocarbons, Zheng et al. investigated the influence of the SFILs on interfacial properties through the combination of experiments and MD simulation [22,27]. The interfacial behaviors are visible from the equilibrated snapshots of different BIL-containing systems, as presented in Fig. 2.16. It can be observed that the longer alkyl chains extend deeper into the reactant phase compared to the shorter ones, while the sulfonic acid groups protrude deeply into the bulk regions of the H_2SO_4 [27].

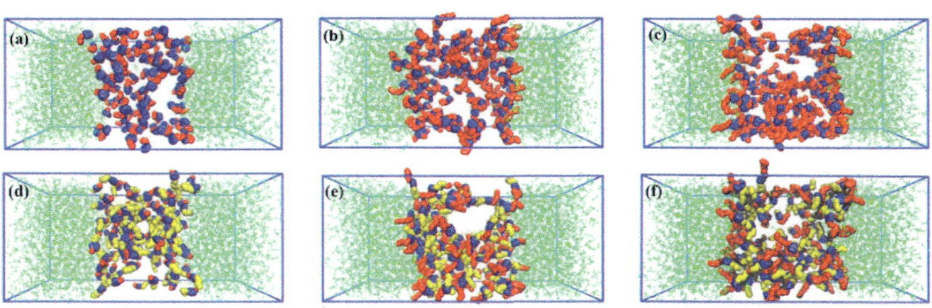

FIG. 2.16 Equilibrated snapshots of different simulation systems: (A) [PMim][HSO_4], (B) [HMim][HSO_4], (C) [OMim][HSO_4], (D) [PSMim][HSO_4], (E) [HPSim][HSO_4], (F) [OPSim][HSO_4]. *Red* represents the alkyl groups; *yellow* represents sulfonic acid groups; *blue* represents the imidazole ring; and *green* represents the reactant mixtures. *From W. Sun, W. Zheng, P. Cao, L. Zhao, Probing interfacial behaviors of Brønsted acidic ionic liquids improved isobutane alkylation with C4 olefin catalyzed by sulfuric acid, Chem. Eng. J. 377 (2019) 119744.*

TABLE 2.7 The interface composition and width of different simulation systems.

	Mole fractions (%)			Interfacial width (nm)	
Interfacial additions	**2-C_4H_8**	***i*-C_4H_{10}**	**Cation**	**Oil phase**	**Acid phase**
None	15.603	12.766	0.000	0.582	0.624
[PMim][HSO_4]	22.170	19.811	4.245	0.652	0.659
[HMim][HSO_4]	16.883	15.584	8.658	0.764	0.803
[OMim][HSO_4]	14.523	12.863	10.373	0.854	0.867
[PSMim][HSO_4]	26.704	24.432	3.409	0.686	0.697
[HPSim][HSO_4]	19.844	18.677	7.782	0.794	0.814
[OPSim][HSO_4]	15.837	13.575	9.502	0.873	0.890

Data from W. Sun, W. Zheng, P. Cao, L. Zhao, Probing interfacial behaviors of Brønsted acidic ionic liquids improved isobutane alkylation with C4 olefin catalyzed by sulfuric acid, Chem. Eng. J. 377 (2019) 119744.

The interfacial composition and interfacial width of different simulation systems are calculated and listed in Table 2.7 The BIL-containing systems show a thicker interfacial thickness compared to the pure H_2SO_4 system, indicating that the BILs can enlarge the interfacial width. With the length of the alkyl chains being prolonged, the interfacial width also becomes thicker regardless of SFILs or non-SFILs [27].

According to the study of the effect of different ratios of SFIL to H_2SO_4 on the SFIL/H_2SO_4-catalyzed C4 alkylation, compared to the pure H_2SO_4, the addition of the SFILs can enhance the catalytic performance of the H_2SO_4, contributing to lower HEs, higher C_8-alkylate selectivity, and higher RON [22], as shown in Fig. 2.17. In the range of a 10wt% SFIL

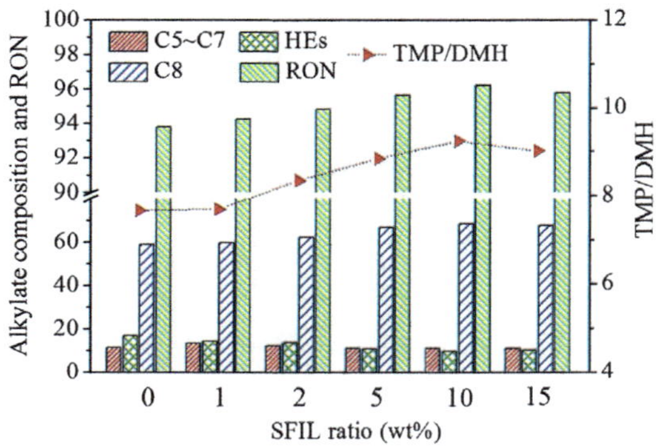

FIG. 2.17 Effect of different ratios of SFIL to H_2SO_4 on the alkylate quality. Reaction conditions: temperature 276.2K, reaction time 15min, stirring rate 3000r/min, ratio of H_2SO_4/hydrocarbon 1.5:1, volume ratio of I/O 8:1. *From W. Zheng, P. Cao, Y. Yuan, C. Huang, Z. Wang, W. Sun, L. Zhao, Experimental and modeling study of isobutane alkylation with C4 olefin catalyzed by Brønsted acidic ionic liquid/sulfuric acid, Chem. Eng. J. 377 (2019) 119578.*

addition ratio, the catalytic performance of SFIL/H_2SO_4 improves with an increased proportion of SFIL, reaching the highest RON of 96.23 at the 10 wt% SFIL concentration.

Moreover, to explain the proton microenvironment and structural features in SFIL/H_2SO_4 binary mixtures, Zheng et al. introduced the sulfonic acid groups (SAG) into ILs, which provided a new interaction site with both HSO_4^- and H_2SO_4 [28]. The introduction of the sulfonic acid groups is observed to provide a strong hydrogen bonding interaction site with both HSO_4^- and H_2SO_4, resulting in an increase of almost two orders of magnitudes than the average numbers of hydrogen bonds between different pairs. The introduction of sulfonic acid groups into the shorter side chains contributes to stronger hydrogen-bonding interactions with both HSO_4^- and H_2SO_4 compared to their introduction into longer side chains.

In addition, the incorporation of [OPSIm][HSO_4] or [Pr_3NPS][HSO_4] as cocatalysts led to enhanced catalytic performance and improved quality of alkylate products (see Fig. 2.18). In-depth analysis through molecular dynamic (MD) simulation reveals intriguing insights. Specifically, [OPSIm][HSO_4] is found to facilitate the dissolution of isobutane molecules at the interface, thereby elevating the I/O ratio from 1.02 to 1.18. On the other hand, [Pr_3NPS][HSO_4] is observed to significantly increase

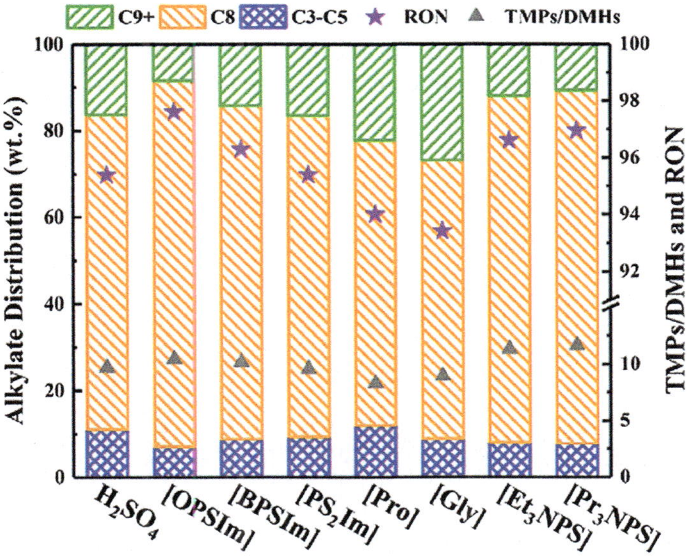

FIG. 2.18 Catalytic performance of different H_2SO_4/ILs for C4 alkylation. Reaction conditions: IL amount, 100 mol/m^3; reaction temperature, 276.15 K; reaction time, 15 min; stirring rate, 3000 rpm; and mass ratio of isobutane to C4 olefins, 87.1:12.9. *From P. Cao, Y. Yuan, C. Huang, W. Sun, L. Zhao, Promoting the sulfuric acid catalyzed isobutane alkylation by quaternary ammonium ionic liquids, AIChE J. 66 (2020) e16979.*

the interfacial width from 0.66 to 0.97 nm and reduce the interfacial tension from 28.49 to 14.62 mN/m. These effects contribute to the enlargement of the reaction area and improvement of ion transfer [23].

To predict and rationally design high-efficiency IL additives for H_2SO_4-catalyzed alkylation, the quantitative structure-activity relationship between 11,000 ILs and interfacial properties of H_2SO_4 alkylation was established by combination of MD simulation, semiempirical density function theory (DFT), and machine learning. The findings reveal that IL additives with longer alkyl chains substantially increase interfacial thickness and reduce interfacial tension (see Fig. 2.19). This effect is primarily driven by both cation- and anion-based descriptors working synergistically. Furthermore, a series of novel IL additives with varying alkyl chain lengths in both cations and anions were prepared. These additives demonstrate the capability to effectively enhance the catalytic performance of H_2SO_4 with only a small addition, resulting in the increase in the RON of the alkylate from 95.87 to 98.08 compared to the pure H_2SO_4 system [18].

Ma et al. further investigated the effect of used H_2SO_4 on the C4 alkylation with the introduction of formyl functional $[N_{1,1,1,1}][C_{10}SO_4]$ IL additives. According to the experimental results, the $[N_{1,1,1,1}][C_{10}SO_4]$ additive can further improve the quality of alkylate with the used H_2SO_4 catalyst (see Fig. 2.20). The optimal mass ratio for the $[N_{1,1,1,1}][C_{10}SO_4]$ additive to the used H_2SO_4 is approximately 0.3 wt%. The reaction time in the formyl functional $[N_{1,1,1,1}][C_{10}SO_4]$-assisting system is shortened compared with the pure used H_2SO_4 system. In addition, they proved used acid with 92.29 wt% H_2SO_4 exhibits better catalytic performance on C4 alkylation

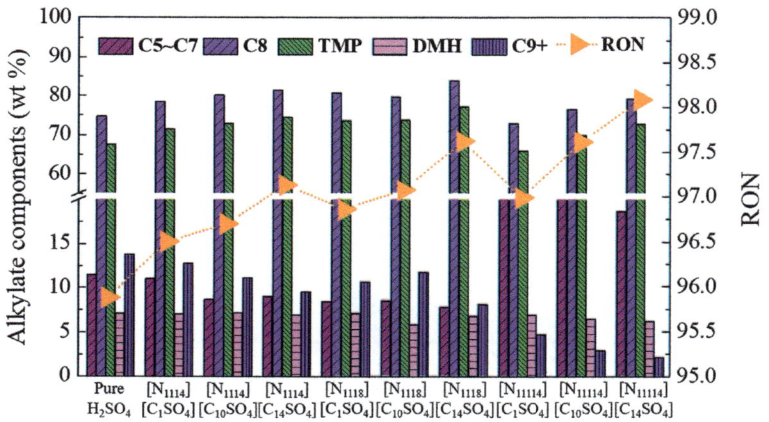

FIG. 2.19 Alkylate components for the IL additives with different combinations of cations and anions. Reaction conditions: temperature 281.2 K, stirring rate 3000 rpm, reaction time 20 min, volume ratio of H_2SO_4/hydrocarbon 1.5:1, volume ratio of I/O 12:1. *From W. Zheng, Z. Ma, W. Sun, L. Zhao, Target high-efficiency ionic liquids to promote H_2SO_4-catalyzed C4 alkylation by machine learning, AICHE J. 68 (2022) e17698.*

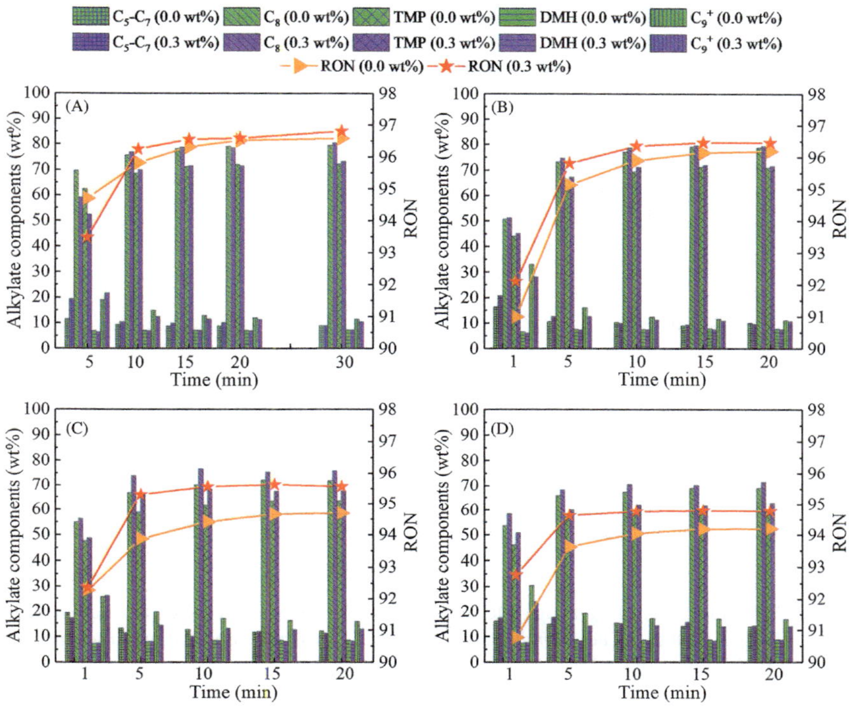

FIG. 2.20 Temperature-dependent alkylate components for the used H_2SO_4-catalyzed alkylation without and with the formyl functional $[N_{1,1,1,1}][C_{10}SO_4]$ additive as a function reaction time. (A) 277.15K; (B) 281.15K; (C) 285.15K; and (D) 289.15K. Reaction conditions: stirring rate 3000rpm/min, formyl functional $[N_{1,1,1,1}][C_{10}SO_4]$ concentration 0.3wt%, volume ratio of H_2SO_4/hydrocarbon 1.5:1, volume ratio of I/O 16:1. *From Z. Ma, W. Zheng, W. Sun, L. Zhao, Enhanced catalytic performance of H_2SO_4-catalyzed C4 alkylation by formyl functional [N1,1,1,1][$C_{10}SO_4$] additive, AICHE J. (2023) e18179.*

than fresh, concentrated H_2SO_4 through experimental and simulation methods. Furthermore, the key rate constant for the formation of TMPs is 1.67kg/(molmin) in used acid catalyzed alkylation, which is larger than that of the fresh, concentrated H_2SO_4 system (1.03kg/(molmin)) [24].

2.1.6.2 *Organic substances as additives*

As early as 1970, Roebuck et al. discovered the effective catalytic properties of the ether-$AlCl_3$ system for C4 alkylation. When aromatic hydrocarbons were introduced to this system, it led to the improved quality of gasoline [29]. Aluminum chloride-ether catalysts demonstrate remarkable activity in C4 alkylation with low-molecular-weight olefins. However, the high activity needs to be regulated for practical purposes. Aromatic hydrocarbons and metal chlorides can modulate the activity in a way that preserves the initial alkylation products and prevents

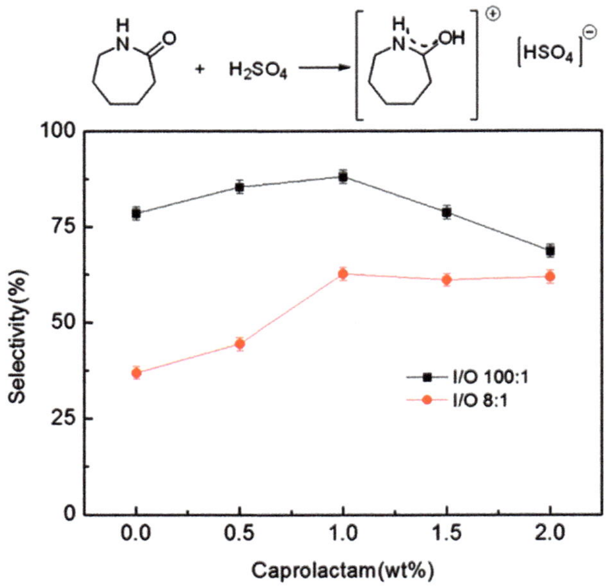

FIG. 2.21 Selectivity of C_8 using different caprolactam amounts in H_2SO_4 as catalyst. *From L. Li, J. Zhang, K. Wang, G. Luo, Caprolactam as a new additive to enhance alkylation of isobutane and butene in H_2SO_4, Ind. Eng. Chem. Res. 55 (2016) 12818–12824.*

subsequent alterations due to isomerization, disproportionation, hydrogen transfer, and cracking. Li et al. utilized caprolactam (CPL) as an additive in the H_2SO_4-catalyzed C4 alkylation. CPL can form a stable caprolactamium hydrogen sulfate complex in H_2SO_4 (see Fig. 2.21) [30]. Initial experiments reveal an increase in C_8 selectivity from 80% to 88% with 1 wt% CPL [30].

The addition of aromatic compounds in H_2SO_4 significantly enhanced C4 alkylation (Fig. 2.22). Specifically, the presence of thiophene in H_2SO_4 leads to a remarkable increase of 23.46% in C_8 selectivity and 5.09 rise in the RON. The proportion of TMPs increases from 58.57% to 84.58%, and the ratio of TMPs to DMHs doubles compared to that with H_2SO_4 alone. This improvement is attributed to the enhanced stability of carbenium ions and acceleration of hydrogen proton release, facilitating by electron-rich π-bond compounds. Moreover, the substitution of thiophene in the alpha site by the sulfonic group further contributes to this effect [31].

Furthermore, Zhang et al. reported the effect of organic substances as additives on the C4 alkylation catalyzed by ILs. In 2007, Zhang et al. found that the introduction of a small amount of benzene into the $[(C_2H_5)_3NH]$ $Cl\text{-}AlCl_3$ IL can increase the TMP content up to 70 wt%. The addition of aromatic compounds alters atom charge distributions, subsequently

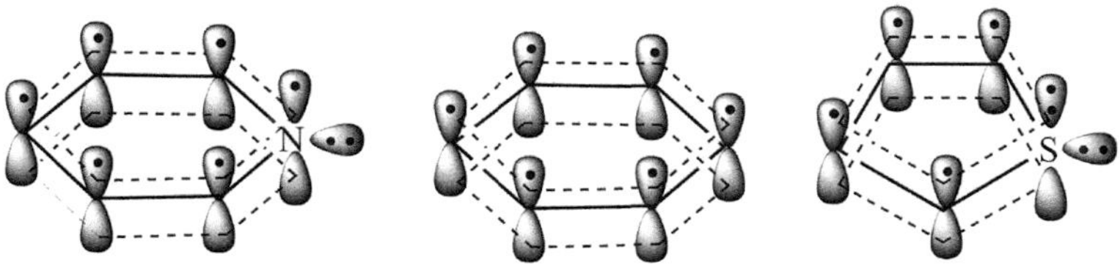

FIG. 2.22 Electronic configuration of different additives: pyridine, benzene, and thiophene. *From H. Zhang, H. Tang, J. Xu, Y. Li, Z. Yang, R. Liu, S. Zhang, Y. Wang, Enhancement of isobutane/butene alkylation by aromatic-compound additives in strong Brønsted acid, Fuel 231 (2018) 224–233.*

adjusting the super acidity of the catalysts. The quality of alkylate is closely reliant on the acidity of the catalysts, even though the amounts of aromatics in the mixtures is relatively low. Table 2.8 characterizes the effects of varying the quantity of benzene added to ILs on their catalytic performance. The addition of 1 wt% is found to yield the optimal alkylate compositions.

Through the experimental methods and theoretical calculations, the roles of aromatics in the C4 alkylation utilizing chloroaluminate ILs (CIL) were investigated. The catalytic system of CILs mixed with small amounts of aromatics exhibits a buffering effect, reducing the overall acidity, modifying the activity of the CIL catalyst, and leading to the inhibition of undesirable by-product formation within the C_{5-7} and C_9^+ range [32]. The catalytic performance of ILs with benzene addition was

TABLE 2.8 Effect of varying the amount of benzene added to the ILs.

Benzene (wt%)	Yield (wt% on olefine)	Composition of alkylate mixtures (wt%)			TMP/DMH
		C_5–C_7	C_8	C_9^+	
0.2	171	24.5	64.0	11.5	6.7
0.4	173	19.2	68.4	12.4	8.2
0.6	173	21.6	66.3	12.1	8.4
1.0	188	13.4	80.9	5.7	8.0
2.0	183	15.9	76.2	7.9	8.7
3.0	184	11.4	77.6	11.0	14.5
4.0	184	14.2	76.8	9.0	15.7

Note: Reaction temperature = 303 K, I/O ratio = 10.
Data from J. Zhang, C. Huang, B. Chen, P. Ren, M. Pu, Isobutane/2-butene alkylation catalyzed by chloroaluminate ionic liquids in the presence of aromatic additives, J. Catal. 249 (2) (2007) 261–268.

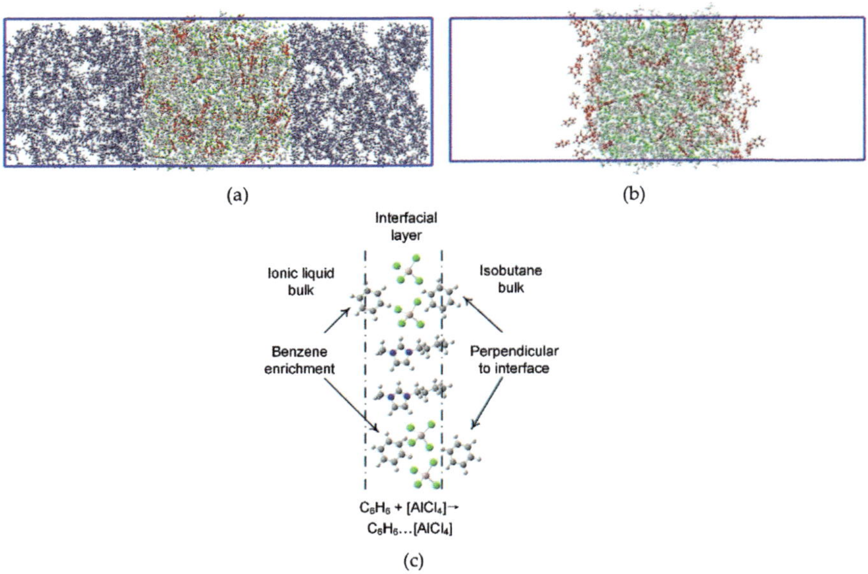

FIG. 2.23 The enrichment of benzene at liquid-liquid interfaces: (A) initial; (B) final of IL/benzene-C4 system snapshots ((B) snapshot without the C4 mixtures); (C) diagram showing benzene, cations, and anions. Color: isobutane/2-butene, *dark blue*; $[Bmim]^+$, *gray*; $[AlCl_4]^-$, *green*; benzene, *red*. *From Y. Liu, G. Wu, R. Hu, G. Gao, Effects of aromatics on ionic liquids for C4 alkylation reaction: Insights from scale-up experiment and molecular dynamics simulation, Chem. Eng. J. 402 (2020) 126252.*

investigated using MD simulations (Fig. 2.23). It has been observed that benzene tends to accumulate within the interfacial layer between the acid and hydrocarbon phases, resulting in an expansion of the interfacial width. This behavior at the interface enhances the dispersion of isobutane in the interfacial layer. The primary factor contributing to the enhanced performance of C4 alkylation is the accumulation of benzene molecules within the acid-hydrocarbon interfacial layer, in which benzene molecules serve to buffer the excessive acidity of anions in the particular region [33].

2.1.6.3 Surfactants as additives

In the alkylation process, the hydrocarbons are the dispersed phase and H_2SO_4 is the continuous phase. It means that surfactants with higher value of hydrophile lipophile balance (HLB) are suitable for alkylation. The first attempts to use surfactant additives in H_2SO_4 alkylation were carried out by the Cities Service Oil Company in 1972, utilizing dodecyl benzenesulfonic acid and *p*-phenylenediamine surfactants. The outcomes of these experiments demonstrate that the addition of surfactants could enhance the quality of the alkylate [34]. In 1975, Texaco, Inc.

evaluated *N,N′,N″*-tri(alkyl)phosphoric triamide surfactants of approximately 0.0005–0.5 wt% as additives, resulting in a slight increase in C_8 selectivity and RON [35].

By 2003, Chen begun to explore the solubility of hydrocarbons in H_2SO_4 using various cationic, anionic, and amphoteric surfactants as additives. The addition of small quantities of the surfactants alters the solubility ratio of isobutane to alkenes, consequently influencing the yield distribution of the alkylate. This effect is attributed to the higher HLB values conferred by the sulfonic and phosphonic groups. Five surfactants of differing natures were employed to evaluate their activity in H_2SO_4 alkylation. Among them, the anionic surfactants amino hydroxynaphthalenedisulfonic acid and naphthalenesulfonic acid with optimal addition amount of 1.0 wt% are found to improve TMP content and reduce the HE contents [36].

Commercial surfactants, including long-alkyl-chain cationic surfactants (hexadecyltrimethylammonium bromide, CATB and dihexadecyldimethylammoniumchloride, DDAB), and an anionic one (sodium dodecyl benzene sulfonate, SDBS), can also enhance the activity and stability of the H_2SO_4 catalyst for C4 alkylation. The addition of surfactant results in the improvement of acid/hydrocarbons interface, including better interfacial dispersion, higher I/O ratio, and lower interfacial tension (Fig. 2.24), which differs significantly from the inherent reaction enhancement achieved through lower temperatures. When SDBS additive is combined with lower temperatures, it leads to a substantial increase in TMP selectivity and higher RON up to 84.20 wt% and 99.07 at 269.2 K, respectively, compared to 60.21 wt% and 94.34 at 281.2 K (Fig. 2.25) [37].

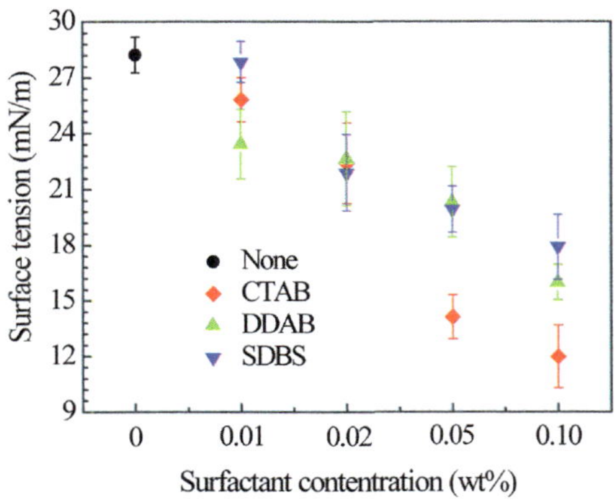

FIG. 2.24 Effect of surfactant concentrations on the interfacial tension of the acid/hydrocarbon interface. *From W. Zheng, Z. Wang, W. Sun, L. Zhao, F. Qian, H_2SO_4-catalyzed isobutane alkylation under low temperatures promoted by long-alkyl-chain surfactant additives, AIChE J. 67 (2021) e17349.*

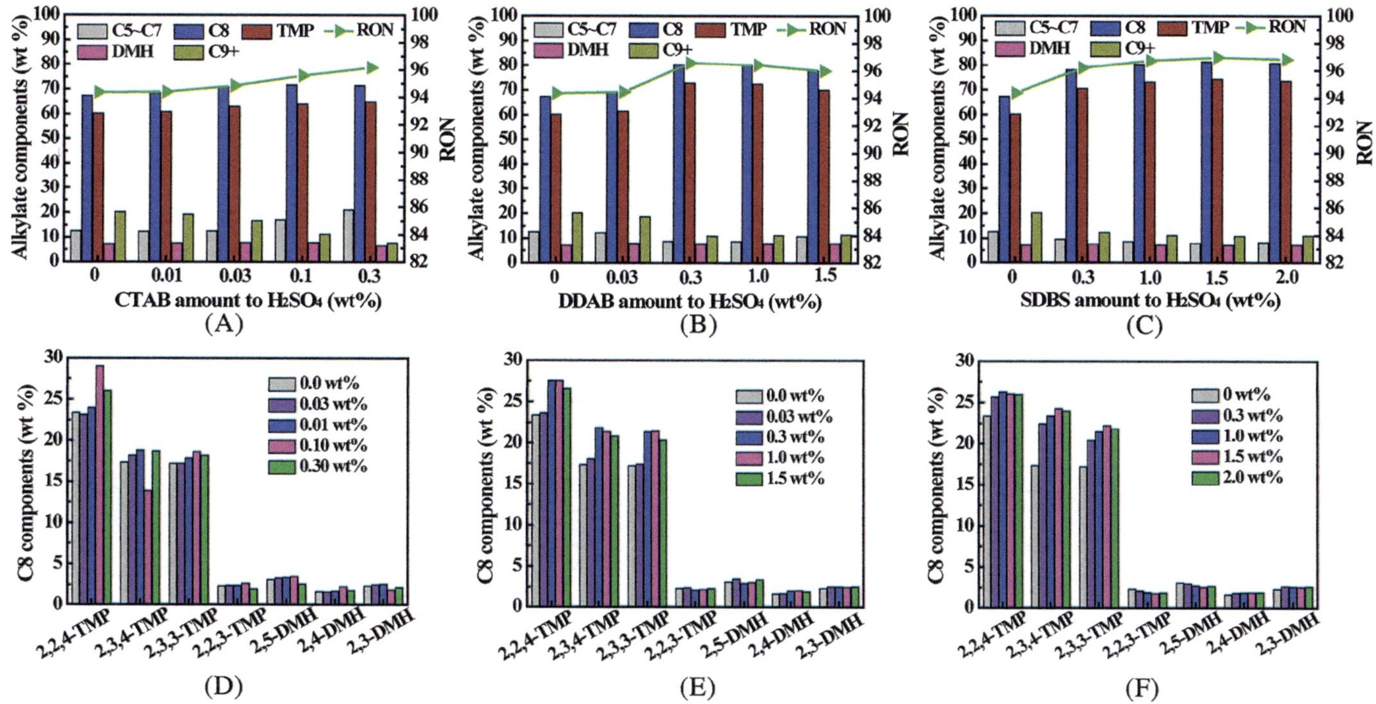

FIG. 2.25 Alkylate components as a function of surfactant concentration. (A) and (D) for the CTAB additive. (B) and (E) for the DDAB additive. (C) and (F) for the SDBS additive. Reaction conditions: temperature 281.2 K, stirring rate 3000 r/min, reaction time 15 min, volume ratio of H_2SO_4/hydrocarbon 1.5:1, volume ratio of I/O 8:1. *CTAB*, hexadecyltrimethylammonium bromide; *DDAB*, dihexadecyldimethylammoniumchloride; *SDBS*, sodium dodecyl benzene sulfonate. *From W. Zheng, Z. Wang, W. Sun, L. Zhao, F. Qian, H_2SO_4-catalyzed isobutane alkylation under low temperatures promoted by long-alkyl-chain surfactant additives, AIChE J. 67 (2021) e17349.*

2.1.6.4 *Deep eutectic solvents as additives*

The deep eutectic solvent (DES) is a molten liquid composed of a hydrogen bond acceptor (HBA) and a hydrogen bond donor (HBD) via a hydrogen-bonding interaction, and thus the melting point is obviously lower than either individual pure component [38,39]. The DES exhibits excellent physical and chemical properties, such as low vapor pressure, relatively wide liquid range, nonflammable, conductivity, and nontoxicity, sharing similar properties as ILs [39]. DES is a new class of ILs' analogs, but have more advantages than ILs, especially given its simple and green preparation process using cheap and widely available raw materials. Therefore, DES has received considerable attention over the past years and has been widely applied in several research fields. For C4 alkylation, the novel DES consisting of trifluoromethanesulfonic acid (TfOH) and taurine (TAU), denoted as $[TfOH]_3[TAU]$, served as the catalyst in the alkylation process for generating high-octane alkylates. To overcome the long-standing issue of poor solubility of volatile and stubborn reactants, such as isobutane and isobutene, polyethylene glycol (PEG) is introduced as a solvent into the catalytic system, and PEG-200 effectively dissolves both $[TfOH]_3[TAU]$ and the reactants, resulting in a catalytic system with high catalytic activity, enhanced C_8-product selectivity as high as 85.54%, and excellent recyclability for the alkylate [40].

However, the utilization of DESs as additives in H_2SO_4-catalyzed C4 alkylation remains infrequent. The influence of DES additives, including choline chlorides-phenol (ChCl-Pho [1:2]), choline chlorides-*p*-toluenesulfonic acid (ChCl-TsOH [1:1]), and choline chlorides-benzoic acid (ChCl-BOA [1:2]), as shown in Fig. 2.26, on the catalytic performance and interfacial properties in H_2SO_4-catalyzed C4 alkylation was extensively examined using a combination of experiments and MD simulations [15]. Interestingly, the introduction of DESs has negligible effects on the reaction time. Among the DESs investigated, ChCl-Pho (1:2) demonstrates the most outstanding catalytic performance, yielding a RON of 97.37 and the C_8 selectivity of 83.64%. Optimal addition ratios for ChCl-Pho (1:2) and

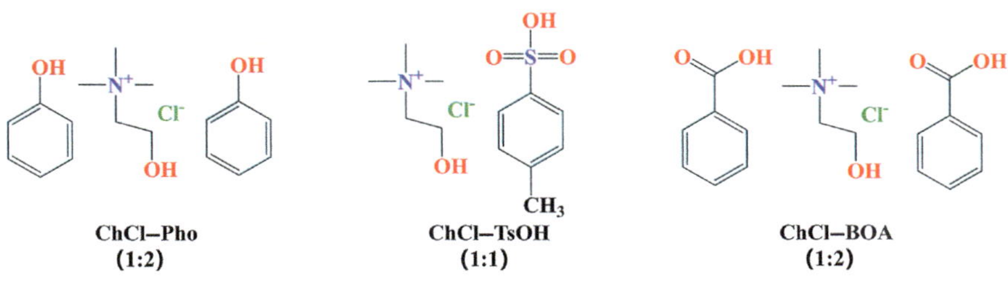

FIG. 2.26 Structures of different DESs. *From Z. Ma, J. Sha, W. Zheng, W. Sun, L. Zhao, Effects of deep eutectic solvents on H_2SO_4-catalyzed alkylation: combining experiment and molecular dynamics simulation, AIChE J. 68 (2022) e17556.*

TABLE 2.9 Temperature-dependent alkylate components for H_2SO_4 alkylation with ChCl-Pho (1:2) as additive.

Additive	Temp (K)	Alkylate composition (wt%)					RON
		$C_5 \sim C_7$	C_8	TMP	C_9^+	DMH	
Pure H_2SO_4	281.2	11.05	74.50	67.56	14.45	6.94	95.80
ChCl-Pho (1:2)	281.2	6.46	83.77	76.98	9.77	6.79	97.39
	277.2	6.07	85.65	79.45	8.28	6.20	98.05
	274.2	5.62	86.07	80.15	8.30	5.92	98.27
	269.2	3.82	90.91	85.83	5.27	5.08	99.43

Data from Z. Ma, J. Sha, W. Zheng, W. Sun, L. Zhao, Effects of deep eutectic solvents on H_2SO-catalyzed alkylation: Combining experiment and molecular dynamics simulation, AIChE J. 68 (2022) e17556.

ChCl-TsOH (1:1) to H_2SO_4 are found to be approximately 1.0wt%, resulting in a significant improvement in C_8 content. In the case of ChCl-Pho (1:2), the C_8 content increases from 74.50wt% in the pure H_2SO_4 system at 281.2K to 83.64wt%, with the corresponding increase in RON from 95.80 to 97.37. However, an excessive addition of DESs led to a reduction in the catalytic enhancement due to decreased H_2SO_4 acidity. ChCl-Pho (1:2) and ChCl-TsOH (1:1) have minimal effects on 2,3,4-TMPs and DMHs but significantly influences the conversion of light ends (LEs) and heavy ends (HEs) into TMPs. The addition of ChCl-Pho (1:2) notably increases the targeted C_8 components and RON at lower reaction temperatures, as listed in Table 2.9. However, the addition of ChCl-Pho (1:2) and ChCl-TsOH (1:1) does not extend the catalytic lifetime of the H_2SO_4 catalyst, although they can improve catalytic performance.

2.1.6.5 *Other additives*

To enhance the solubility of isobutane in H_2SO_4, trifluoroacetic acid (TFA) is used as a more efficient additive due to its favorable chemical properties, including excellent solubility in organic compounds, lower viscosity compared to H_2SO_4, and its ability to act as a buffer against strong inorganic acids [41]. Table 2.10 provides a comparison between TFA/H_2SO_4 and other composite catalysts with highlighted differences in reaction time required to achieve high C_8 selectivity. The reduction in reaction time with the TFA additive can be attributed to significantly increasing the solubility of isobutane and reducing the viscosity of the catalyst. Interestingly, the reaction could have achieved even better results at lower temperatures. The optimal conditions for the best outcome were identified as follows: temperature of 12°C, isobutane-to-olefin (I/O) ratio of 8mol/mol, hydrogen-to-carbon (H/C) ratio of 1.5v/v,

TABLE 2.10 Comparison of Li's work and the work of other researchers.

Catalyst	I/O	H/C	Temperature (°C)	Stirring speed (r/min)	Reaction time (min)	Selectivity of C_8 (%)
H_2SO_4	5	0.8	5	1000	10	68.4
H_2SO_4-[Bmim][PF_6]	5	0.8	5	1000	10	68.0
HFAU Zeolite	40		50		96	44.0
$C_{32}H_{68}P$-$2AlCl_3$	110		80		90	68.9
$H_2SO_4^+$ Caprolacatam	10	1.5	12	1000	12.5	39.8
H_2SO_4/TFSA	5.12		10		31.8	86.4
H_2SO_4/TFA	8	1.5	12	1000	3.5	90

Data from L. Li, J. Zhang, C. Du, G. Luo, Intensification of the sulfuric acid alkylation process with trifluoroacetic acid, AIChE J. 65 (2019) 113–119.

stirring rate of 1000 rpm, and the addition of 20 wt% TFA for a reaction duration of 3.5 min.

Graphene is a two-dimensional (2D) carbon allotrope, which can be viewed as both a solid and a macromolecule with molecular weights of more than 10^6–10^7 g/mol. In natural graphite, the graphene layers stick together through very pronounced π-π stacking interactions [42].

A series of carbon-based materials were examined as effective additives to enhance the product distribution in C4 alkylation catalyzed by H_2SO_4. Notably, among these materials, graphene oxide (GO) emerges as the most efficient additive, as shown in Fig. 2.27. The inclusion of 0.1 wt% of GO sheets results in an approximately 11% increase in the weight percentage of C_8 in the product and improves the RON of the alkylate by around 2 units. Interestingly, the introduction of GO into H_2SO_4 does not appreciably alter its acidity [43]. The enhancement in product distribution can be ascribed to the improved emulsification and better dispersion of the reactants within the acid medium. Furthermore, the lifespan of the catalytic system extends from 110 cycles (without any additive) to 140 cycles when the additive is present. This efficient and environmentally friendly emulsifier has the potential to be a particularly promising addition to the current H_2SO_4 alkylation process [43].

2.2 HF catalyst

HF is more dangerous since toxic aerosol clouds can be formed once it is released. For the HF process, several methods to reduce dangers have now been reported. Conoco-Phillips and UOP have developed the additives for HF in some of their units to reduce the dangers. For this purpose, Conoco-Phillips recommends sulfolane, whereas UOP has recommended pyridine or picolene. The additives can reduce the amount of HF in the aerosol clouds by 60%–90%. Despite the potential dangers, HF processes have several advantages compared to H_2SO_4 processes. The temperature employed in HF reactors is significantly higher, usually 30–40°C vs 5–10°C in H_2SO_4 processes. For HF processes, the costs to remove the reaction heats plus energy supplied by agitation are much lower. Cooling water can be used, whereas refrigeration is required for H_2SO_4 processes. Much more rapid kinetics occur in HF-type units. The Conoco-Phillips reactor is simply a vertical tube that has a residence time estimated to be about 10 s. For H_2SO_4 processes, residence time is in the range of approximately 20–30 min. Thus, the size and costs of the reactors for HF are much lower. In addition, distillation is employed in the refinery to regenerate used HF. Operating costs are relatively low. Used H_2SO_4 is usually shipped, however, to a nearby plant where sulfuric acid is produced. The costs of regenerating H_2SO_4 acid have often been as high as 25%–30% of the total operating costs.

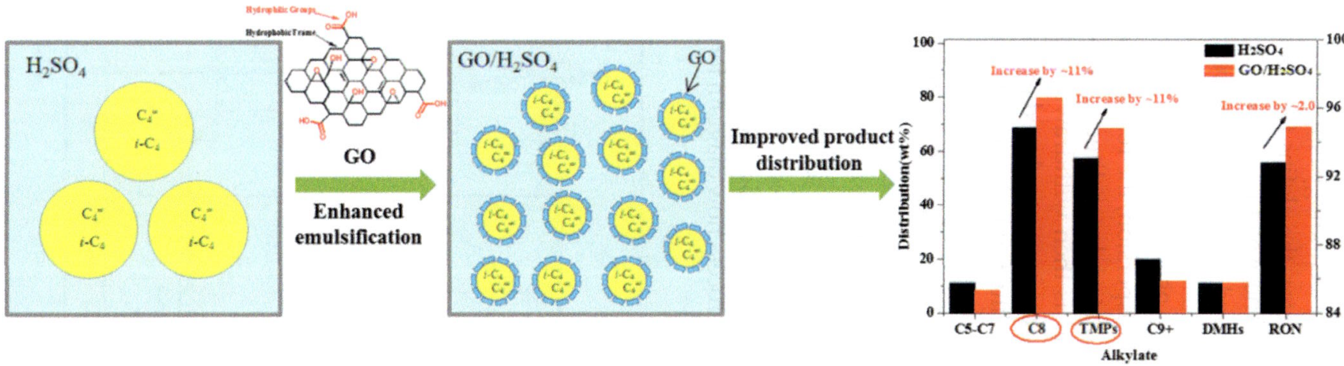

FIG. 2.27 The abstract graph of Meng's work. *From X. Meng, P. Prawang, Z. Li, H. Wang, S. Zhang, Carbon-based materials enhanced emulsification to improve product distribution in isobutane/butene alkylation catalyzed by sulfuric acid, Ind. Eng. Chem. Res. 56 (2017) 7700–7707.*

2.2.1 Conoco-Phillips reduced volatility process

Conoco-Phillips alkylation technology employs a noncooled riser reactor. Prior to entering the reactor, the olefins feed undergo dehydration to eliminate any excess water. These olefins are premixed with iso-paraffins, and feed nozzles are employed to introduce the reactant mixture into the reactor, both at the bottom and along its length. The complete process flow diagram is depicted in Fig. 2.28.

To ensure effective acid dispersion, the reactor is equipped with perforated trays. The produced materials are directed into the settler for hydrocarbon separation. The acid is then withdrawn and cooled at the lower section of the reactor before being reintroduced to control the temperature within the reactor. Fig. 2.29 illustrates the reactor utilized in this process. The standard operating temperature is approximately 24°C. The acid strength is typically maintained within the range of 89–92 wt%.

2.2.2 UOP Alkyplus HF alkylation process

UOP Alkyplus HF alkylation process is based on a concept similar to the Stratco contactor reactor, with the primary distinction lying in the orientation of the reactor. UOP employs a vertical reactor-heat exchanger configuration. Fig. 2.30 depicts the process flow diagram for the Alkyplus process.

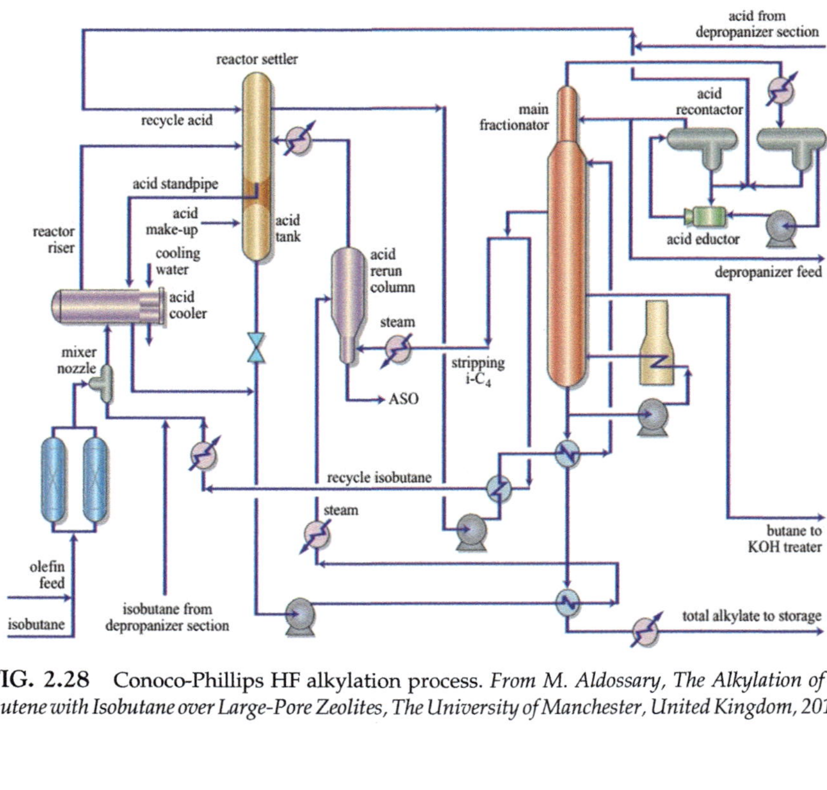

FIG. 2.28 Conoco-Phillips HF alkylation process. *From M. Aldossary, The Alkylation of 2-Butene with Isobutane over Large-Pore Zeolites, The University of Manchester, United Kingdom, 2018.*

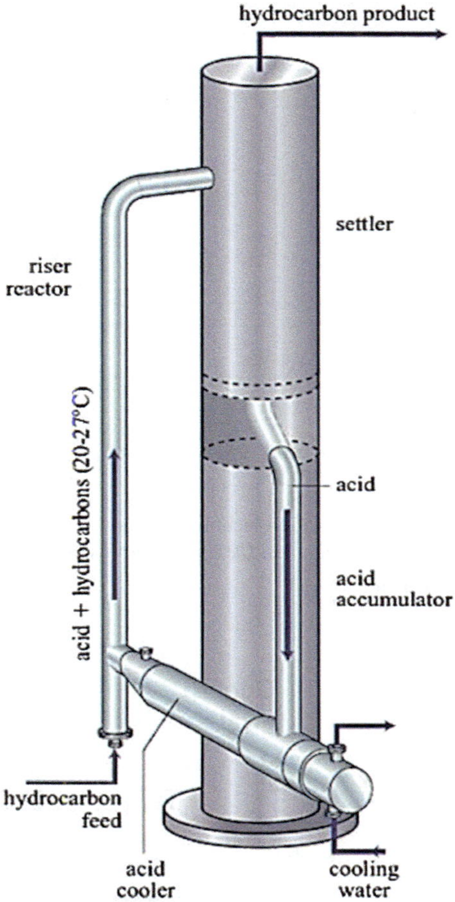

FIG. 2.29 Conoco-Phillips HF alkylation reactor. *From M. Aldossary, The Alkylation of 2-Butene with Isobutane over Large-Pore Zeolites, The University of Manchester, United Kingdom, 2018.*

The olefins and isobutane are mixed and the mixture enters a drying bed to remove excess water. The reactants enter the reactor shell through feed nozzles. The acid is introduced from the reactor bottom. Cooling water flows through coils inside the reactor to control the temperature and remove any excess heat. The reactor schematic is shown in Fig. 2.31.

The reactor effluent proceeds into a settler where the emulsion is separated. Acid is extracted from the settler and cycled back to the reactor. Subsequently, the hydrocarbons are directed to the main fractionator, where alkylates are collected at the bottom. A side stream, containing isobutane and HF, is withdrawn from the fractionator and returned to the reactor. No external acid regeneration is necessary; the acid is distilled along with the product during routine operations.

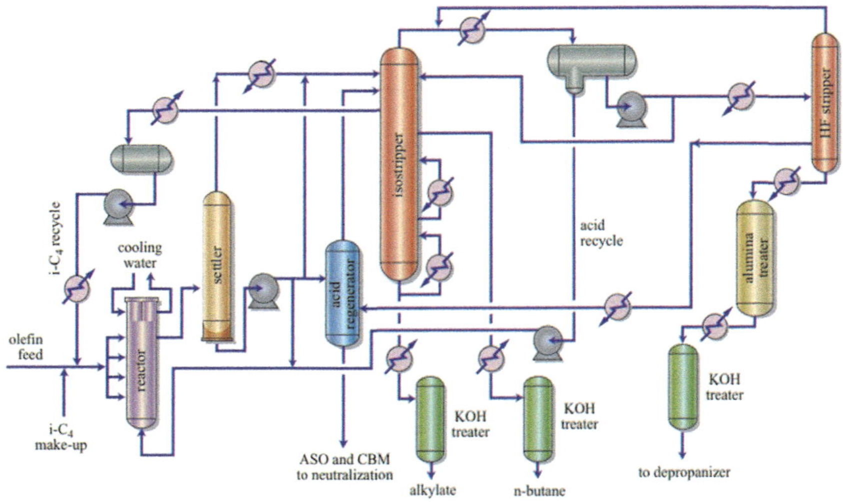

FIG. 2.30 UOP Alkyplus HF alkylation technology. *From M. Aldossary, The Alkylation of 2-Butene with Isobutane over Large-Pore Zeolites, The University of Manchester, United Kingdom, 2018.*

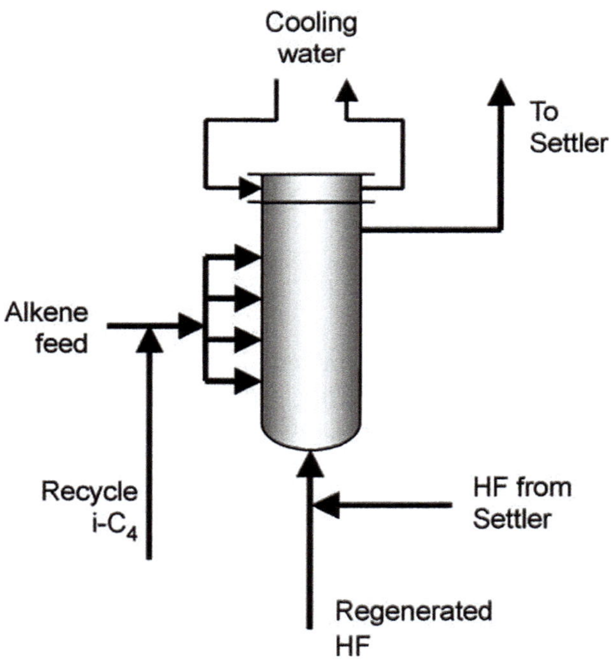

FIG. 2.31 UOP Alkyplus HF alkylation reactor. *From M. Aldossary, The Alkylation of 2-Butene with Isobutane over Large-Pore Zeolites, The University of Manchester, United Kingdom, 2018.*

The HF alkylation process necessitates the installation of costly mitigation systems. These systems encompass intricate acid leak detection systems and water spray towers. Some technology licensors show some process enhancements, such as the UOP/Chevron ALKAD process, which employs an additive based on the amine salts of HF to create complexes. Additionally, Phillips/ExxonMobil proposes the use of sulfones to reduce the generation of HF aerosols during potential leaks.

2.2.3 Operating parameters

2.2.3.1 Effect of olefin type

The HF alkylation utilizes isobutane and light olefins as the primary raw materials. Isopentane alkylation is not commonly employed in industry due to the low octane number of the alkylate product compared to the isopentane itself (RON = 93). The light olefins employed in alkylation are primarily composed of C_3–C_5 olefins. Mixed butenes are predominantly utilized in industrial processes. It should be noted that the obtained products from the alkylation reaction involving different olefins and isobutane display significant variations in the octane numbers under specific reaction conditions. The temperature of 25–30°C and alkane/olefin ratio of 10–20, as demonstrated in Table 2.11, were used.

From Table 2.12, the alkylation of isobutane and propylene produces significant amounts of TMPs under varying alkane/olefin ratios, with even large excesses of isobutane. The increase in alkane/olefin ratio results in higher RON, but a sharp increase in isobutane consumption indicates that a significant amount of isobutane converted to isobutene and participated in the reaction. The elevated alkane/olefin ratio increases the probability of hydrogen ion provided by isobutane and decreases the chances of polymerization, resulting in lower contents of HEs in the reaction product and lower dry point of the alkylate. For instance, as the

TABLE 2.11 The RON alkylate obtained from different olefins.

Olefins	RON
Propene	91–93
Mixed butenes	94–96
Pentene	91–92
Mixed hexenes	81–83
Mixed propene and butenes	93–94

Data from B.L. Gao, High Octane Gasoline Component Production Technology, China Petrochemical Press, Beijing, 2006.

TABLE 2.12 The impact of I/O ratio on the properties of propylene alkylate.

$iC_4/C_3^=$	5	20	70
Components of alkylate (%)			
iC_5	5.0	4.0	2.3
iC_7	54.5	49.3	43.8
TMPs	20.0	34.0	47.5
DMHs	5	3.2	3.2
C_9^+	15.5	9.5	3.2
Yield (volume ratio of propylene)	1.66	1.72	1.88
Consumption of isobutane (volume ratio of propylene)	1.20	1.36	1.88
Properties of alkylate RON	88.7	91.6	93.2
Density (20°C) (kg/L)	0.700	0.7008	0.700
Dry point (°C)	213	179	164
Relative molecular weight	107.5	107.5	107.3

Data from B.L. Gao, High Octane Gasoline Component Production Technology, China Petrochemical Press, Beijing, 2006.

alkane/olefin ratio increases from 5 to 70, the TMP contents increase from 20% to 47.5% and HE components decrease from 15.5% to 3.2%.

2.2.3.2 *Effect of butene type*

Butene is the major olefinic feedstock for the alkylation process, which is an important route to produce high RON gasoline blending components. Isobutene, 1-butene, *cis*-2-butene, and *trans*-2-butene are four isomers of butene that can also be used as olefinic feedstocks for alkylation. Under high I/O ratio conditions, the composition of alkylate obtained from different butenes is shown in Table 2.13. These results provide valuable information for optimizing the alkylation process and controlling the quality of the alkylate product.

From Table 2.13, the alkylate generated through the alkylation of *cis*- and *trans*-2-butene exhibits a higher content of TMPs, a lower proportion of HEs, and a lower dry point, followed by isobutene, while 1-butene is the least desirable. Consequently, for HF alkylate processes, it is advisable to maximize the utilization of the 2-butene component present in the available resources. To enhance the quality of the alkylate, a selective hydrogenation process can be employed to remove butadiene in the C4 feedstock. Simultaneously, the majority of 1-butene can be isomerized into 2-butene. This strategic approach allows for an improvement in the quality of the alkylate.

TABLE 2.13 Compositions of alkylate from different olefinic feedstock.

Butene	1-Butene	*cis*-2-Butene	*trans*-2-Butene	Isobutene
I/O ratio	11	12	14	13
Alkylate components (wt%)				
Isopentane	3.31	1.84	1.79	5.44
N-pentane	0.01	0.07	0	0.05
2-Methylpentane	1.47	1.37	1.39	2.88
3-Methylpentane	0.18	0.12	0.13	0.23
2,4-Dimethylpentane	1.42	1.30	1.30	2.39
2,2,3-Triptane	0.03	0	0	0
2-Methylhexane	0.13	0.14	0.05	0.13
2,3-Dimethylpentane	0.72	0.63	0.68	1.18
3-Methylhexane	0.08	0.20	0.05	0
2,2,4-TMP	41.30	48.71	48.93	50.21
2,5-DMH	3.75	2.09	2.10	3.26
2,4-DMH	6.41	2.87	2.88	3.56
2,2,3-TMP	1.38	1.13	1.13	1.07
2,3,4-TMP	12.77	21.39	20.93	10.69
2,3,3-TMP	10.38	12.92	12.86	9.35
2,3-DMH	7.61	2.14	2.20	1.78
2-Methylheptane	0.18	0	0	0
3,4-DMH	1.01	0.15	0	0.21
2,2,5-Trimethylhexane	0.66	0.10	0.05	1.31
Heavy ends	7.02	2.83	2.75	6.26
Properties of alkylate RON	94.4	97.8	97.6	95.6
Dry point (°C)	180.6	164.4	163.9	177.2

Data from B.L. Gao, High Octane Gasoline Component Production Technology, China Petrochemical Press, Beijing, 2006.

2.2.3.3 *Effect of impurities*

The quality of the feedstock is of paramount importance in HF alkylation, requiring strict control over impurity levels. These impurities encompass water, sulfur compounds, diolefins, and more. If there exists an upstream MTBE unit, the impurities can also include compounds such as tert-butanol and dimethyl ether. Empirical evidence suggests that the majority of issues encountered during the operation of alkylation units can be traced back to the quality of the feedstock.

(1) H_2O

The water content in the feedstock serves as a critical control parameter in HF alkylation. Under typical conditions, HF exists in a complexed state, while water is polar. When introduced into HF, water facilitates its dissociation, thereby augmenting the catalytic activity of HF. It is imperative to recognize that in the absence of water, HF lacks acidity, which leads to a corresponding absence of catalytic activity.

As the water content in the acid increases, the catalytic activity of HF gradually intensifies. However, once the water content reaches a certain threshold, the catalytic activity of HF declines [44]. When the water content reaches 10%, olefins react with HF to form organic fluorides, and even in the presence of a significant amount of isobutane, the alkylation cannot occur. Additionally, when the water content exceeds 5%, it exacerbates the corrosion of the equipment, resulting in frequent shutdowns and maintenance of the unit [6,45]. The influence of water content in HF on the composition of alkylate is outlined in Table 2.14.

TABLE 2.14 Effect of water content on the composition of alkylate.

	Composition of alkylate (%)				
Mass fraction of water (%)	C_5	C_6	C_7	C_8	C_9^+
0.25	10.39	4.94	17.76	48.75	18.16
0.43	5.5	3.83	19.43	58.30	12.87
0.6	4.57	2.89	19.16	60.60	12.79
0.9	3.62	2.63	24.29	60.39	9.67
1.4	2.89	2.35	23.29	62.83	8.63
2.0	3.51	2.54	36.35	60.10	7.50
2.8	3.72	2.88	25.48	61.43	6.49
4.4	4.28	3.44	28.40	56.56	7.32

The reaction raw materials are propylene 42.5%, butene 57.2%, and pentene 0.3%.
Data from B.L. Gao, High Octane Gasoline Component Production Technology, China Petrochemical Press, Beijing, 2006.

From Table 2.14, as the water content increases, the content of low-octane C_9^+ heavy fractions significantly decrease. The lowest C_9^+ heavy fraction content is observed at a water content of 2.8%, while the ideal component C8 content is highest at a water content of 1.4%. Generally, the ideal range for water content in HF is considered to be between 1% and 3%, within which the octane number of the alkylate undergoes minimal variation.

To mitigate the introduction of water-bearing components into the alkylation feedstock, HF alkylation units are equipped with drying processes [2,46]. For the assurance of consistent and dependable drying, it is customary to maintain two sets of dryers, filled with 3A and 4A molecular sieves, or alumina as desiccants. One set operates while the other stands by for regeneration. In cases where the raw material contains an excessive amount of water, it is permissible to temporarily connect the two dryer sets in series. Typically, the water content in the feedstock is controlled to be less than 10 μg/g after the drying process.

Molecular sieve desiccants yield higher levels of drying effectiveness, despite the requirement of higher regeneration temperatures (>220°C). In contrast, alumina desiccants are more cost-effective and require lower regeneration temperatures (>170°C), although their adsorption capacity and dehydration efficiency do not match that of molecular sieves [6]. The regeneration cycle for desiccants typically ranges from 1 to 5 days. In order to assess desiccant performance and ensure the dehydration efficacy of the alkylation feedstock entering the reaction section, the installation of online moisture analyzer postdryers is recommended. Real time monitoring via an online moisture analyzer enables the extension of the desiccant regeneration cycle and a reduction in regeneration-related consumption. The indication for replacing desiccants with fresh ones becomes apparent after approximately 6–7 days of desiccant usage, characterized by a notable increase in pressure drop within the dryer due to impurity blockage and other factors.

(2) Sulfides

Sulfides represent a prevalent class of impurities within alkylation feedstocks. The sulfide content within the raw material exerts a significant influence on the alkylation process, particularly regarding the production of acid-soluble oils and the consequential elevation in acid consumption [46]. The relationship between sulfur content in the feedstock and acid consumption is elucidated in Table 2.15. Notably, when the sulfur content in the alkylation feedstock remains below 20 μg/g, acid consumption by the unit remains within acceptable parameters. However, surpassing a sulfur content threshold of 50 μg/g in the raw material leads to a sharp and pronounced escalation

TABLE 2.15 The relationship between sulfur content in the feedstock and acid consumption.

Sulfur content of raw materials (μg/g)	Acid consumption (kg)
10	0.404
20	0.608
50	2.025
100	4.05
250	80% of Hydrofluoric acid in the device will be consumed in the short term

Data from B.L. Gao, High Octane Gasoline Component Production Technology, China Petrochemical Press, Beijing, 2006.

in acid consumption. When the sulfur content in the feedstock reaches 100μg/g, it becomes imperative to halt feedstock intake into the unit [6]. Apart from its impact on acid consumption, the presence of sulfides in the raw material can introduce undesirable qualities to the alkylate, including discoloration, odorous attributes, and the potential for foaming phenomena.

Therefore, it is imperative to operate the liquefied gas desulfurization and desulfurization alcohol removal units effectively to ensure that the sulfur content in liquefied gas remains below 20μg/g. In cases where issues arise within the desulfurization system, it is essential to divert inadequately desulfurized liquefied gas out of the system. If, inadvertently, this highly sulfur-containing portion of liquefied gas is introduced into the alkylation unit, it can result in the production of a substantial quantity of acid-soluble oil.

(3) Butadiene

The butadiene content in catalytic cracking liquefied gas typically resides at approximately 0.5%. However, factors such as elevated reaction temperatures in the catalytic cracking unit or the incorporation of substantial quantities of heavy residues into the catalytic cracking feedstock, can elevate the butadiene content. Elevated butadiene levels can detrimentally affect the economic feasibility of the unit, primarily because 1kg of butadiene yields 0.7–1kg of ASO, resulting in an additional HF consumption ranging from 0.5 to 20kg per 1kg of ASO [47]. Furthermore, an increased presence of butadiene in the raw material can lead to a rise in the dry point of the alkylate oil and an augmentation of the organic fluorides content within the circulating isobutane, propane, and butane products in the unit. Presently, both domestic and international practices predominantly employ selective hydrogenation as the method of choice for butadiene removal. This process concurrently isomerizes a significant proportion of 1-butene into 2-butene, a transformation that significantly enhances the octane number of the alkylate.

(4) Oxygen-containing compounds

In the presence of an MTBE unit upstream of the alkylation facility, the prospect of introducing novel impurities into the alkylation process, notably methanol, dimethyl ether (DME), and tert-butanol, must be carefully considered [6]. Methanol, within the alkylation unit, has the capacity to generate methane or water. Dimethyl ether, conversely, contributes to the production of lightweight acid-soluble hydrocarbons. The challenge lies in the separation of these lightweight acid-soluble hydrocarbons from the bottom of the acid regeneration tower, thereby resulting in a precipitous deterioration in the quality of the circulating acid.

To mitigate these concerns, the MTBE units should be engineered to operate with processes characterized by an alcohol-to-olefin ratio equal to or less than 1 [6]. This choice optimizes methanol utilization and forestalls the undue introduction of excess methanol into the alkylation unit. Moreover, it is imperative to meticulously manage the distillation system within the MTBE unit to curtail the ingress of diverse oxygen-containing compounds and impurities into the alkylation feedstock.

In the quest to diminish acid consumption and augment the yield of alkylate oil, conventional distillation techniques are often employed for the removal of dimethyl ether. The typical operational conditions and compositional variations for the dimethyl ether removal tower are comprehensively delineated in Tables 2.16 and 2.17.

2.2.3.4 *Effect of temperature*

The typical reaction temperature for HF alkylation ranges from 15°C to 50°C. The temperature of the circulating cooling water at the equipment is commonly employed as the reaction temperature. With the increase in reaction temperature, the reaction rate accelerates. However, it also leads to an increase in the formation of polymers and heavy components beyond C_8 components, resulting in an elevated dry point and a decreased RON of the product.

The reduction of the reaction temperature contributes to an enhanced RON of the alkylate, suggesting that the alkylating reaction at lower temperatures promotes the generation of higher octane-value components. Simultaneously, it reduces the occurrence of side reactions, inhibits the formation of ASO, and lowers the dry point of the alkylate. Excessively low reaction temperatures, such as below 15°C, can facilitate the

TABLE 2.16 The operating parameters of dimethyl ether removal tower.

Project	Design value	Calibration value
Tower top pressure (gauge pressure) (MPa)	0.65	0.66
Tower top temperature (°C)	50	54
Tower kettle temperature (°C)	56	59
Feed rate (kg/h)	7100	8270
Tower top discharge volume (kg/h)	76	214
Tower kettle discharge volume (kg/h)	7024	8056
C_4 yield (%)	99	98
Energy consumption (MJ/t)	280.2	511.2

Data from B.L. Gao, High Octane Gasoline Component Production Technology, China Petrochemical Press, Beijing, 2006.

TABLE 2.17 Composition changes before and after dimethyl ether removal.

Component	Before removing dimethyl ether	After removing dimethyl ether
$\omega(C_3)$%	0.23	0.13
$\omega(C_4)$%	97.78	99.37
Methanol (μg/g)	65	–
Dimethyl ether (μg/g)	1617	35

Data from B.L. Gao, High Octane Gasoline Component Production Technology, China Petrochemical Press, Beijing, 2006.

production of organic fluorides, which is undesirable. Generally, the reaction temperature is controlled at approximately 30°C for optimal results.

2.2.3.5 *Effect of ratio of acid to hydrocarbon*

It is commonly accepted that the C4 alkylation takes place at the interface between the acid and hydrocarbon phases. Therefore, ensuring a sufficient supply of HF and promoting thorough dispersion of hydrocarbons in the acid phase are crucial for establishing a sufficient acid-hydrocarbon interface. Typically, the acid acts as the continuous phase, while the hydrocarbon serves as the dispersed phase in this system. To achieve effective contact between the acid and hydrocarbon, it is essential to maintain the acid as the continuous phase. This is particularly important when the hydrocarbon is fully atomized. To accomplish this, a minimum requirement for the acid-hydrocarbon ratio should be met. If the acid-hydrocarbon ratio falls below 4:1, insufficient contact between the acid and hydrocarbon leads to decreased product quality and increased formation of by-products.

2.3 Ionic liquid catalyst

ILs are regarded as environmentally friendly catalysts, offering a greener alternative for producing alkylate, which serves as an efficient blending component to promote cleaner oil compositions. ILs have attracted significant research attention due to their tunable physicochemical properties. Concerns about safety and environmental impact associated with traditional acid catalysts, such as anhydrous HF and large amounts of H_2SO_4, in commercial C4 alkylation processes have prompted extensive exploration of IL catalysts. IL catalysts can be classified into different types, including: (1) metal-based Lewis acidic ILs, often chloroaluminates, (2) metal-based Brønsted-Lewis (bifunctional) acidic ILs, (3) nonmetal-based Brønsted acidic ILs, and (4) immobilized/supported ILs (Fig. 2.32). Among them, chloroaluminate ILs exhibit distinctive chemical properties due to their anionic speciation, which can be finely controlled by adjusting the mole fraction, resulting in unique properties such as low melting point, viscosity, acidity, and high conductivity. Their exceptional properties and tunability make them attractive candidates for

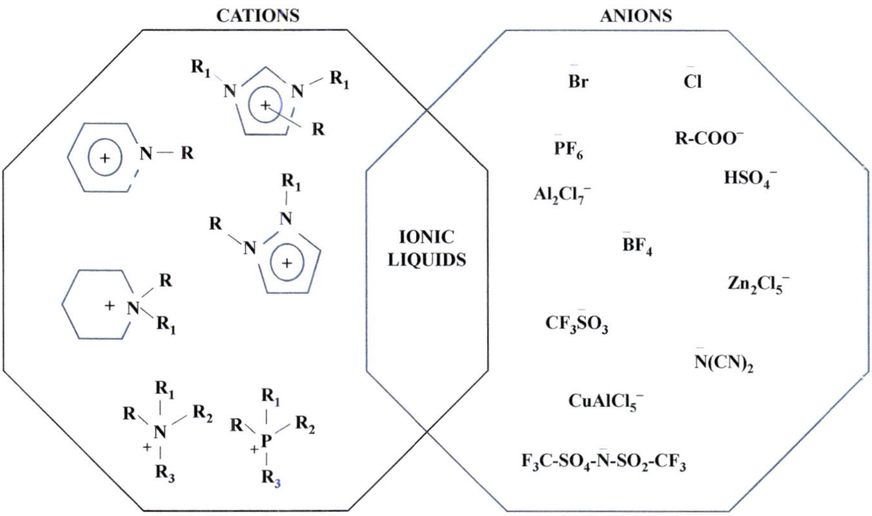

FIG. 2.32 Common cations and anions used in ILs. *Modified from D. Guzmán-Lucero, J. Guzmán-Pantoja, H.D. Velázquez, N.V. Likhanova, M. Bazaldua-Domínguez, A. Vega-Paz, R. Martínez-Palou, Isobutane/butene alkylation reaction using ionic liquids as catalysts. Toward a sustainable industry, Mol. Catal. 515 (2021) 111892.*

addressing safety and environmental concerns associated with conventional acid catalysts.

Significant progress has been achieved in the application of chloroaluminate ILs at room temperature, encompassing a wide range of processes such as catalysis, extractions, biocatalysis, and polymerization. The successful implementation of chloroaluminate ILs at room temperature opens up new possibilities for greener and more sustainable alkylation reactions.

Chauvin et al. firstly investigated the effective catalysis of C4 alkylation using 1-butyl-3-methylimidazolium chloroaluminate ([C_4mim]Cl-$AlCl_3$). Both the acidity and catalytic activity of ILs can be fine-tuned by adjusting the molar fraction of $AlCl_3$ (x). An increase in x from 0.55 to 0.60 results in a notable rise in the primary product, trimethylpentanes (TMPs), from 29.0% to 76.8%. Consequently, this leads to an enhancement in the RON of the alkylate [48]. However, further increase in the $AlCl_3$ molar fraction to 0.65 results in reduced TMP production and an increase in isopentane and DMHs. When $x = 0.65$, the acidity of catalyst becomes overly high with cracking reactions becoming predominant and an increase in the content of LEs. Therefore, to achieve desirable selectivity, the acidity of the catalyst needed to be carefully controlled. Another advantage of the IL is the limited solubility of the alkylate in the catalyst, simplifying the separation of the catalyst from the product [48].

Subsequently, Yoo et al. conducted an evaluation of the catalytic activity involving 1-alkyl-3-methylimidazolium halides ([$C_n$mim]$_X$)-aluminum chloride with different alkyl groups (butyl, hexyl, and octyl, corresponding to $n = 4, 6, 8$) and halides (Cl, Br, I) in the C4 alkylation [49]. The ILs with longer alkyl chain on the cation (i.e., [C_8mim]$^+$) exhibit relatively higher butene conversion compared to one with a smaller alkyl group (e.g., [C_4mim]$^+$ or [C_6mim]$^+$) with the same anion. This difference is attributed to the increased solubility of reactants in the ILs with longer alkyl chains. A critical criterion for evaluating alkylation catalyst performance is the TMPs/DMHs ratio. A higher value is desirable as it typically corresponds to higher alkylate quality. With the same anion, reactions catalyzed by [C_4mim]Br-$AlCl_3$, [C_6mim]Br-$AlCl_3$, and [C_8mim]Br-$AlCl_3$ yields comparable TMPs/DMHs ratios regardless of the alkyl chain length in cation. Furthermore, the alkyl chain length of the cation has a relatively minor impact on the weight percentage of C_8 products in the alkylate. The choice of anion also influences the catalytic performance. Among ILs with different anions, those containing bromide, such as [C_8mim]Br-$AlCl_3$, demonstrate higher catalytic activity and produce a greater amount of the desired TMPs due to their higher inherent acidity, as confirmed by Fourier transform-infrared spectroscopy (FT-IR) analysis. Modifying the molar fraction of $AlCl_3$ in the catalyst (x) leads to variations in the acidity of ILs. ILs with more acidic anions, like [C_8mim]Br-$AlCl_3$ with $x=0.58$, exhibit higher initial catalytic activity. As x increases, both the butene conversion and the TMPs/DMHs ratio also increase. However, over time, the catalytic activity gradually decreases (butene conversion drops from 70.4% at 10 min to 35.1% at 300 min). This deactivation is attributed to the moisture sensitivity and decomposition of the IL [49].

Encouraged by the fact that ether compounds exhibit higher solubility of isobutane, potentially preventing undesired side reactions like butene oligomerization during C4 alkylation, Zhang and colleagues synthesized a range of chloroaluminate ILs incorporating ether groups for catalyzing C4 alkylation, including 1-methoxyethyl 3-methylimidazolium bromide ([MOEmim]Br)-$AlCl_3$, 1-methoxyethyl-3-methyl-imidazolium bis(trifluoromethylsulfonyl) imide ([MOEmim][NTf_2])-$AlCl_3$, and *N*-methoxyethyl *N*,*N*,*N*-trimethylammonium bis(trifluoromethylsulfonyl) imide ([(MOE)Me_3N][NTf_2])-$AlCl_3$ [50]. The solubility of isobutane in these ILs is notably higher. These ILs with ether groups exhibit enhanced catalytic activity compared to those without ether bonds, which is attributed to the presence of suitable acidic sites for C_8 compound formation and the improved transfer of protons from isobutane to C_8^+ species, facilitated by the increased isobutane solubility in the catalyst. In terms of catalytic efficacy, ILs with ether groups followed the order in the alkylation process: [MOEmim][NTf_2]-$AlCl_3$ < [(MOE)Me_3N][NTf_2]-$AlCl_3$ < [MOEmim]Br-$AlCl_3$ ($x=0.75$). Under optimal reaction conditions, the resulting product achieves a weight percentage of 66.6% for C_8 content and RON of 93.0, which is comparable to the performance of H_2SO_4 (65.0% for C_8 content and 92.3 for RON). Furthermore, [MOEmim]Br-$AlCl_3$ can be recycled and reused. However, after the fifth run, the weight percentages of C_8 and RON decrease to 62.2% and 88.0, respectively. This reduction in catalytic activity can be attributed to the formation of acid-soluble compounds and the decomposition of the IL when exposed to air during the separation process of oil from the catalyst [50].

Liu et al. made a study regarding the influence of CuCl content in the catalytic system. When the molar ratio of CuCl/$AlCl_3$ reaches 0.29, there is a significant enhancement in the production of TMPs in comparison to the ILs without CuCl or the IL with the CuCl/$AlCl_3$ ratio of 0.23. This notable improvement can be primarily attributed to the presence of $[AlCl_4CuCl]^-$, as evidenced by FT-IR, electrospray ionization-mass spectroscopy (ESI-MS), and ^{27}Al nuclear magnetic resonance (^{27}Al NMR) analyses, within the catalytic system. Importantly, this compound is absent in the [Et_3NH]Cl-$AlCl_3$/CuCl system when CuCl/$AlCl_3$ is at 0.23. Consequently, the composite anion plays a pivotal role in enhancing the quality of the alkylate. Under the most favorable reaction conditions, the proportion of TMPs in the resulting product can be elevated to 92.0%, and the RON of the alkylate reaches an impressive 99.5. Furthermore, this composite IL system exhibits higher selectivity for TMPs compared to H_2SO_4 under the same reaction conditions [51].

Zheng et al. focused on the liquid/liquid interfacial behaviors between the binary reactant mixtures (equimolar trans-2-butene and isobutane, also called C4 mixtures) and ILs with various alkyl groups on the cation and different types of anions. According to the MD simulation, the ILs with longer chains on cations exhibit a significant density enrichment of alkyl chains at the interface and tend to orient themselves with alkyl chains perpendicular to the interface and protruding into the reactant phase, in good agreement with the van der Waals energy between the reactants and cations of the ILs [52]. As shown in Table 2.18, the interface width calculated from the H_2SO_4 or IL phase density profiles and the oil phase (the reactant mixture phase) density profiles is presented. Compared to H_2SO_4, ILs with longer chains could obviously improve the interfacial width, solubility, and diffusion of reactants at the interface, in which reactants are much easier to disperse into the IL phase with a lower interface tension.

With the investigation of ILs with various anions corresponding to different acidity using MD simulations, the equilibrated snapshots of MD simulations for IL-binary hydrocarbon mixture systems are presented in Fig. 2.33. The well-defined interface can be found between both phases. As expected, the hydrocarbon mixture phase can be noticed to be dissolved in the IL phase. Nevertheless, ILs display negligible solubility in the hydrocarbon mixture phase. Additionally, the influence of the anion of ILs on the mixing of both phases is noteworthy. In contrast to the

TABLE 2.18 The interface composition and width of different simulation systems.

	Mass fractions (%)				Width (oil) (nm)	Width (acid) (nm)
System	Cation	Anion	2-C_4H_8	i-C_4H_{10}		
H_2SO_4-C4 mixtures	–	–	9.84	8.16	0.587	0.688
[EMim][$AlCl_4$]-C4 mixtures	27.73	43.43	15.59	13.25	0.703	0.636
[BMim][$AlCl_4$]-C4 mixtures	32.52	35.71	17.33	14.44	0.797	0.661
[HMim][$AlCl_4$]-C4 mixtures	33.80	32.94	18.05	15.22	0.851	0.721
[OMim][$AlCl_4$]-C4 mixtures	35.11	26.08	20.53	18.28	0.984	0.821

Data from W. Zheng, H. Wang, W. Xie, L. Zhao, W. Sun, Understanding interfacial behaviors of isobutane alkylation with C4 olefin catalyzed by sulfuric acid or ionic liquids, AIChE J. 64 (2018) 950–960.

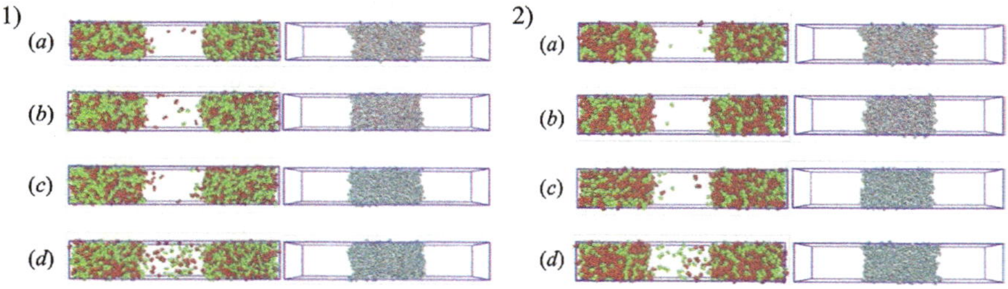

FIG. 2.33 The equilibrated snapshots of MD simulation for different systems. (a) [Bmim][PF_6], (b) [Bmim][BF_4], (c) [Bmim][$AlCl_4$], and (d) [Bmim][Al_2Cl_7]. For clarity, the IL molecules are removed from the left simulation box, and hydrocarbon molecules are removed from the right simulation box. (1) The IL-C4 mixture systems. (2) The IL-C8 mixture systems. 2-butene or isooctane molecules are shown in *red*, and isobutane molecules are shown in *green*. IL molecules are shown on the right side. *From W. Zheng, W. Xie, W. Sun, L. Zhao, Modeling of the interfacial behaviors for the isobutane alkylation with C4 olefin using ionic liquid as catalyst, Chem. Eng. Sci. 166 (2017) 42–52.*

neutral ILs, acidic ones facilitated a more efficient dispersion of binary hydrocarbon mixtures into the IL phase, exhibiting a mole fraction approximately 4 times higher in the bulk and approximately 1.2 times higher at the interface. The variation in interfacial properties among different hydrocarbon-IL systems aligns well with the interaction energy involving both the hydrocarbons and the cations and anions of the ILs [53].

Furthermore, with the deeper investigation of the solubility and diffusion in the 1-alkyl-3-methylimidazolium ILs with various anions, Zheng et al. predicted isobutane solubility in ILs follows the order of [BMIm][Al_2Cl_7] > [BMIm][$AlCl_4$] > [BMIm][Tf_2N] > [BMIm][PF_6] > [BMIm][BF_4] for ILs with different anions, and [OMIm][$AlCl_4$] > [HMIm][$AlCl_4$] > [BMIm][$AlCl_4$] for ILs with different alkyl chain lengths, as shown in Fig. 2.34. The increased solubility of isobutane can be attributed to a combination of factors, including weaker cation-anion interactions, lower solubility parameter, and larger free volume. The diffusion of isobutane follows a consistent pattern with solubility for ILs featuring different anions, but exhibits an opposing trend for ILs with varying alkyl chain lengths.

In addition, Zheng et al. further conducted multiscale modeling of C4 alkylation using composite ILs as catalyst [54]. The kinetics of C4 alkylation catalyzed by composite ionic liquids (CILs) was examined, and the rate constant associated with the formation of TMPs in CIL alkylation is more than two orders of magnitude greater than that in H_2SO_4 alkylation. The improvement in the rate constants for CIL alkylation at the macroscopic scale can be attributed to the enhanced solvation and diffusion of isobutane in the CIL at the microscopic scale, which is supported by the

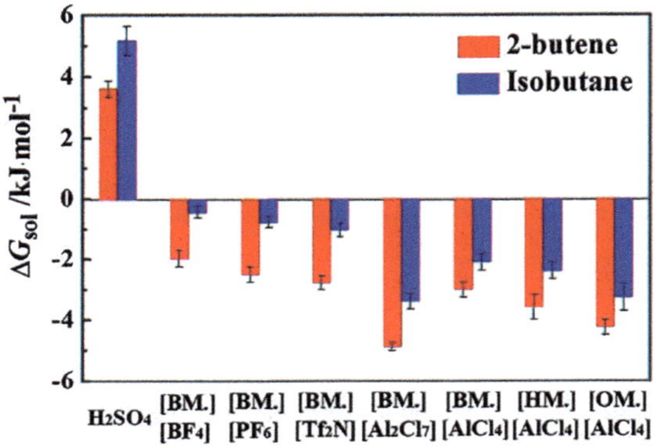

FIG. 2.34 Simulated solvation free energy of isobutane and 2-butene in ILs at 300 K and 1 bar. *From W. Zheng, L. Zheng, W. Sun, L. Zhao, Screening of imidazolium ionic liquids for the isobutane alkylation based on molecular dynamic simulation, Chem. Eng. Sci. 183 (2018) 115–122.*

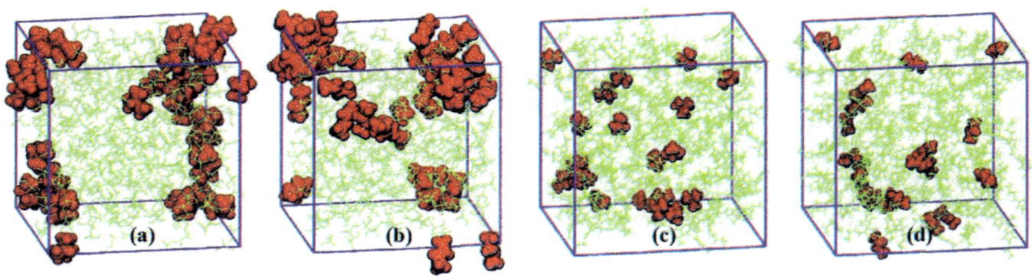

FIG. 2.35 Different equilibration snapshots for the diffusion systems. (A) H_2SO_4-isobutane system; (B) H_2SO_4-2-butene system; (C) IL-isobutane system; and (D) IL-2-butene system. C4 hydrocarbons are in *red*. ILs and H_2SO_4 are in *green*. *From W. Zheng, D. Li, W. Sun, L. Zhao, Multi-scale modeling of isobutane alkylation with 2-butene using composite ionic liquids as catalyst, Chem. Eng. Sci. 186 (2018) 209–218.*

calculation of solvation free energy and diffusion coefficient using MD simulation (shown in Fig. 2.35).

Recently, Zhang et al. reported that an aluminum chloride-based IL medium can promote the breakdown of these polyolefin plastics into liquid fuels at temperatures below 100°C [55]. Polyethylene and polypropylene exhibit commendable properties, yet their resistance to chemical degradation poses a formidable challenge in the context of waste management. Particularly, the exigency of subjecting these polymers to elevated temperatures for the purpose of carbon-carbon bond cleavage results in a pronounced energy burden. The method compensates for the energy directed toward bond scission with energy released from alkylation reactions among the resulting intermediates (Fig. 2.36).

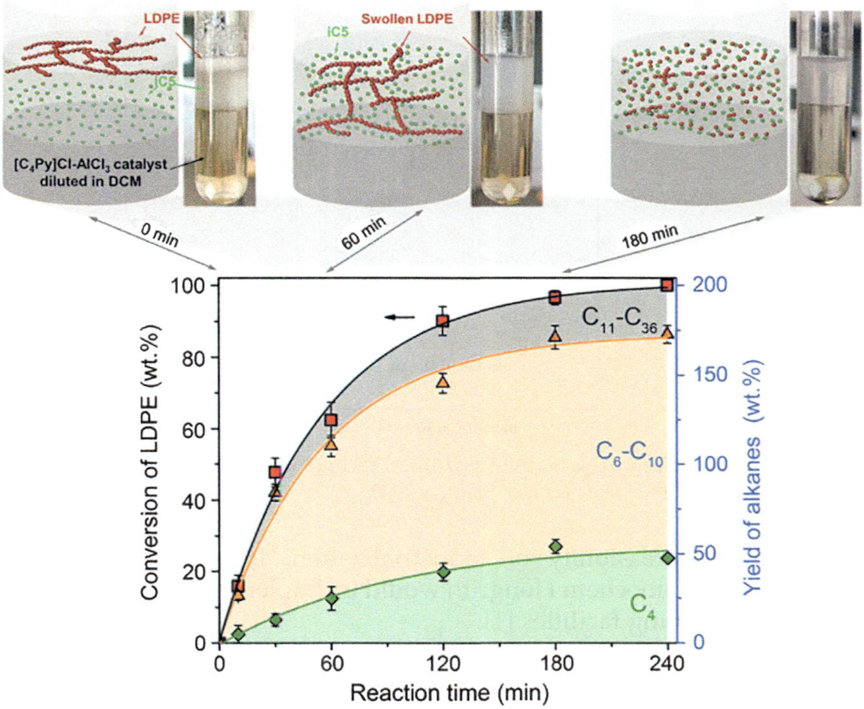

FIG. 2.36 The change in product distribution as a function of reaction time. *From W. Zhang, S. Kim, L. Wahl, R. Khare, L. Hale, J. Hu, D.M. Camaioni, O.Y. Gutiérrez, Y. Liu, J.A. Lercher, Low-temperature upcycling of polyolefins into liquid alkanes via tandem cracking-alkylation, Science, 379 (2023) 807–811.*

2.3.1 CIL alkylation process

The world's first industrial-scale production facility (the CILA process) for the C4 alkylation using composite ILs was established with a capacity of 10 wt/a [1]. The alkylate produced in this process has an octane number exceeding 97, and the catalyst equivalent consumption is as low as 5 kg/ton of alkylate. The energy consumption is 157 kg of standard oil per ton of alkylate. CILA technology employs a composite IL system consisting of acidic chloroaluminate ILs and metal additives (Fig. 2.37). This system effectively suppresses isomerization and side reactions, enhancing the yield and selectivity of the alkylate. Compared to current HF and H_2SO_4 alkylation processes, the IL alkylation process produces products with higher RON.

In 2016, UOP purchased the ISOALKY IL alkylation technology. Prior to this, Chevron conducted 5-year operational trial of small-scale demonstration unit for this process at its refinery in Salt Lake City, Utah,

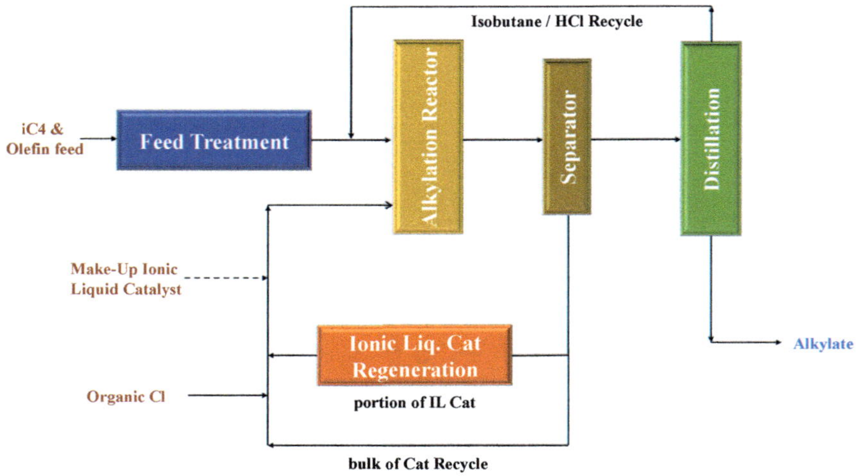

FIG. 2.37 Overall process scheme of ISOALKY Technology. *Modified from M.B. Shiflett, Commercial Applications of Ionic Liquids, Springer 2020.*

confirming the feasibility of industrialization. In April 2019, UOP announced that Sinochem Hongrun would be implementing this technology in their refining facilities [1].

2.3.2 Operating parameters

2.3.2.1 *Effect of impurities in feed*

Various impurities found in hydrocarbon feeds can influence the alkylation process. Common hydrocarbons used for alkylates include C4 raffinate from MTBE production, typically containing trace amounts of contaminants like water, methanol, MTBE, and *n*-butyl mercaptan. In this study, the impact of these four contaminants on the performance of the $[N_{222}H][CF_3SO_3(CF_3SO_3H)_2]$ catalyst was examined, as depicted in Fig. 2.38. As illustrated in Fig. 2.38, water, when present as an impurity, has a negligible effect on the catalytic activity of the $[N_{222}H][CF_3SO_3(CF_3SO_3H)_2]$ system, as long as its weight fraction remains below 10 mg/g. In fact, even trace amounts of water (<1 mg/g) appear to enhance catalyst activity, possibly by improving its protonation capability. However, when the weight fraction of water exceeds 10 mg/g, the RON and C8 selectivity sharply decrease, while the C_9^+ fraction significantly increases. Clearly, the water content of 10 mg/g marks the turning point for catalyst performance of the $[N_{222}H][CF_3SO_3(CF_3SO_3H)_2]$ and is therefore considered the upper limit for water accumulation in the catalyst. Similarly, the upper accumulation limits for methanol (CH_3OH), MTBE, and *n*-butyl mercaptan (n-C_4H_9SH) in the catalyst are 5, 20, and 5 mg/g, respectively [56] (Table 2.19).

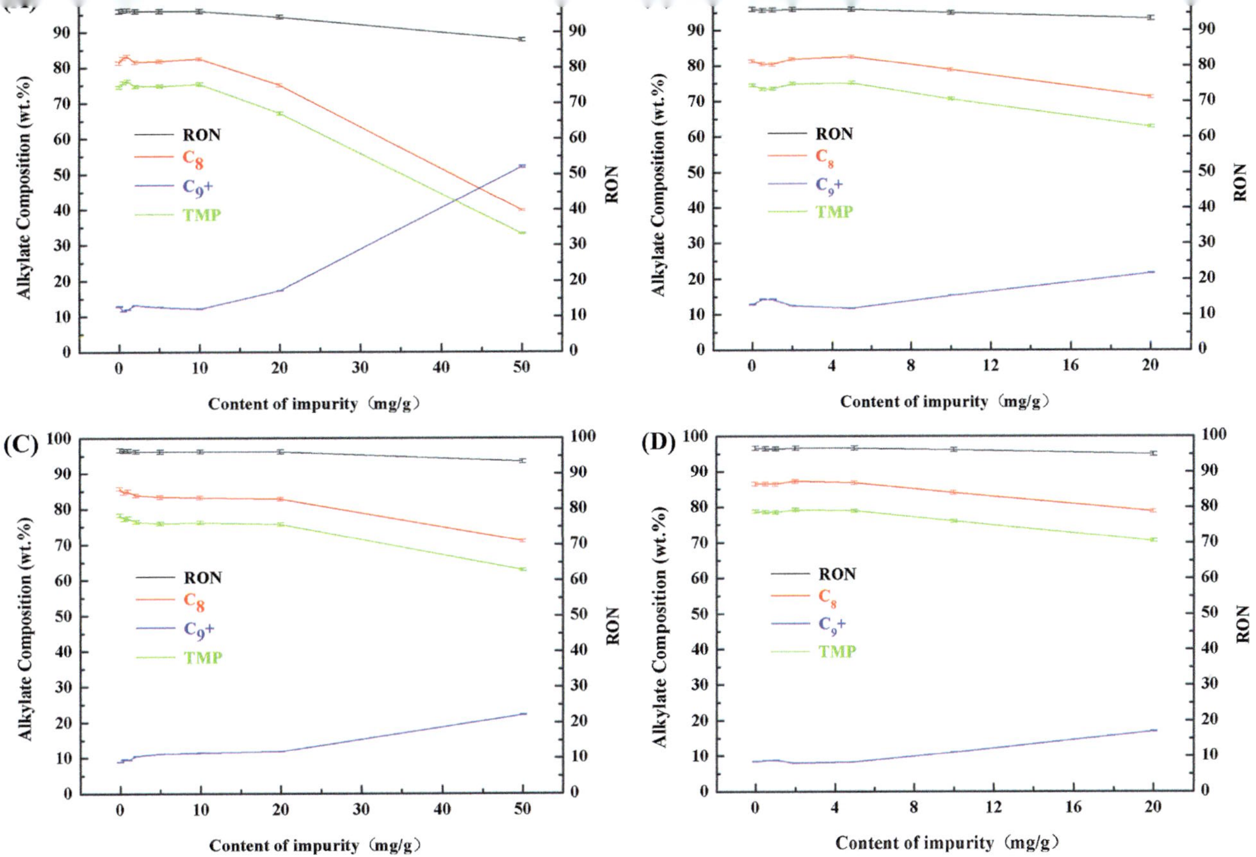

FIG. 2.38 Effect of H_2O contents (A), methanol contents (B), MTBE contents (C), and *n*-butyl mercaptan contents (D) in the catalyst on the alkylation reaction. Reaction conditions: catalyst 40 mL, feed rate 500 mL/h, χTfOH 0.69, acid/hydrocarbon 30/50, 5°C, 10 min. *From A. Wang, G. Zhao, F. Liu, L. Ullah, S. Zhang, A. Zheng, Anionic clusters enhanced catalytic performance of protic acid ionic liquids for isobutane alkylation, Ind. Eng. Chem. Res. 55 (2016) 8271–8280.*

TABLE 2.19 Threshold value of impurities in the catalyst.

Impurities	Quantity (mg/g)
H_2O	<5
CH_3OH	<5
MTBE	≤20
n-C_4H_9SH	<5

Data from A. Wang, G. Zhao, F. Liu, L. Ullah, S. Zhang, A. Zheng, Anionic clusters enhanced catalytic performance of protic acid ionic liquids for isobutane alkylation, Ind. Eng. Chem. Res. 55 (2016) 8271–8280.

2.3.2.2 Effect of reaction temperature

Increased reaction temperature has a negative impact on the selectivity of TMPs and the TMP/DMH ratio. Chauvin et al. found the optimized temperature is in the range of −5°C to 15°C, although the product distribution at −15°C is slightly inferior compared to that at −5°C [57]. This discrepancy can be attributed to the increased viscosity of the IL at −15°C. In the presence of the chloroaluminate IL catalyst and 1.0 wt% benzene additive, Zhang and colleagues have found that the TMP/DMH ratio in the alkylate increases with decreasing reaction temperature. C4 alkylation is an exothermic process, and therefore, lower temperatures promote a balanced distribution of TMPs. Moreover, lower temperatures inhibit side reactions, favoring the formation of C_8 and TMPs. Both the product distribution and the RON of the alkylate exhibit favorable characteristics at 0°C [48].

In order to study systemically the effect of temperature on C4 alkylation, Zhang et al. first calculated the Gibbs free energy of the reaction and equilibrium constants under 0.4 MPa, at different temperatures [58]. The six key alkylation reactions were chosen for simplifying the thermodynamic calculation and analysis:

$$i - C4^{\circ} + C4^{=} \rightarrow 2,2,4 - \mathrm{TMP} \tag{2.2}$$

$$i - C4^{\circ} + C4^{=} \rightarrow 2,2,3 - \mathrm{TMP} \tag{2.3}$$

$$i - C4^{\circ} + C4^{=} \rightarrow 2,3,3 - \mathrm{TMP} \tag{2.4}$$

$$i - C4^{\circ} + C4^{=} \rightarrow 2,3,4 - \mathrm{TMP} \tag{2.5}$$

$$i - C4^{\circ} + C4^{=} \rightarrow 2,4 - \mathrm{DMH} \tag{2.6}$$

$$i - C4^{\circ} + C4^{=} \rightarrow 3 - \mathrm{MH} \tag{2.7}$$

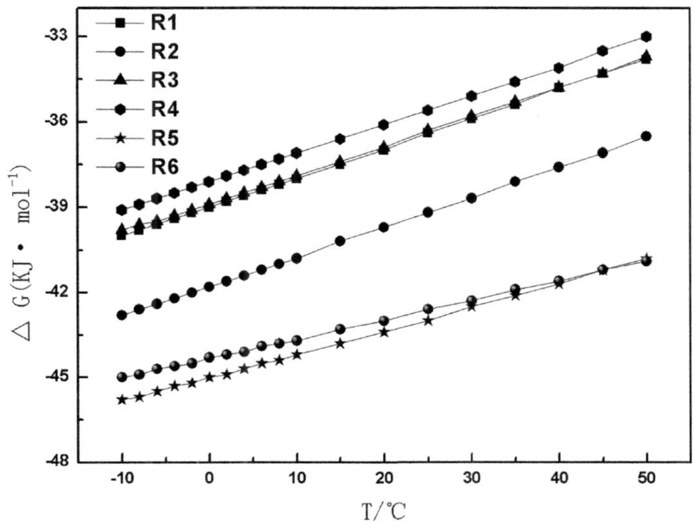

FIG. 2.39 Gibbs free energy change trend of alkylation at different temperatures under 0.4MPa. *From A. Wang, G. Zhao, F. Liu, L. Ullah, S. Zhang, A. Zheng, Anionic clusters enhanced catalytic performance of protic acid ionic liquids for isobutane alkylation, Ind. Eng. Chem. Res. 55 (2016) 8271–8280.*

where TMP represents trimethylpentane, DMH represents dimethylhexane, and MH represents methylheptane. The Gibbs free energy data of all pure components at different temperatures and 0.4MPa were collected from the Aspen properties with the Aspen Plus program.

Based on the data presented in Fig. 2.39, the calculated ΔG for each independent reaction at varying temperatures is negative. This negativity indicates that the alkylation reactions occur spontaneously within the investigated temperature range. However, as the temperature increases, ΔG exhibits a rising trend. This phenomenon implies that elevating the reaction temperature reduces the spontaneity of the alkylation reactions and the formation of TMP and DMH. From a thermodynamic perspective, a similar conclusion was reached that lower reaction temperature is favorable for the production of C_8 components.

2.3.2.3 *Effect of reaction time*

Compared to H_2SO_4 alkylation, the key difference is that C4 alkylation catalyzed by ILs occurs much faster, with reactions taking place on a timescale of seconds. Consequently, the increased solubility and diffusion of isobutane in the ILs lead to a more rapid hydride transfer from isobutane to C_8^+ ions. This, in turn, results in a shorter induction period and a reduced overall reaction time for C4 alkylation [59]. Fig. 2.40 illustrates the evolution of alkylate components and RON with respect to reaction

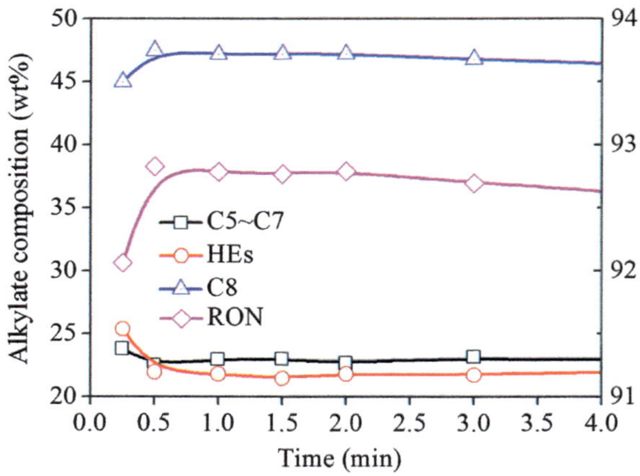

FIG. 2.40 Effect of reaction time on the alkylate components and RON. Reaction conditions: stirring rate 3000r/min, volume ratio of IL/hydrocarbon 1:1, volume ratio of I/O 8:1, temperature 288.2K. *From W. Zheng, D. Li, W. Sun, L. Zhao, Multi-scale modeling of isobutane alkylation with 2-butene using composite ionic liquids as catalyst, Chem. Eng. Sci. 186 (2018) 209–218.*

time. Evidently, the yields of C_5–C_7 and HE components experience a significant decline before 0.5min, after which they exhibit a slight increase. Conversely, the yield of C_8 components follows the opposite trend. Furthermore, the RON of the alkylate reaches its peak value before 0.5min, after which it slightly decreases with prolonged reaction time. This suggests that the optimal reaction time for catalytic C4 alkylation using ILs falls within the range of 30s [59,60].

2.3.2.4 *Effect of acid strength*

The acidity of ILs plays a crucial role in alkylation. The widely accepted mechanism of catalytic alkylation involves the formation of carbonium ions initiated by Brønsted acids. It is believed that the Brønsted acid species originate from the interaction of Lewis acid species with HCl, which is generated through the hydrolysis of chloroaluminate in the presence of trace amounts of water in the reactant. The nature of the Lewis acid centers influences the properties of the Brønsted acid sites. Qiao and Deng conducted experimental studies that confirmed the considerable acidity of chloroaluminate ILs. These specific ILs intrinsically exhibit characteristics of Lewis acids. The emergence of Brønsted acid properties occurs through the interaction between the Lewis acid components and either HCl or the proton from the Et_3NH^+ cation. The acidity of the resulting Brønsted acid within the IL is, in turn, affected by the properties of the Lewis acid components present [61].

As illustrated in Fig. 2.41, the ideal acidity range for $[N_{222}H][CF_3SO_3(CF_3SO_3H)_x]$ to achieve optimal alkylate production falls between

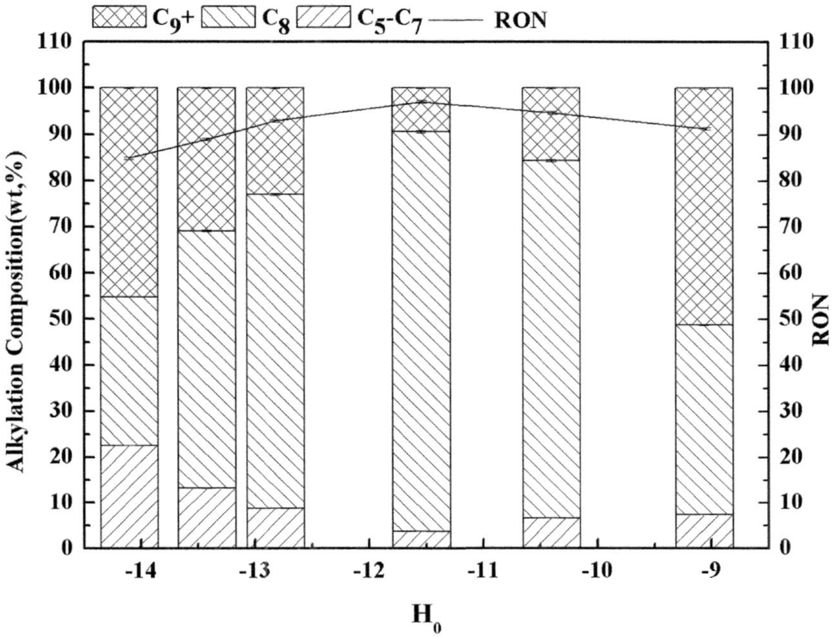

FIG. 2.41 Alkylate composition and RON varied with the acid strength of the catalyst system. Reaction conditions: reaction time 10 min, catalyst 40 mL, feed rate 500 mL/h, acid/hydrocarbon 30/50, 5°C. *From A. Wang, G. Zhao, F. Liu, L. Ullah, S. Zhang, A. Zheng, Anionic clusters enhanced catalytic performance of protic acid ionic liquids for isobutane alkylation, Ind. Eng. Chem. Res. 55 (2016) 8271–8280.*

−13.77 ($\chi_{TfOH}=0.90$) and −10.4 ($\chi_{TfOH}=0.49$) on the H_0 scale. This range stands in contrast to other catalysts like H_2SO_4 (−11.6 to −10.4) [17], HF (−9.0 to −8.0) [62], and even $[N_{222}H]HSO_4$/TfOH. Within this range, the maximum C_8 selectivity reaches 86.23%, the highest RON peaks at 97.3, and the lowest levels of by-products (C_5–C_7, 3.8% and C_9^+, 9.44%) are obtained at −11.52. In comparison, for TfOH/water, $[N_{222}H]HSO_4$/TfOH, and CF_3CH_2OH/TfOH, the respective optimal acidity points are −10.7 [62], −9.2 [63], and −9.6 [64].

Below the optimum acidity range, there is an increase in C_9^+ fraction in the product stream with a reduction in the catalyst acidity, leading to a lower quality product. Conversely, when H_0 is −13.77 and −14.1, there are significantly more C_5–C_7 and C_9^+ components in the alkylate product than C_8. This disparity occurs because their acidity exceeds the upper limit of the optimal acidity range. These findings collectively underscore that not only acid strength but also the composition and microstructure of the catalyst play a crucial role in its catalytic performance [16,64].

2.4 Zeolite catalyst

Zeolite catalysts play an extremely important role in various petroleum refining processes, including isomerization, catalytic cracking, and hydrocracking. However, their application in the C4 alkylation reaction with light olefins remains unsuccessful due to rapid deactivation. To date, liquid acid catalysts such as H_2SO_4 and HF have been the primary commercial catalysts for C4 alkylation. Nonetheless, the strong corrosiveness of H_2SO_4 and the volatile nature of toxic HF pose significant environmental and human health risks. Consequently, zeolite is considered as the top candidate for future C4 alkylation catalysts due to its noncorrosive nature and structural stability. Additionally, as early as 2006, the economic characteristics of zeolite catalysts for petroleum refining were clearly outlined in Table 2.20 [46]. Zeolite-catalyzed C4 alkylation processes have been extensively studied for decades.

2.4.1 Alkylation process

Four process developments in the C4 alkylation area deserve special mention: Alkylene from UOP, FBA processes from Haldor Topsøe, Alky-Clean from CB&I, and K-SAAT from KBR.

TABLE 2.20 Estimated US refining catalysts market in 1991.

Process	Capacity millions (bpsd[a])	Catalyst consumption (millions lb)	Catalyst cost ($/lb)	Market (millions $)
Catalyst cracking	5.41	410	0.800	330
Hydrotreating	9.55	38	2.400	91
Hydrocracking	1.29	8	9.000	72
Catalyst reforming	3.91	5	6.500	33
Isomerization	0.46	0.3	6.000	2
Alkylation, H_2SO_4	0.63	4.550	0.055	250
Alkylation, HF	0.40	9	0.800	7
Total				785

[a] *Barrel per stream day.*

Data from A. Corma, A. Martínez, Chemistry, catalysts, and processes for isoparaffin–olefin alkylation: actual situation and future trends, Catal. Rev. 35 (1993) 483–570.

2.4.1.1 AlkyClean technology

AlkyClean is currently the only solid catalyst-based technology that has been successfully commercialized for use in alkylation processes in the petroleum industry, which was developed by McDermott using the solid catalyst of Albemarle, AlkyStar [65], as shown in Fig. 2.42. The AlkyClean solid acid alkylation process employs a straightforward fixed-bed design, and the entire unit is constructed using carbon steel materials. The AlkyClean process consists of four main components: feedstock pretreatment, reaction system, catalyst regeneration, and product separation system. The key to this process lies in the reaction system, which comprises multiple fixed-bed rotating reactors, and the catalyst regeneration technology [66].

The reaction temperature is maintained between 50°C and 90°C, eliminating the need for cooling costs associated with H_2SO_4 and HF processes that operate at temperatures less than 50°C [66]. To ensure high production efficiency and catalyst activity, three parallel reactors are used. At any given time, two reactors are in production while one is undergoing mild regeneration with hydrogen, with each cycle taking 1–3h. One reactor is periodically taken offline and subjected to high-temperature regeneration at 250°C. The key features of the AlkyClean process are minimizing olefin

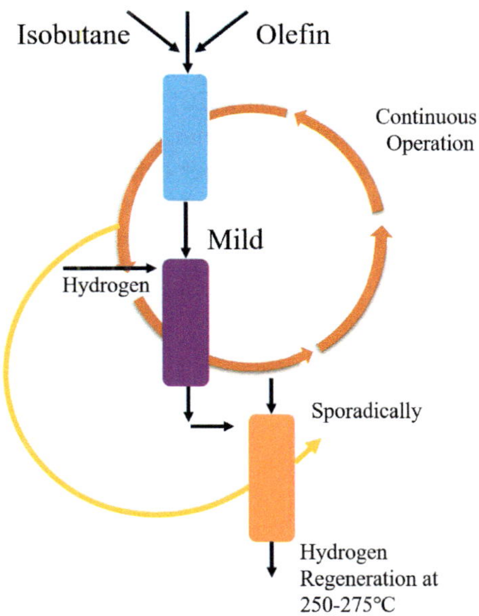

FIG. 2.42 AlkyClean process. *Modified from H. Diaz Velazquez, N. Likhanova, N. Aljammal, F. Verpoort, R. Martinez-Palou, New insights into the progress on the isobutane/butene alkylation reaction and related processes for high-quality fuel production. A critical review, Energ. Fuel 34 (2020) 15525–15556.*

concentrations and maximizing the I/O ratio in the reaction zone, which is achieved through dispersed injection of olefin feed and a high degree of mixing. Isobutane reacts with olefins at temperatures ranging from 20°C to 90°C with I/O of 8:1 to 10:1, resulting in alkylate with RON exceeding 95 [66].

The AlkyClean technology was implemented in 2015 at Shandong Wonfull Petrochemical in China, producing the alkylate product with high octane number (95–98) and sulfur content of less than 1 ppm [65]. The process employs the zeolite-supported metal catalyst, which can be regenerated with hydrogen at moderate temperatures [67,68]. The process allows for the reaction of C3 to C5 olefins with isobutane, producing predominantly C8 fraction alkylate, the catalyst is regenerated online once deactivation happens. The cost of an AlkyClean unit is expected to be 10%–12% lower than the corresponding H_2SO_4 unit and similar to the HF unit because the AlkyClean unit does not use liquid acid catalysts, resulting in lower equipment maintenance costs, significantly lower hazards, and no waste acid handling issues. The estimated production costs are approximately 3% lower compared to H_2SO_4 catalyzed systems, making it competitive with HF catalyzed systems [66].

2.4.1.2 K-SAAT technology

KBR is commercially promoting the K-SAAT (Solid Acid Alkylation Technology), which is based on the previously developed ExSact process by EXELUS Inc. The K-SAAT technology employs two reactors with multiple sections. As shown in Fig. 2.43 [69,70], one of the reactors operates in the alkylation mode, and the other regenerates the solid acid catalyst. The key to K-SAAT technology is the ExSact catalyst, which overcomes the limitations of rapid deactivation of common solid acid catalysts and exhibits excellent alkylation activity, stability, and environmental

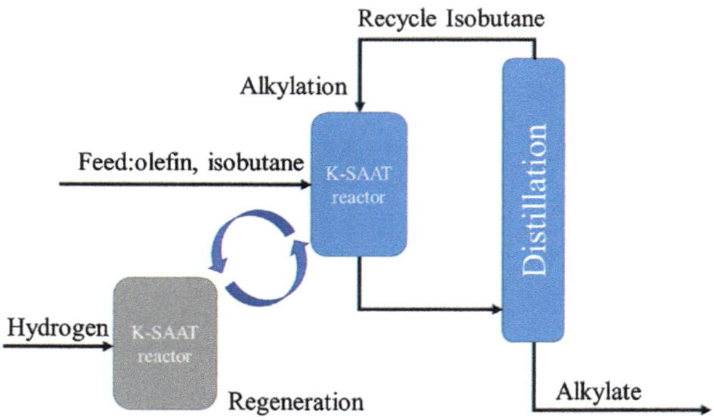

FIG. 2.43 A K-SAAT process. *Modified from American Fuel & Petrochemical Manufacturers 2018 Operations & Process Technology Summit.*

friendliness [66]. As compared to other liquid and solid acid alkylation technologies, the K-SAAT offers excellent RON, higher alkylate yield, and greater olefin consumption efficiency. KBR has licensed the K-SAAT unit to Haike Ruilin Petrochemical Co. Ltd. and Luoyang Aiyou Chemical Co. Ltd., both in China [65].

2.4.1.3 *Alkylene technology*

The Alkylene technology is developed by UOP and employs the Pt-KCl-$AlCl_3/Al_2O_3$ catalyst. The process, as shown in Fig. 2.44, is similar to the Fluid Catalytic Cracking (FCC) unit [71]. Liquid olefin feedstock undergoes pretreatment and is combined with recycled isobutane and regenerated catalyst. Then, they enter the up-flow reactor where the alkylation reaction takes place. At the top of the reactor, the catalyst is separated from the hydrocarbons. The postreaction catalyst enters the catalyst regeneration area. The hydrocarbons go through a separation system and are divided into alkylate, isobutane, and light hydrocarbons. After cooling step, isobutane is reintroduced into the reactor to increase the ratio of alkene to alkane inside the reactor. In the regeneration area, the postreaction catalyst flows counter currently with isobutane and hydrogen to wash away some soft coke, partially restoring the catalyst activity of alkylation. After some time, the completely deactivated catalyst is carried into the regenerator, where it undergoes full regeneration under high temperature with hydrogen atmosphere. The reaction temperature for the Alkylene process maintains at 10–40°C, and the I/O ranges from 6 to 15/1 [71,72].

2.4.1.4 *FBA technology*

The Fixed Bed Alkylation (FBA) technology was developed by Haldor Topsøe A/S and employs a supported liquid phase (SLP) catalyst where a liquid superacid is loaded on a porous carrier material, which has been

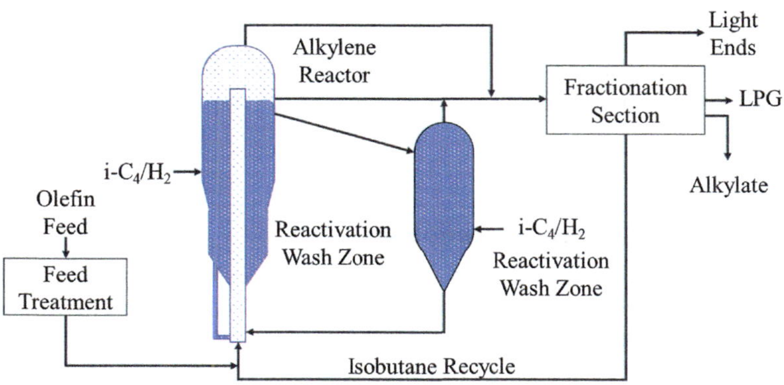

FIG. 2.44 The Alkylene process. *Modified from S.I. Hommeltoft, Isobutane alkylation: recent developments and future perspectives, Appl. Catal. A Gen. 221 (2001) 42–428.*

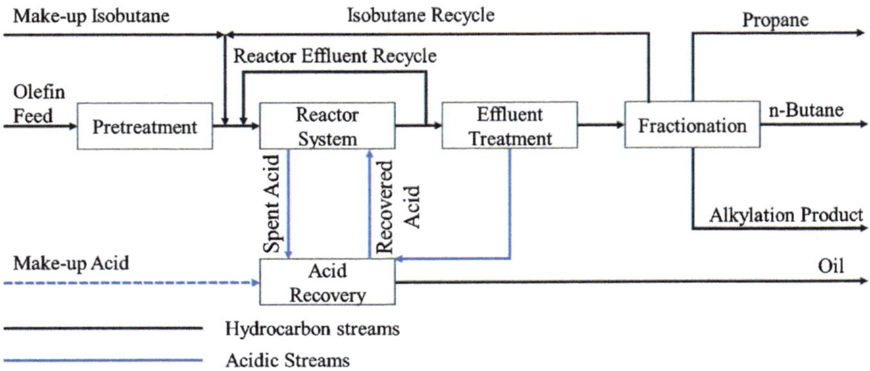

FIG. 2.45 Simplified process flow diagram for the FBA process. *From S.I. Hommeltoft, Isobutane alkylation: recent developments and future perspectives, Appl. Catal. A Gen. 221 (2001) 421–428.*

extensively validated in Denmark using 0.5 barrel per day pilot unit [71]. It is reported that cumulative pilot experience exceeds 20,000 h, of which 8000 h used to convert various raw materials from refineries worldwide. The reactor operates adiabatically, and the reaction temperature is controlled by the cooled reactor effluent loop. Waste acid is recovered using proprietary technology, which does not consume any chemicals and produces a small amount of light oil as the only by-product.

The process includes waste treatment based on proprietary technology to remove trace acids from the product stream leaving the reactor section, as shown in Fig. 2.45 [71]. When the product stream leaves waste treatment and enters the fractionator, the stream contained no acid, because the acid removed from the reactor is recycled back into the alkylation process. FBA technology has been proven to handle variations in feed composition. Moreover, the catalysts in the FBA unit have mobile acid sites, exhibit more effective catalytic activity than truly solid catalysts with fixed acid sites [73]. The catalyst system can process most typical alkylation raw materials without pretreatment. Nevertheless, the presence of impurities in the raw materials increases the requirements for acid recovery capabilities.

2.4.1.5 Other alkylation technologies

In addition to the solid acid alkylation processes mentioned above, Lurgi developed the Eurofuel alkylation process [74,75]. The Eurofuel process employs reactive distillation technology and consists of two main distillation columns: the reaction column and the product separation column. In the production process, the catalyst is suspended in isobutane and introduced from the top into the reaction column. The feedstocks enter the reaction column from various sidelines, and the alkylation reaction occurs on the tray in the tower. At the bottom of the tower, a portion of isobutane evaporates, generating upflowing vapor that keeps the catalyst suspended and thoroughly mixed with the reactants. After condensation at the top of the tower, the mixtures return to the tray as a diluent. Another portion of isobutane, along with the product, enters the separation column. Alkylate is drawn from the sidelines, while isobutane condenses at the top of the separation column and mixes with the catalyst at the bottom of the column, forming a catalyst suspension that reenters the reaction column. Deactivated catalysts undergo high temperature hydrogen regeneration outside the reactor. This process utilizes molecular sieve catalysts with hydrogenation functionality, operates at temperatures ranging from 50°C to 100°C, and maintains an olefin-to-alkane ratio of 6-12/1.

Additionally, SINOPEC Co. LTD. developed the ZCA-1 solid acid alkylation technology using the AlB-2 alkylation catalyst [72,76]. The catalyst exhibits high activity, good selectivity, and high stability at low temperature. The process contains a small amount of regenerating aids and can be repeatedly regenerated and reused under certain conditions. With the AlB-2 catalyst and a dual fixed bed reaction/regeneration switching operation mode, the process has low operating temperature, high product quality, low cost, and energy consumption similar to liquid acid processes. The process does not have issues related to waste disposal or corrosion, making it environmentally friendly and inherently safe. Moreover, the process has been polited at 100 tons/year at Yanshan Petrochemical Company in China, with the resulting alkylate with an RON of 95–98.

2.4.2 Effect of zeolite structure

The efficiency of current oil refining and petrochemicals largely depends on the use of highly active, selective, and durable catalysts; zeolites are typical examples of solid acids and the large surface area supports the development of environmentally friendly processes. The outstanding level of development in refining and petrochemicals is the result of a combination of intensive research and innovation over many years [77].

Zeolite is a kind of porous crystalline aluminosilicate, which has the advantages of high stability, clear pore structure, high catalytic activity, strong adsorption capacity, and strong ion exchange capacity [78]. The special properies of zeolites makes them incredibly versatile and useful in adsorption of the separation process, gas storage, and oil refining catalyst agent and petrochemical industries, including catalytic cracking, hydrocracking, and alkylation [79–82]. For C4 alkylation to produce alkylate oil of high quality, zeolites with different topologies, including HY, USY, HX, beta, mordenite, ZSM-12, LTL, and MCM-22 zeolite, have been widely used as solid acid catalysts in recent decades, and some zeolites show good catalytic performance [83–87]. However, zeolite catalysts often suffer from rapid deactivation due to carbonaceous deposits and difficult regeneration, which limits their industrial application [88–90]. An important reason for their wide application and limitations is the balance of their chemical and structural aspects and the reaction environments. It is worth noting that conceptually distinguishing the effects of zeolite chemistry and structure helps to understand the reaction processes, but cannot isolate the effects of the two properties [91]. The complexity arises because the chemistry of a particular molecular sieve is related to its framework density [91].

According to Feller et al., only large-pore zeolites exhibit sufficient activity and selectivity for the alkylation reaction [91]. The effect on the topologies with different pore sizes was studied in detail by Chu and Chester [92]. Specially, the reaction of isobutane and butene were examined over two different zeolite catalyst systems, ZSM-5 and REHY. The results show that ZSM-5 is inactive for the alkylation due to its constrained pore size, while the desirable TMPs selectivity appeared to be controlled by the diffusion in the medium-pore ZSM-5 zeolite and large-pore REHY zeolite. Mota et al. [93] studied the effect of different dimensional channels on the C4 alkylation, and found that beta-zeolite with uniform channels is superior to USY zeolite in TMPs selectivity and catalytic lifetime because of the isobutane diffusion restriction in micropores. Corma et al. [46] also considered that relatively large pores should be required for the C4 alkylation reaction. Indeed, not only is the initiating alkyl chain a bulky one, but easy diffusion of TMP should also occur. Thus, they selected 12-member-ring Y zeolites and 10-member-ring ZSM-5 zeolite to study the influence of pore size and pore dimension on the C4 alkylation. The results show that the larger pore size zeolites are better than intermediate pore-size zeolites, as shown in Table 2.21. Moreover, three-dimensional (3D) pore structure zeolites are more conducive to diffuse than one-dimensional (1D) pore structure zeolites, which propagates the alkylation.

Yoo et al. [86] studied the effect of structural characteristics on the alkylation activities of five 12-member-ring zeolites with different dimensional channels, including beta, USY, LTL, mordenite, and ZSM-12 zeolites under under extremely low olefin feed (I/O molar ratio of 1000, olefin WHSV of $0.01\,h^{-1}$) to keep real alkylation without undesirable oligomerizations. They found that 3D beta-zeolite and 1D ZSM-12 zeolite with linear channels and without connecting the huge supercage have less steric resistance than 3D USY and LTL zeolites, which promotes the formation of huge TMPs (2,2,4-TMP, 2,2,3-TMP) and inhibits the production of heavy hydrocarbons (C_9^+). Hence, the adsorption and diffusion of C4 reactants (isobutane and butene) are dependent on the topologies of zeolites to great extent, which plays a crucial role in determining the

TABLE 2.21 Comparison of zeolite activities.

Catalyst	REHY	Deal REHY	REUSY	USY	ZSM-5
Temperature (°F)	193	200	193	190	207
Pressure (psig)	335	375	335	320	375
WHSV (1-$C_4^=$)	7.2	7.2	3.7	1.9	1.0
C_5^+ yield (g/g 1-$C_4^=$ charged)	0.5	0.4	0.2	0.2	0
C_5^+ selectivity (g/g 1-$C_4^=$ converted)	1.0	0.6	0.9	0.9	–
$C_4^=$ conversion (wt%)	44	57	74	74	–
RON	78	76	74	74	–

Data from A. Corma, A. Martínez, Chemistry, catalysts, and processes for isoparaffin–olefin alkylation: actual situation and future trends, Catal. Rev. 35 (1993) 483–570.

catalytic performance and lifetime of zeolites. Li et al. investigated the adsorption and diffusion behaviors of C4 reactants on zeolites with different topologies by configuration bias Monte Carlo (CBMC) and MD simulations under experimental conditions at 348 K and 2.5 MPa [94]. It is revealed that surface area and pore size are the most relevant structural parameters for zeolites to determine the adsorption and diffusion of C4 reactants.

Since zeolites are classified as microporous materials, which are crystalline aluminosilicates of alkaline elements, rare-earth elements (REEs) or other monovalent or multivalent metals [95]. Because of their unique properties, such as large specific surface area, high ion-exchange ability, high thermal stability, presence of strong acidic centers, precisely defined system of micropores, and channels which enable conducting shape-selective catalytic reactions, they are applied in catalysis for distinguishing reagents, products, and transition states in a molecular scale [95]. Nevertheless, the diffusion of branched molecules is limited in the micropores, and the transport of reagents with size similar to the size of the micropores is difficult, as shown in Fig. 2.46. Intracrystalline diffusion of adsorbed molecules in zeolite crystals is critically important in relation to the application of zeolites as catalysts. The diffusion in zeolites, intracrystalline diffusion, also known as configurational diffusion, is an activated process, the energy of activation arising largely from steric hindrance [96]. The diffusional activation energy and therefore the diffusivity, for a given sorbate, are therefore directly related to the pore size, as shown schematically in Fig. 2.46 [97].

Over the past decades, developing the improved experimental methods for measuring adsorption and diffusion within zeolite crystals to increase the understanding of transport mechanisms has attracted

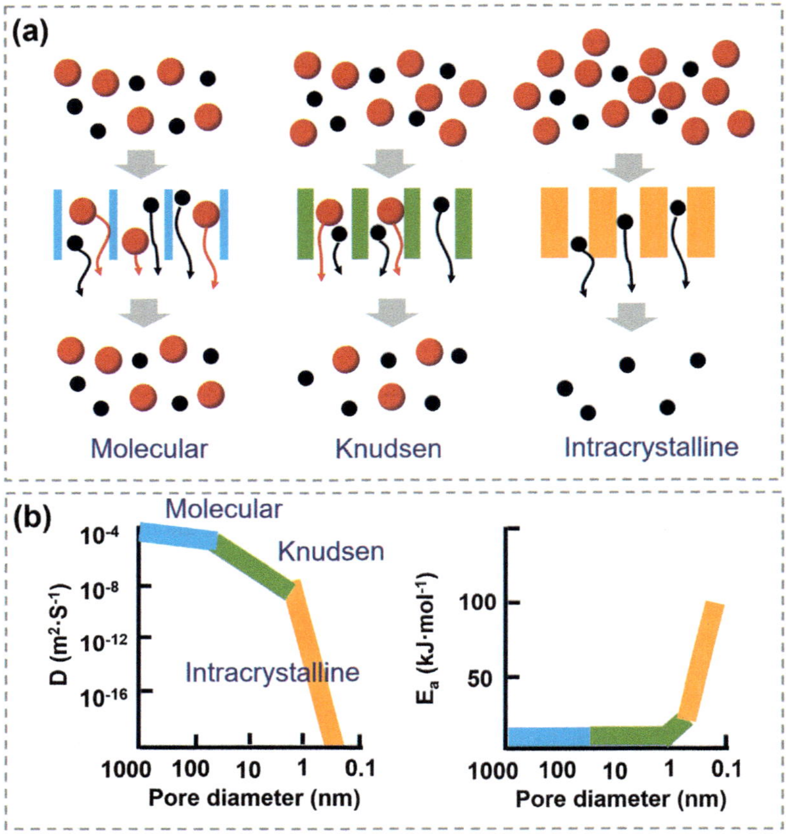

FIG. 2.46 (A) Illustration of the diffusion of large *(red)* and small *(black)* molecules within macropores (in *blue*), mesopores (in *green*), and micropores (in *yellow*). (B) Magnitudes of diffusion coefficients (left) and activation energy (right) in different diffusion regimes as a function of pore sizes. The *blue*, *green*, and *yellow* regimes correspond to diffusion in macropores, mesopores, and micropores, respectively. *Modified from M. Hartmann, A.G. Machoke, W. Schwieger, Catalytic test reactions for the evaluation of hierarchical zeolites, Chem. Soc. Rev. 45 (2016) 3313–3330.*

considerable researches [96]. The development of hierarchical catalysts in particular based on microporous zeolites has facilitated the study of intracrystalline diffusion. Hierarchical zeolites include all microporous materials with at least two classes of pores [97]. According to the International Union of Pure and Applied Chemistry (IUPAC), pores are divided into three groups, according to their internal pore width: micropores of widths less than 2 nm, mesopores of widths between 2 and 50 nm and macropores of widths larger than 50 nm. The diffusion in zeolites changes with the pore size. In micropores, the configuration/activation-diffusion mechanism dominates due to the tight fit of molecules and pore sizes, which is common in solids and is characterized by low diffusivity, perhaps 2–3 orders of magnitude smaller than that observed in Knudsen and molecular mechanisms, as shown in Fig. 2.46 [97]. Knudsen and molecular diffusion are commonly observed in solid materials containing mesopores and macropores, as shown in Fig. 2.46. The low diffusivity in the micropores significantly reduces the transport of different reactants into and out of the active sites. The delayed transport of reactants often favors the further conversion of these molecules into unwanted by-products that can also act as coke precursors, such as butene and C_9^+ molecules in C4 alkylation, leading to a build-up of coke, which in turn plugs the micropores and deactivates the zeolite catalyst [98]. As a result, many zeolite-catalyzed reactions are deactivated by coke formation, which significantly shortens the catalyst lifetime [98]. It is particularly important to promote the diffusion of butene and products to inhibit the formation of coke in zeolites pores for C4 alkylation.

In separation and catalysis applications, adsorption and diffusion are normally considered mutually exclusive. That is, rapid diffusion is generally accompanied by weak adsorption and vice versa. Liu et al. analyzed an anomalous loading-dependent mechanism of *p*-xylene diffusion in a newly developed zeolite called SCM-15, as shown in Fig. 2.47 [99]. Their

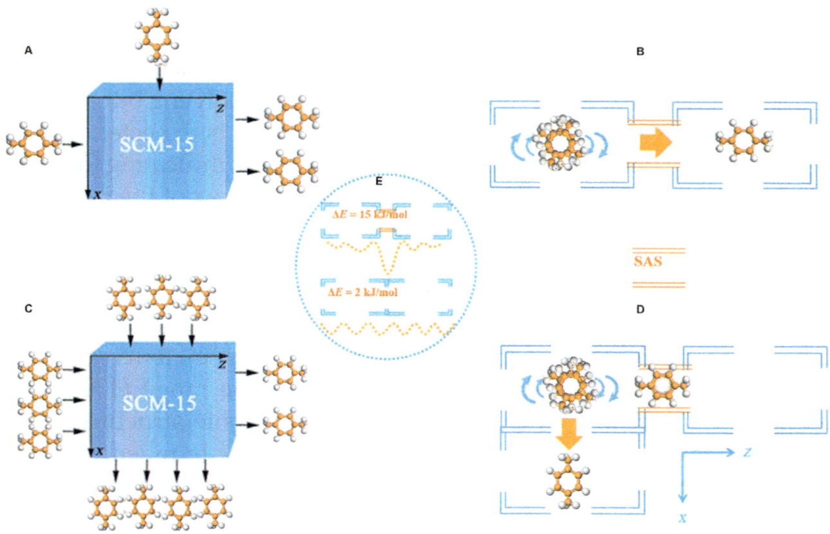

FIG. 2.47 Mechanism of *p*-xylene diffusion in SCM-15 zeolite [99]. (A and B) Diffusion mechanism at low loading. The presence of intersections and strong adsorption sites (SAS) along the Z direction facilitates the rotation of molecules for self-chosen different directions. (C and D) Diffusion mechanism at high loading. Rotation is efficiently suppressed at high loading because of the preferential occupancy of *p*-xylene along Z. Therefore, *p*-xylene molecules tend to migrate to the channels along X, where they move fast because of the low diffusion barrier at continuum intersections (E). *From Z. Liu, J. Yuan, J.M. van Baten, J. Zhou, X. Tang, C. Zhao, W. Chen, X. Yi, R. Krishna, G. Sastre, A. Zheng, Synergistically enhance confined diffusion by continuum intersecting channels in zeolites, Sci. Adv. 7 2021 eabf0775.*

result was obtained that the unique system of "continuum intersecting channels" plays a key role in the diffusion process for the molecule-selective pathways. At low pressure, the presence of strong adsorption sites and intersections provide space for molecule rotation, facilitating the diffusion of *p*-xylene along the *z* direction. Upon increasing the molecular uptake, the adsorbates move faster along the *z* direction because of the effect of continuum intersections in reducing the diffusion barriers and thus maintaining the large diffusion coefficient of the diffusing compound. This mechanism synergistically improves the diffusion in zeolites with continuum intersecting channels.

2.4.3 Effect of acidity

In liquid acid-catalyzed C4 alkylation systems, the acidic strength of the liquid acid catalyst is one of the key factors to decide the quality of the alkylate. Acidity of the zeolites is also an important for C4 alkylation. Zeolite acidity comprises two distinct facets: the acid amount, i.e., the number (mole) of acidic sites sensed in a portion (kg) of material, and the acid strength, i.e., the energy (kJ) involved during the interaction between such acid sites and the basic molecules used to sense them [100]. The acid amount as well as the acid strength depends strongly on the characteristics of the basic molecule used in the determination. In systems such as aqueous solutions of weak acids, the acid amount and the acid strength are inherently referred to water, which not only acts as a solvent, but as a base. If the solvent is changed, for example to acetone, it is expected that the energetics and extents of dissociation be also affected, indicating that acidity is a relative concept at all levels, and solid acids are not the exception. The acid sites that can be found on the zeolite structure and in solids, in general, are Brønsted (protonic) or Lewis (nonprotonic) in nature. Zeolite acidity is derived from the tetrahedral aluminum bonding in the aluminosilicate structure; the aluminum and silicon atoms (T atoms) of the zeolite lattice are chemically bound together by sharing oxygen atoms at the corners; T–O–T (inter-tetrahedrons) varies over a wide range (130°–180°), leading to an increase in the number of known zeolite structures [101]. The introduction of trivalent aluminum atoms in the tetrahedral framework brings negative charges, which are balanced by cations outside the framework (Fig. 2.48), such as Na^+, K^+, Mg^{2+}, and NH_4^+ [100]. Under certain conditions, the charge-compensating cations can be easily replaced by cation exchange with either di- or trivalent cations, as well as NH_4^+, which after thermal decomposition leads to the formation of acidic hydroxyl groups in the zeolite, which behave as Brønsted acid sites (BASs).

The content of framework aluminum informs about the highest possible abundance of BASs. However, aluminum and acid amounts hardly match

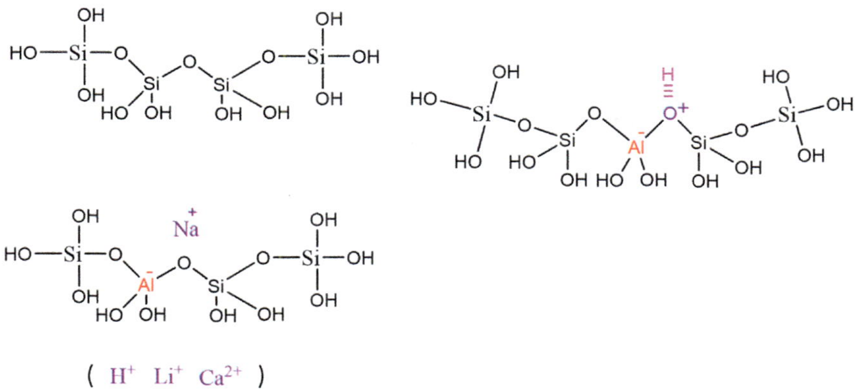

FIG. 2.48 Top left: Fragment of the tetrahedral silica lattice. Bottom left: The isomorphic insertion of aluminum in the lattice gives rise to negative charges compensated by extra-framework cations. Right: Structure of the BAS. *From L.-E. Sandoval-Díaz, J.-A. González-Amaya, C.-A. Trujillo, General aspects of zeolite acidity characterization, Microporous Mesoporous Mater. 215 (2015) 229–243.*

because zeolites easily release aluminum during thermal or chemical treatments. Y zeolite, one of the most important commercial zeolites, is easy to dealuminate. Additionally, the ion exchange is an equilibrium process that cannot be taken to 100%, and BASs are used to dehydroxylate at high temperature. Released framework aluminum accumulates inside zeolite porosity to produce what is known as EFAL, the EFAL acts as a neutralizing agent since it can experience ion exchange with BASs and may cause blockage at pores, reducing further the acid amount. These EFAL species often act as Lewis acid sites (LAS). The balanced alkali cations change into alumina species (AlO^+, $Al(OH)_2{}^+$, $AlOH^{2+}$, AlOOH, or $Al(OH)_3$, etc.), which are called Lewis acid sites. The main group or transition metal elements, such as Ga, B, Ti, and Sn, can be introduced onto the zeolites as extra-framework species or into the zeolite framework by the isomorphous substitution of Al^{3+} or Si^{4+}, forming metal LAS, as shown in Fig. 2.49 [102].

The cation exchange also affects the BAS. As mentioned above, Na^+ is replaced by a proton to form an acid, and the smaller the amount of substitution, the lower the amount of acid. Nivarthy et al. reported the influence of the concentration of acid sites for isobutane/*n*-butene alkylation on zeolite BEA with varying degrees of Na^+ ion exchange [103]. The results showed that the probability of side reactions, such as multiple alkylation, enhanced with the increase degree of Na^+ exchange, causes a decrease of the catalysts' lifetime and a decrease of the integral turnovers per acid site in line with the above proposal. The Na^+ cations acting as weak Lewis acid sites may also enhance the local olefin concentrations, thereby increasing the probability of multiple alkylation. Cationic species

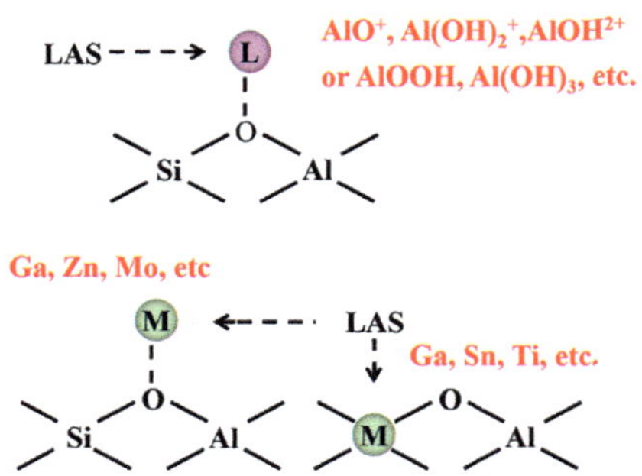

FIG. 2.49 Lewis acid sites (LAS) on zeolites. *From X. Zhao, J. Xu, F. Deng, Solid-state NMR for metal-containing zeolites: from active sites to reaction mechanism, Front. Chem. Sci. 14 (2020) 159–187.*

on the zeolites also is a factor affecting the BAS. Ward et al. reported, in 1968, on the effect of alkali and alkaline earth ion exchange forms on the acidity of zeolites using pyridine-adsorbing infrared spectra [104]. Gerzeliev et al. reported on the effect of a series of cations on the FAU-type zeolites for C4 alkylation and obtained the similar results [105–107]. Ward studied the influence of the alkali and alkaline earth ion on the acid sites of the FAU type zeolites [104]. The results showed that the population of BAS increases with decreasing cation radius and increasing electrostatic potential and field. The divalent cation forms also have structural hydroxyl groups. The electrostatic potential or field due to the divalent cation dissociates adsorbed water, producing MOH^+ and structural hydroxyl groups. The results also showed that calcination temperature affects the acid type of zeolites. None of the zeolites studied exhibits Lewis acidity if the calcination temperature is less than 500°C. At 650°C, calcination temperature, Lewis acidity is observed. Then, the influence of calcination temperature on the structural hydroxyl groups and acidity of stabilized hydrogen Y zeolite was developed by Ward et al. [108]. They found that the concentration of LAS increases with increasing calcination temperatures and decreasing hydroxyl content. Moreover, the concentration of BAS plus twice the concentration of LAS remains constant over a wide range of temperatures. At the same time, they studied the effect of a small amount of water on the acidity of several alkali, alkaline earth, hydrogen, and mixed cationic zeolites by observing the infrared spectral changes of chemically adsorbed pyridine [109]. Here, the pyridinium ion, characterized by an infrared band at 1545 cm^{-1}, is a measure of the Brønsted acidity and coordinately bound pyridine, with an absorption band at 1451 cm^{-1}, is a measure of Lewis acidity. The presence of trace water could generate some BAS, which convert LAS after the elimination of water. Besides, from 1969 to 1971, Ward et al. investigated a series of the nature of active sites on zeolites, especially, the catalytic activity of rare-earth Y zeolite for *o*-xylene isomerization [110]. The rare-earth Y zeolite contains several types of structural hydroxyl groups. The BAS are hydroxyl groups similar to those in hydrogen Y zeolite but the hydroxyl groups attached to the rare-earth cations are nonacidic. The acidity and catalytic activity are intermediate between those of magnesium and hydrogen Y zeolites. The aluminum content is just a rough estimate of zeolite potential. The acidities of zeolite catalysts depend on the coordination of aluminum and on the chemical nature of its neighbors. Ravi et al. reviewed the effect of different Lewis acid types on the acidity of zeolites [111]. Different calcination temperature and base solution in zeolite synthesis could result in different LAS. Fig. 2.50 summarizes the current Lewis acidic aluminum in zeolite [111].

When no Brønsted acidity is present in the zeolite, no alkylation occurs [46]. However, as soon as acid hydroxyl groups are formed by the methods mentioned above, alkylation is observed on these catalysts.

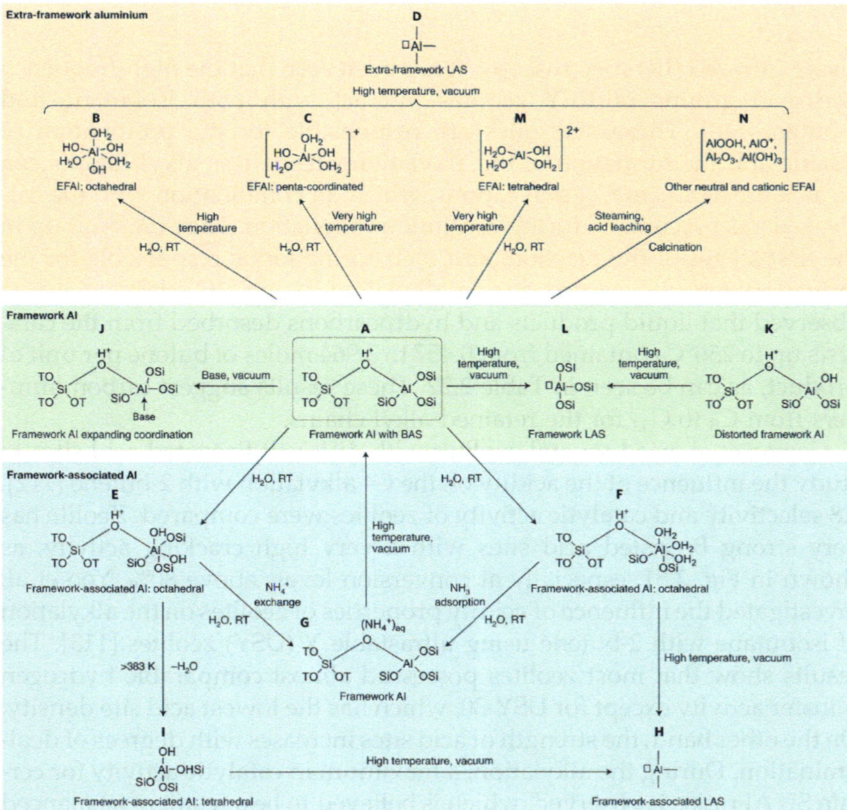

FIG. 2.50 A summary of the current understanding of Lewis acidic aluminum in zeolite. *From M. Ravi, V.L. Sushkevich, J.A. van Bokhoven, Towards a better understanding of Lewis acidic aluminium in zeolites, Nat. Mater. 19 (2020) 1047–1056.*

TABLE 2.22 Butene-1-^{14}C applied to isobutane alkylation with zeolite catalyst.

	Specific activities of fractions	
Name	**μC/g**	**OCI/P[a]**
Liquid product	0.5525	0.457
Stripped product to:		
100°C	2.2683	1.878
100–150°C	1.8430	1.526
150–200°C	2.5800	2.136
200–250°C	4.7872	3.964

[a] *Moles olefin charge incorporated/molar unit of product. A molar unit is the equivalent of 56.10 g product. Carbon number of product depends on the isoparaffin incorporated.*
Data from A. Corma, A. Martínez, Chemistry, catalysts, and processes for isoparaffin–olefin alkylation: actual situation and future trends, Catal. Rev. 35 (1993) 483–570.

Using infrared (IR) spectroscopy, it has been seen that the high-frequency hydroxyl groups of REY zeolites interact with both isobutane and 1-butene [46]. These acid sites are responsible for the protonation of butene and the formation of the alkylation sites. These alkyl cations can be as large as C_{19} to C_{23} polycations, which, in combination with the catalyst, act as precursors during continuous alkylation. They are built up in the first stages of the reaction, and their formation is responsible for the induction period observed during alkylation. Using ^{14}C-1-butene, it was observed that liquid products and hydrocarbons desorbed from the catalysts up to 250°C contained from 0.457 to 3.964 moles of butene per unit of product, as can be seen in Table 2.22. These results suggest carbon numbers from C_8 to C_{32} for the retained alkyl chains.

Corma et al. used several zeolites with different Brønsted acid sites to study the influence of the acidity on the C4 alkylation with 2-butene [112]. C8 selectivity and catalytic activity of zeolites were compared. Zeolite has very strong Brønsted acid sites with a very high cracking activity, as shown in Fig. 2.51, especially at conversion levels above 80%. Yoo et al. investigated the influence of acidity properties of zeolites on the alkylation of isobutane with 2-butene using ultrastable Y (USY) zeolites [113]. The results show that most zeolites possessed almost comparable hydrogen transfer activity except for USY-30, which has the lowest acid site density. On the other hand, the strength of acid sites increases with degrees of dealumination. During the alkylation, a maximum in catalytic activity for certain Si/Al ratios is observed, which is believed to be a result of a balanced level of acidic properties between the concentration and strength of Brønsted acid sites. It can prohibit the formation of heavy carbonaceous materials that deactivate the zeolites.

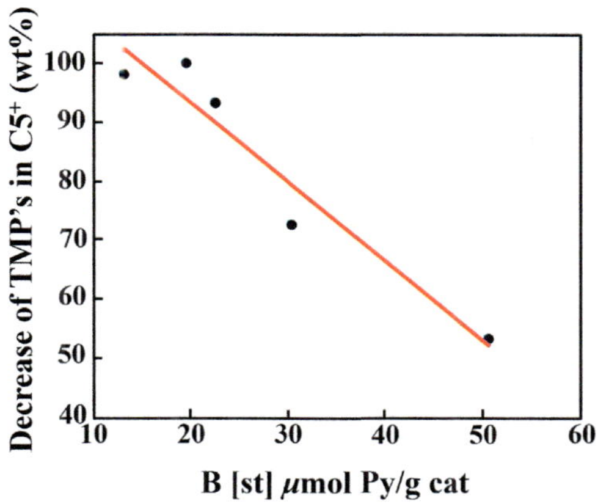

FIG. 2.51 Correlation between the relative decrease of TMP after 15 min TOS for different zeolite structures and the concentration of strong Brønsted acid sites (400°C pyridine desorption temperature) in the "fresh" zeolite. *Modified from A. Corma, A. Martinez, C. Martinez, Isobutane/2-butene alkylation on MCM-22 catalyst. Influence of zeolite structure and acidity on activity and selectivity. Catal. Lett. 28 (1994) 187–201.*

Besides, Corma et al. also studied the influence of the framework and extra-framework composition of zeolite beta on its activity, selectivity, and deactivation behavior during the alkylation of isobutane with 2-butene by submitting a H-beta sample to different postsynthesis treatments [114]. They found that the interaction of dispersed cationic extra-framework aluminum (EFAL) species in the zeolite channels with framework hydroxyls to form Brønsted acid sites enhances acidity. However, the formation of condensed-type EFAL that migrates to the zeolite surface decreased the accessibility of reactants to the acid sites. At the same time, Corma with his coworkers also studied the role of EFAL species in USY catalysts during the liquid phase alkylation of isobutane with 2-butene [115]. Here, the dealuminated USY zeolites, which contained the EFAL generated during steam stabilization, were treated with $(NH_4)_2SiF_6$ (AHF) to selectively remove the EFAL. They found that the elimination of EFAL in mildly dealuminated USY samples causes a decrease of the initial cracking and alkylation activity, and increases the deactivation rate of the zeolite. However, both the alkylation activity and the catalyst life are improved after elimination of EFAL from the highly dealuminated catalysts. These results have been interpreted in terms of changes in Brønsted acid site density and strength distribution produced after the AHF treatment on the different USY catalysts. It also has been reported that the use of large-pore zeolites, in combination with Lewis acids, greatly increases the activity and selectivity of the zeolite, thus affecting alkylation with high olefin space velocity and low isoparaffin-to-olefin ratio [46].

2.4.4 Modification of zeolites

Modification of zeolites is a method used to adjust the properties of zeolites to enhance their acidity or diffusion efficiency so that the alkylation selectivity and catalyst lifetime can be improved. Therefore, many investigations about zeolite modification on C4 alkylation have been reported, including metal catalyst combination, pore structure control, surface modification, ion exchange, and hydrophobization.

Rare-earth metal ion exchange method is a common method since LaY is applied in the cracking process [90,116–118]. rare-earth exchanged faujasites show a good ability to catalyze hydride transfer in alkylation reaction among large pore zeolites [119], which is related to two factors, high aluminum and high strong Brønsted acid sites concentrations. The high aluminum content in the zeolite framework results in optimal bond strength between the zeolite oxygen and the secondary or tertiary carbon atom of the alkoxy groups, which is the ground state of the carbenium ions in the transition state of olefin addition or hydride transfer. The high concentration of strong Brønsted acid sites associated with high framework Al concentration allows to generate a high concentration of alkoxy groups or carbenium ions, increasing the probability of hydride transfer rather than olefin addition for the individual alkoxy group. Both factors help to compensate the higher energy barrier required for the hydride transfer in solid catalysts compared to liquid acids [119]. It has been proposed that the rare earth ions exchanged in X or Y zeolites hydrolyze upon calcination, donating protons to generate Brønsted acid sites [120]: $[La(H_2O)_n]^{3+} = [La(OH)(H_2O)_{n-1}]^{2+} + H^+$. Ward obtained the presence of O–H stretching bands in IR spectra that are attributed to –OH groups of the aluminosilicate framework attached to the exchanged metal cation, indicating that rare earth ions caused the hydrolysis [110]. Lee and Rees investigated the effect of the calcination temperature on the migration of lanthanum ions from the supercages to the sodalite cages [121]. The authors reported that, at temperatures above 333 K, the hydrated La^{3+} cations begin to hydrolyze water and migrate into sodalite cages, the hydrolysis of the rare earth cations produces both Brønsted acid sites and Lewis acid sites. However, it is hard to understand under which conditions water may lead to the partial or full hydrolysis of framework aluminum atoms and what role the rare-earth metal cations play in this process. Guzman et al. explored a series of La^{3+}-exchanged X zeolites on C4 alkylation with cis-2-butene [119]. The activity in isobutane/cis-2-butene alkylation is mainly controlled by the fraction of strong Brønsted acid sites and the space velocity of the olefin in relation to the absolute concentration of these strong Brønsted acid sites. Catalysts with similar concentration of strong Brønsted acid sites and lower concentration of weak Brønsted sites showed a higher lifetime. Adding a certain amount of Ca^{2+} cations to faujasite-type zeolites

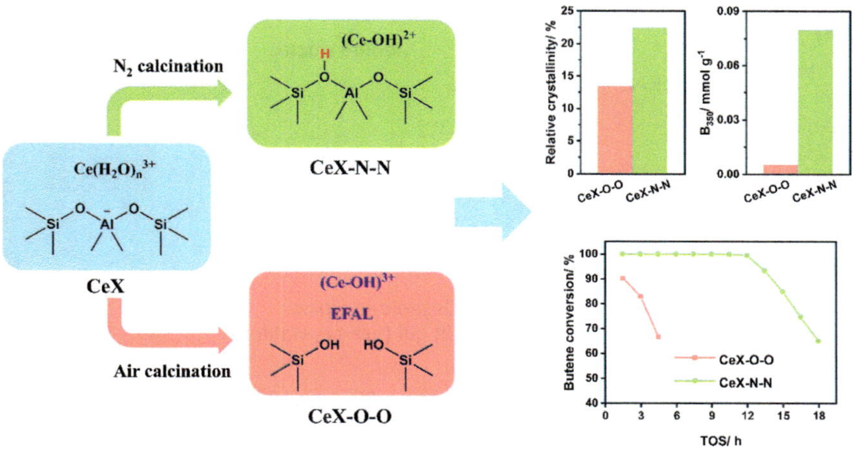

FIG. 2.52 The influencing mechanism of catalyst calcination atmospheres on the crystallinity, acidity, and alkylation stability of CeX zeolites. *From Z. Yang, R. Zhang, H. Zhang, H. Tang, R. Liu, S. Zhang, Comparative catalytic study on butene/isobutane alkylation over LaX and CeX zeolites: the influence of calcination atmosphere, Chin. J. Chem. Eng. 46 (2022) 173–183.*

containing REEs can stabilize the catalytic activity of *n*-butene mixtures for C4 alkylation [116].

The value of the Ce element could also affect the C4 alkylation results with olefin. Yang et al. prepared a series of Ce ion-exchanged X zeolites with different calcination atmospheres, as shown in Fig. 2.52 [117]. It is found that the Ce^{IV} contents in CeX zeolites are adjusted and their impacts on physicochemical properties and catalytic performance in C4 alkylation established. The crystallinity of CeX zeolite is found to be negatively correlated with the Ce^{IV} content. This is believed to be due to the water formed during the oxidation of Ce^{III}, which facilitates the framework dealumination. CeX zeolite calcined in oxygen produced only one-sixteenth the concentration of strong Brønsted acid and deactivated at the beginning of the alkylation process, whereas CeX calcined in nitrogen had a catalytic lifetime of 12 h. Therefore, cerium-containing zeolites should be prepared carefully to avoid cerium oxidation.

Hierarchical zeolites have already been applied in the alkylation, isomerization, transformation of methanol to hydrocarbons (MTH), aromatization, condensation, or catalytic cracking [95]. Choi et al. [122] synthesized a hierarchical BEA zeolite with trimodal pores, including micropores, mesopores, and macropores, for isobutane/2-butene alkylation. Specifically, it can be divided into three steps. First, micropore zeolites (MZ) are synthesized, then mesopores are introduced by structure-directing agents (micro-/mesoporosity, 2MZ), and finally macropores are introduced (micro-/meso-/macroporosity, 3MZ), as

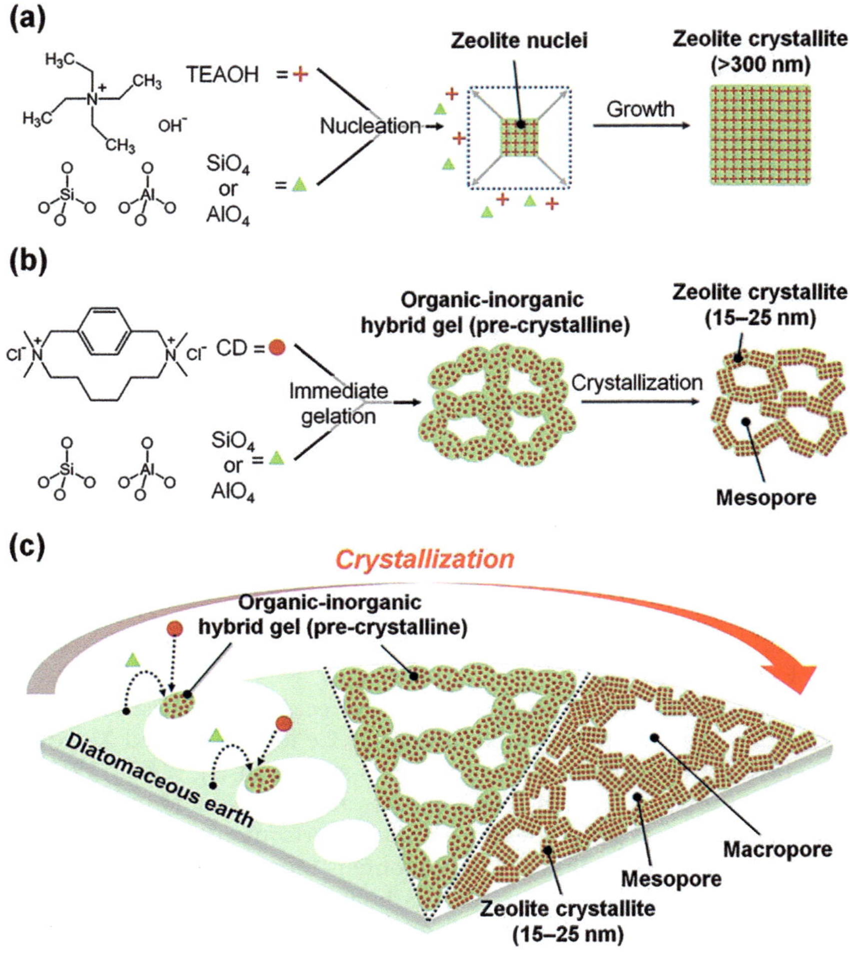

FIG. 2.53 Schematic illustration of the preparation of (A) MZ, (B) 2MZ, and (C) 3MZ samples. *From M. Jin, M. Oh, M. Choi, Hierarchical BEA zeolite with trimodal micro-/meso-/macroporosity as a selective and long-lived catalyst for isobutane/2-butene alkylation, ACS Catal. 12 (2022) 4067–4077.*

shown in Fig. 2.53. Then, a series of BEA zeolites with similar acid site distributions but different hierarchical porosities were used as catalysts in C4 alkylation. It is shown that the hierarchically porous BEA zeolite with trimodal micro-/meso-/macroporosity, synthesized by the pseudo-solid-state crystallization of diatomaceous earth, exhibited remarkably improved alkylation selectivity to TMPs and lifetime. Furthermore, the hierarchically porous BEA zeolite required 4 times less frequent hydrogenative regeneration than the zeolite containing only micropores. Such excellent catalytic properties could be attributed to the considerably facilitated molecular transport of hydride donors (e.g., isobutane) and bulky alkylation products to or from the zeolite micropores owing to the highly interconnected secondary macro- and mesoporous structures.

2.4.5 Operating parameters

In addition to the physicochemical properties of the catalysts, operating parameters such as reaction temperature, feed ratio of isobutane to olefin, olefin space velocity, and olefin type also determine the catalytic activity of zeolites in the C4 alkylation process. On the premise of ensuring a liquid-solid reaction, reaction pressure has a lower impact on the alkylation reaction. Typical process conditions for alkylation reactions catalyzed by different catalysts are shown in Table 2.23 [72].

2.4.5.1 *Effect of olefin types*

Ethylene, propylene, 1-butene, 2-butene, isobutene, pentene, and isopentene can all serve as feedstocks for alkylation reactions, and their reactivity varies for different catalysts. Guisnet and Gnep [123] found variations in the reactivity of 2-butene and propylene on USY zeolite catalysts. They discovered that in the isobutane/propylene system, dimerization reactions produce C6 hydrocarbons, alkylation reactions produce C7 hydrocarbons, and self-alkylation reactions produce C8 hydrocarbons. The rate of isobutane/propylene alkylation is twice than that of self-alkylation, 4 times than that of propylene dimerization, and propylene conversion is much lower than that of 2-butene. Nivarthy and others [124] compared the alkylation performance of ethylene, propylene, and 2-butene on an H-BEA zeolite. The reactivity of the olefins follows this order: 2-butene > propylene > ethylene. The products can also be categorized as dimerization reactions, alkylation reactions, and self-alkylation reactions. Ethylene more readily generates 1-butene through dimerization reactions, and 1-butene subsequently continues to react to

TABLE 2.23 These results suggest carbon numbers from C_8 to C_{32} for the retained alkyl chains.

	HF	**H_2SO_4**	**Zeolite**
Reaction temperature (°C)	10–40	5–15	50–100
Feedstock alkene ratio (mol/mol)	11–14	7–10	6–150
Olefin space velocity (h^{-1})	0.1–0.6	0.03–0.2	0.01–1.0
Exhaust acid concentration (wt%)	83–92	89–93	–

Data from S. Li, Molecular simulation study of structure-performance relationship of zeolite catalyst in C4 alkylation process, China University of Petroleum (East China) 2020.

form octanes, resulting in C8 product distribution almost identical to that of 2-butene. In the case of C4 olefins, isobutene is more prone to polymerization compared to linear butenes. Therefore, isobutene alkylation systems may cause the catalyst to deactivate rapidly, with higher acid consumption. In contrast, the alkylation performance of linear butenes is better than that of other olefins [125,126], which is why most alkylation studies use linear butenes as raw materials.

2.4.5.2 *Effect of temperature*

The reaction temperature not only affects the activation energy barrier of elementary reactions but also has a significant impact on the solubility of reactants and products, adsorption and desorption processes, and diffusion. In the sulfuric acid-catalyzed alkylation system, when the temperature exceeds 18°C, secondary reactions are enhanced, leading to increased acid consumption and a reduction in the octane rating of the products. However, the temperature should not be too low, because at lower temperatures, the viscosity of sulfuric acid significantly increases, which hinders the effective contact between the acid and the hydrocarbons. Hydrofluoric acid, on the other hand, has lower viscosity and higher solubility in isobutane, which promotes the conversion rate of isobutane. Therefore, hydrofluoric acid catalytic systems can operate at higher temperatures. In contrast to liquid acids, molecular sieve catalytic systems typically require higher temperatures. This is primarily due to the weaker acidity of molecular sieves or the smaller contact area between the solid catalyst and the liquid phase, which necessitates higher activation energy for the reaction steps and effective diffusion of larger molecules within the catalyst pores at elevated temperatures [27].

Generally, the suitable temperature range for zeolite catalyzed alkylation reactions is between 50°C and 100°C. At lower temperatures, polymerization reactions predominantly occur, and the reduced temperature hampers diffusion. This results in poor catalyst stability, and the main reason for catalyst deactivation is pore blockage by large hydrocarbon molecules. At higher temperatures, despite enhanced hydrogen transfer activity and diffusion capabilities, catalyst stability is higher. However, severe cracking occurs at elevated temperatures, leading to an increase in C5–C7 content and a decrease in C8 content in the products. The primary reason for catalyst deactivation at high temperatures is the irreversible adsorption of generated aromatic compounds occupying the active sites.

Studies have shown that the suitable temperature range for alkylation using H-BEA molecular sieves is between 75°C and 90°C [127]. The stability of LaX molecular sieves is best in the range of 70–80°C [128]. For H-BEA, the optimal reaction temperature is around 75°C [129].

Furthermore, changing the catalyst properties can also lead to corresponding changes in the optimal reaction temperature. Kirsch and others found that, for catalysts containing 0.2 wt% Na^+, the suitable reaction temperature is 40°C, while for catalysts with 1.0 wt% Na^+, the appropriate reaction temperature is 80°C. Recent literature typically evaluates catalyst reaction activity at around 80°C [126].

The above discussion reveals that temperature significantly influences the catalytic performance of molecular sieve catalysts in alkylation. Cracking reactions have a higher activation energy and are endothermic; thus, raising the temperature is most favorable for cracking reactions. As the temperature increases, the alkylation reaction rate increases. Therefore, catalytic activity increases with higher temperatures, but the selectivity and stability of alkylation reactions initially increase and then decrease with rising temperatures.

2.4.5.3 *Effect of inlet of I/O ratio and olefin space velocity*

The I/O ratio can be adjusted to change the reaction rate by altering the concentration term in the rate equation. Increasing the I/O ratio leads to higher content of isobutane and increases the hydrogen transfer rate. Conversely, an increase in butene content results in a higher polymerization reaction rate. The relative rates of hydrogen transfer and polymerization reactions significantly affect the selectivity and stability of the catalyst.

Yoo and colleagues [130] found that higher I/O ratios are favorable for molecular sieve catalyst stability and selectivity. At the same time, the effect of the silicon-to-aluminum ratio on alkylation performance is reduced. While a higher I/O ratio is beneficial for alkylation reactions, excessively high I/O ratios can lead to excessive recycling of isobutane within the unit, increasing operational costs. Therefore, industrial units need to optimize and select appropriate feedstock I/O ratio. The alkene space velocity determines the production rate of the alkylation unit. De Jong and others [131] found through kinetic modeling that increasing the I/O ratio and reducing alkene space velocity contribute to extending the catalyst lifespan. Taylor and Sherwood also arrived at similar conclusions. However, Feller and colleagues [128] discovered that alkene space velocity has almost no impact on the productivity of alkylation. The catalytic reaction at acid sites occurs a fixed number of times and is independent of alkene space velocity.

Solid acid alkylation processes require suitable reaction temperatures, feedstock alkene ratios, olefin space velocities, and feedstock compositions. In addition to these factors, researchers have also investigated the influence of the phase of the reactants and the reactor type on the catalytic activity of solid acid catalysts in alkylation. In terms of the phase of the reactants, liquid phase conditions are most favorable for alkylation reactions, while gas phase and supercritical conditions are less conducive to alkylation reactions. Regarding reactor types, due to the enhanced diffusion of substances, full mixed flow reactors exhibit superior alkylation catalytic performance compared to tubular reactors.

From another perspective, zeolite catalysts can be viewed as microreactors, and the impact of the I/O ratio within molecular sieve pores has a more direct influence on the catalytic performance of alkylation. The mentioned process conditions also have a significant impact on the I/O ratio inside the zeolite pores, which in turn may affect the catalytic performance of the catalyst. Research on this issue is valuable for gaining a new perspective on the alkylation process under zeolite catalysis and establishing an evaluation criterion during the development of novel, porous, solid acid alkylation catalysts. This evaluation criterion can help predict the alkylation performance of materials and guide the efficient design and optimization of high-performance porous materials for alkylation catalysts.

2.4.6 Other solid acids

In addition to zeolite catalysts, there are other solid acid catalysts such as aluminum chloride, sulfonic resins, heteropolyacids, etc. Strong Lewis acid catalysts are highly active in alkylation reactions. Aluminum chloride, in the presence of hydrogen chloride, was the first catalyst used for this purpose. Hydrogen fluoride salts, such as boron trifluoride (BF_3) and antimony pentafluoride (SbF_5), are also common catalysts for alkylation reactions [73]. Other solid acid catalysts explored by researchers for alkylation reactions include zeolites [83,132,133], sulfated zirconia [134–136], mixed oxides, and heteropolyacids [71,137,138], as well as acidic resins [139,140]. In general, oxide promoted by sulfate is considered a good solid acid catalytic material. Sulfated zirconia achieves its activity due to its high acidity ($H_o \leq -16.04$). Although sulfated alumina has low activity due to its low acidity ($H_o \leq -14.5$) [141–143], it has advantages over sulfated zirconia because it is widely available, heat-resistant, inexpensive, and phase-change resistant during preparation. The amount of sulfur plays a crucial role in the activity of sulfate-promoted oxides. The optimal sulfur content includes only the amount required to cover the single-layer surface of the substrate; any accumulation of sulfate ions beyond this point leads to decreased catalytic activity and faster deactivation. An increase in sulfur content reduces pore size distribution and pore volume, which is attributed to poorly dispersed sulfate ions block pores, leading to a transition from mesopores to micropores [144]. It is worth noting that the type of sulfate precursor also determines the texture and characteristics of the sample. For example, when sulfuric acid is used as the precursor, the carrier dissolves, and then sulfate salts deposit during the drying stage, whereas interaction occurs during the calcination stage when ammonium sulfate is used.

The active sites in sulfate-promoted oxides consist of single sulfate or polymeric sulfate species bonded to the oxide surface [141,145,146]. Catalytic activity occurs when the sulfur content exceeds half of the monolayer concentration. Compared to zeolite catalysts, sulfated zirconia exhibits higher activity and stability in conversions [147]. Nafion (sulfonated tetrafluoroethylene fluoropolymer copolymer) is a synthetic polymer and a choice for butene and isobutene alkylation [148–150]. Due to the presence of adjacent perfluorinated carbon groups in its structure with electron-withdrawing tendencies, Nafion is inherently a strong acid [151]; however, it has a small surface area. Its low surface area results in relatively poor catalytic activity, though its catalytic activity can be improved by loading it onto high-surface-area supports [152,153]. It has a Hammett acidity value of ≥ -12 and a thermal stability range of 180–200°C [154]. Nafion resin loaded with mesoporous structured SBA-15 modified with end-capped OH groups has high surface area ($400\,m^2/g$) and narrow pore size distribution. The end-capping of surface –OH groups reduces surface silanol density and provides a hydrophobic environment for isobutene/1-butene reactions [155]. Triflic acid (trifluoromethanesulfonic acid) developed by Haldor Topsoe A/S appears promising for use in solid-liquid phase reaction processes [73,156]. Heteropolyacids, with their unique molecular properties and strong acidity, are suitable candidates for C4 alkylation [134,135,157]. Compared to solid phases with low-density strong Brønsted sites, the porous alkaline salt of 12-tungstophosphoric acid significantly increases the ratio of isobutene alkylation to oligomerization when using high-density supercritical isobutene [158]. However, heteropolyacids are mostly liquid phase catalysts, which have problems such as difficulty in recycling, environmental pollution, and equipment corrosion. Moreover, the specific surface area of heteropolyacids is small; so, single heteropolyacid is not conducive to fully exerting catalytic activity and needs to be supported on a porous solid carrier. In addition, its selectivity toward target products, especially trimethylpentane, is very low [159].

References

[1] D.X. Cao, X.N. Yang, G.Q. Cao, C4 alkylation technology development trend, Pet. Petrochem. Today 28 (2020) 29–37.

[2] M. Aldossary, The Alkylation of 2-Butene With Isobutane Over Large-Pore Zeolites, The University of Manchester, United Kingdom, 2018.

[3] H.X. Ma, F. Zhou, K. Kai, Advances in liquid acid alkylation technology, Chem. Ind. Eng. Prog. 33 (2014) 32–40.

[4] M.H. Dong, B.N. Zong, SINOALKY Sulfuric acid alkylation process technology and its industrial application, Pet. Process. Petrochem. 50 (2019) 29–32.

[5] P. Cao, L. Zheng, W. Sun, L. Zhao, Multiscale modeling of isobutane alkylation with mixed C4 olefins using sulfuric acid as catalyst, Ind. Eng. Chem. Res. 58 (2019) 6340–6349.

[6] B.L. Gao, High Octane Gasoline Component Production Technology, China Petrochemical Press, Beijing, 2006.
[7] K. Li, R.E. Eckert, L.F. Albright, Alkylation of isobutane with light olefins using sulfuric acid. Operating variables affecting both chemical and physical phenomena, Ind. Eng. Chem. Process. Des. Dev. 9 (1970) 441–446.
[8] J.G. Speight, Handbook of Industrial Hydrocarbon Processes, Gulf Professional Publishing, 2019.
[9] Y. Xin, Y. Hu, M. Li, K. Chi, S. Zhang, F. Gao, S. Jiang, Y. Wang, C. Ren, G. Li, Isobutane alkylation catalyzed by H_2SO_4: effect of H_2SO_4 acid impurities on alkylate distribution, Energy Fuel 35 (2021) 1664–1676.
[10] Y. Xin, Y. Hu, M. Li, K. Chi, S. Zhang, F. Gao, S. Jiang, Y. Wang, C. Ren, G. Li, Isobutane alkylation catalyzed by H_2SO_4: effect of H_2SO_4 acid impurities on alkylate distribution, Energ. Fuel 35 (2020) 1664–1676.
[11] D.L. Minnick, R.R. Kore, C.J. Lyon, B. Subramaniam, M.B. Shiflett, A.M. Scurto, Understanding sulfur content in alkylate from sulfuric acid-catalyzed C3/C4 alkylations, Energ. Fuel 33 (2019) 4659–4670.
[12] J.R. Rostrup-Nielsen, Innovation and the catalytic process industry—the science and the challenge, Chem. Eng. Sci. 50 (1995) 4061–4071.
[13] J.C. Kuo, Slurry Fischer-Tropsch/Mobil Two Stage Process of Converting SYNGAS to High Octane Gasoline, Final report, Mobil Research and Development Corp, Paulsboro, NJ, 1983.
[14] G. Parkinson, Revving up for alkylation, Chem. Eng. 108 (2001) 27.
[15] Z. Ma, J. Sha, W. Zheng, W. Sun, L. Zhao, Effects of deep eutectic solvents on H_2SO_4-catalyzed alkylation: Combining experiment and molecular dynamics simulation, AIChE J. 68 (2022) e17556.
[16] S. Tang, A.M. Scurto, B. Subramaniam, Improved 1-butene/isobutane alkylation with acidic ionic liquids and tunable acid/ionic liquid mixtures, J. Catal. 268 (2009) 243–250.
[17] Q. Huang, G. Zhao, S. Zhang, F. Yang, Improved catalytic lifetime of H_2SO_4 for isobutane alkylation with trace amount of ionic liquids buffer, Ind. Eng. Chem. Res. 54 (2015) 1464–1469.
[18] W. Zheng, Z. Ma, W. Sun, L. Zhao, Target high-efficiency ionic liquids to promote H_2SO_4-catalyzed C4 alkylation by machine learning, AICHE J. 68 (2022) e17698.
[19] Y. Zhao, T. Li, X. Meng, H. Wang, Y. Zhang, H. Wang, S. Zhang, Improvement of product distribution through enhanced mass transfer in isobutane/butene alkylation, Chem. Eng. Res. Des. 143 (2019) 190–200.
[20] H. Zhang, R. Liu, Z. Yang, F. Huo, R. Zhang, Z. Li, S. Zhang, Y. Wang, Alkylation of isobutane/butene promoted by fluoride-containing ionic liquids, Fuel 211 (2018) 233–240.
[21] Y. Liu, R. Li, H. Sun, R. Hu, Effects of catalyst composition on the ionic liquid catalyzed isobutane/2-butene alkylation, J Mol. Catal. A Chem. 398 (2015) 133–139.
[22] W. Zheng, P. Cao, Y. Yuan, C. Huang, Z. Wang, W. Sun, L. Zhao, Experimental and modeling study of isobutane alkylation with C4 olefin catalyzed by Brønsted acidic ionic liquid/sulfuric acid, Chem. Eng. J. 377 (2019) 119578.
[23] P. Cao, Y. Yuan, C. Huang, W. Sun, L. Zhao, Promoting the sulfuric acid catalyzed isobutane alkylation by quaternary ammonium ionic liquids, AIChE J. 66 (2020) e16979.
[24] Z. Ma, W. Zheng, W. Sun, L. Zhao, Enhanced catalytic performance of H_2SO_4-catalyzed C4 alkylation by formyl functional $[N_{1,1,1,1}][C_{10}SO_4]$ additive, AICHE J. (2023) e18179.
[25] Y. Zhao, Ionic Liquid-Enhanced Concentrated Sulfuric Acid-Catalyzed Isobutane/Butene Alkylation Reaction, Zhengzhou University, 2018.
[26] W. Sun, W. Zheng, W. Xie, L. Zhao, Understanding structure–property relationship of SO_3H-functionalized ionic liquids together with sulfuric acid in catalyzing isobutane alkylation with C4 olefin, Ind. Eng. Chem. Res. 57 (2018) 15310–15318.

[27] W. Sun, W. Zheng, P. Cao, L. Zhao, Probing interfacial behaviors of Brønsted acidic ionic liquids improved isobutane alkylation with C4 olefin catalyzed by sulfuric acid, Chem. Eng. J. 377 (2019) 119744.
[28] W. Zheng, P. Cao, W. Sun, L. Zhao, Towards an understanding of the microstructure and interfacial properties of the ionic liquid/sulfuric acid catalyst in liquid-liquid reactions, Chem. Eng. Sci. 205 (2019) 287–298.
[29] A. Roebuck, B. Evering, Isobutane-olefin alklation with inhibited aluminum chloride catalysts, Ind. Eng. Chem. Process. Des. Dev. 9 (1970) 76–82.
[30] L. Li, J. Zhang, K. Wang, G. Luo, Caprolactam as a new additive to enhance alkylation of isobutane and butene in H_2SO_4, Ind. Eng. Chem. Res. 55 (2016) 12818–12824.
[31] H. Zhang, H. Tang, J. Xu, Y. Li, Z. Yang, R. Liu, S. Zhang, Y. Wang, Enhancement of isobutane/butene alkylation by aromatic-compound additives in strong Brønsted acid, Fuel 231 (2018) 224–233.
[32] G. Wu, Y. Liu, G. Liu, R. Hu, G. Gao, Role of aromatics in isobutane alkylation of chloroaluminate ionic liquids: insights from aromatic–ion interaction, J. Catal. 396 (2021) 54–64.
[33] Y. Liu, G. Wu, R. Hu, G. Gao, Effects of aromatics on ionic liquids for C4 alkylation reaction: Insights from scale-up experiment and molecular dynamics simulation, Chem. Eng. J. 402 (2020) 126252.
[34] M.S. Rakow, W.H. Lockwood Jr., Dodecylbenzenesulfonic Acid Addition in Sulfuric Acid Alkylation, US Patent 3,655,807, 1972.
[35] F.C. Mccoy, E.L. Cole, Alkylation Catalyst Additive, US Patent 3,865,896, 1975.
[36] W.-S. Chen, Solubility measurements of isobutane/alkenes in sulfuric acid: applications to alkylation, Appl. Catal. A Gen. 255 (2003) 231–237.
[37] W. Zheng, Z. Wang, W. Sun, L. Zhao, F. Qian, H_2SO_4-catalyzed isobutane alkylation under low temperatures promoted by long-alkyl-chain surfactant additives, AIChE J. 67 (2021) e17349.
[38] Q. Zhang, K.D.O. Vigier, S. Royer, F. Jérôme, Deep eutectic solvents: syntheses, properties and applications, Chem. Soc. Rev. 41 (2012) 7108–7146.
[39] A.P. Abbott, D. Boothby, G. Capper, D.L. Davies, R.K. Rasheed, Deep eutectic solvents formed between choline chloride and carboxylic acids: versatile alternatives to ionic liquids, J. Am. Chem. Soc. 126 (2004) 9142–9147.
[40] F.-L. Yu, Y.-L. Gu, X. Gao, Q.-C. Liu, C.-X. Xie, S.-T. Yu, Alkylation of isobutane and isobutene catalyzed by trifluoromethanesulfonic acid-taurine deep eutectic solvents in polyethylene glycol, Chem. Commun. 55 (2019) 4833–4836.
[41] L. Li, J. Zhang, C. Du, G. Luo, Intensification of the sulfuric acid alkylation process with trifluoroacetic acid, AIChE J. 65 (2019) 113–119.
[42] S. Eigler, A. Hirsch, Chemistry with graphene and graphene oxide—challenges for synthetic chemists, Angew. Chem. Int. Edit. 53 (2014) 7720–7738.
[43] X. Meng, P. Prawang, Z. Li, H. Wang, S. Zhang, Carbon-based materials enhanced emulsification to improve product distribution in isobutane/butene alkylation catalyzed by sulfuric acid, Ind. Eng. Chem. Res. 56 (2017) 7700–7707.
[44] P. Esteves, C. Araújo, B. Horta, L. Alvarez, C. Zicovich-Wilson, A. Ramírez-Solís, The isobutylene–isobutane alkylation process in liquid HF revisited, J. Phys. Chem. B 109 (2005) 12946–12955.
[45] A. Eastman, B. Randolph, W. Moore, R. Heald, Consider online monitoring of HF acid when optimizing alkylation operations, Hydrocarb. Process. 80 (2001) 95.
[46] A. Corma, A. Martinez, Chemistry, catalysts, and processes for isoparaffin–olefin alkylation: actual situation and future trends, Catal. Rev. 35 (1993) 483–570.
[47] L.F. Albright, Alkylation of isobutane with C3–C5 olefins: feedstock consumption, acid usage, and alkylate quality for different processes, Ind. Eng. Chem. Res. 41 (2002) 5627–5631.

[48] Y. Chauvin, A. Hirschauer, H. Olivier, Alkylation of isobutane with 2-butene using 1-butyl-3-methylimidazolium chloride—aluminium chloride molten salts as catalysts, J. Mol. Catal. A Chem. 92 (1994) 155–165.
[49] K. Yoo, V.V. Namboodiri, R.S. Varma, P.G. Smirniotis, Ionic liquid-catalyzed alkylation of isobutane with 2-butene, J. Catal. 222 (2004) 511–519.
[50] D. Lu, G. Zhao, B. Ren, Isobutane alkylation catalyzed by ether functionalized ionic liquids, Chem. Ind. Eng. Soc. China J. 66 (2015) 2481–2487.
[51] Y. Liu, R. Hu, C. Xu, H. Su, Alkylation of isobutene with 2-butene using composite ionic liquid catalysts, Appl. Catal. A Gen. 346 (2008) 189–193.
[52] W. Zheng, H. Wang, W. Xie, L. Zhao, W. Sun, Understanding interfacial behaviors of isobutane alkylation with C4 olefin catalyzed by sulfuric acid or ionic liquids, AIChE J. 64 (2018) 950–960.
[53] W. Zheng, W. Xie, W. Sun, L. Zhao, Modeling of the interfacial behaviors for the isobutane alkylation with C4 olefin using ionic liquid as catalyst, Chem. Eng. Sci. 166 (2017) 42–52.
[54] W. Zheng, D. Li, W. Sun, L. Zhao, Multi-scale modeling of isobutane alkylation with 2-butene using composite ionic liquids as catalyst, Chem. Eng. Sci. 186 (2018) 209–218.
[55] W. Zhang, S. Kim, L. Wahl, R. Khare, L. Hale, J. Hu, D.M. Camaioni, O.Y. Gutiérrez, Y. Liu, J.A. Lercher, Low-temperature upcycling of polyolefins into liquid alkanes via tandem cracking-alkylation, Science 379 (2023) 807–811.
[56] A. Wang, G. Zhao, F. Liu, L. Ullah, S. Zhang, A. Zheng, Anionic clusters enhanced catalytic performance of protic acid ionic liquids for isobutane alkylation, Ind. Eng. Chem. Res. 55 (2016) 8271–8280.
[57] Y. Chauvin, S. Einloft, H. Olivier, Catalytic dimerization of propene by nickel-phosphine complexes in 1-butyl-3-methylimidazolium chloride/$AlEtxCl_{3-x}$ (x= 0, 1) ionic liquids, Ind. Eng. Chem. Res. 34 (1995) 1149–1155.
[58] Y. He, Z. Man, Thermodynamic analysis on i-C4/$C4^{=}$ alkylation, J. Fuel Chem. Technol. 34 (2006) 591.
[59] Y. Liu, L. Wang, R. Li, R. Hu, Reaction mechanism of ionic liquid catalyzed alkylation: alkylation of 2-butene with deuterated isobutene, J. Mol. Catal. A Chem. 421 (2016) 29–36.
[60] Z. Liu, X. Meng, R. Zhang, C. Xu, H. Dong, Y. Hu, Reaction performance of isobutane alkylation catalyzed by a composite ionic liquid at a short contact time, AICHE J. 60 (2014) 2244–2253.
[61] K. Qiao, Y. Deng, Alkylations of benzene in room temperature ionic liquids modified with HCl, J. Mol. Catal. A Chem. 171 (2001) 81–84.
[62] G.A. Olah, P. Batamack, D. Deffieux, B. Török, Q. Wang, Á. Molnár, G.S. Prakash, Acidity dependence of the trifluoromethanesulfonic acid catalyzed isobutane-isobutylene alkylation modified with trifluoroacetic acid or water, Appl. Catal. A Gen. 146 (1996) 107–117.
[63] P. Cui, G. Zhao, H. Ren, J. Huang, S. Zhang, Ionic liquid enhanced alkylation of isobutane and 1-butene, Catal. Today 200 (2013) 30–35.
[64] H. Ren, G. Zhao, S. Zhang, P. Cui, J. Huang, Triflic acid catalyzed isobutane alkylation with trifluoroethanol as a promoter, Catal. Commun. 18 (2012) 85–88.
[65] H. Diaz Velazquez, N. Likhanova, N. Aljammal, F. Verpoort, R. Martinez-Palou, New insights into the progress on the isobutane/butene alkylation reaction and related processes for high-quality fuel production. A critical review, Energ. Fuel 34 (2020) 15525–15556.
[66] T.G. Xu, X.J. Wu, G.M. Wen, W.C. Zhang, F.M. Sun, C4 alkylation process technology research and application progress, Refin. Chem. Ind. 27 (2016) 1–3.
[67] G. Rothenberg, Catalysis: Concepts and Green Applications, John Wiley & Sons, 2017.

[68] E.H. Van Broekhoven, F.R.M. Cabre, P. Bogaard, G. Klaver, M. Vonhof, Process for Alkylating Hydrocarbons, US Patent 5,986,158, 1999.

[69] S. Singhal, S. Agarwal, S. Arora, N. Singhal, A. Kumar, Solid acids: potential catalysts for alkene–isoalkane alkylation, Catal. Sci. Technol. 7 (2017) 5810–5819.

[70] A. Badakhshan, M. Kazemeini, F. Khorasheh, S. Sahebdelfar, Solid acid catalytic alkylation: a means for gasoline-aromatics reduction, in: World Petroleum Congress, WPC 2002, pp. WPC-32253, 2002.

[71] S.I. Hommeltoft, Isobutane alkylation: recent developments and future perspectives, Appl. Catal. A Gen. 221 (2001) 421–428.

[72] S. Li, Molecular simulation study of structure-performance relationship of zeolite catalyst in C4 alkylation process, China University of Petroleum (East China), 2020.

[73] S.I. Hommeltoft, O. Ekelund, J. Zavilla, Role of ester intermediates in isobutane alkylation and its consequence for the choice of catalyst system, Ind. Eng. Chem. Res. 36 (1997) 3491–3497.

[74] J.G. Bi, Advances in alkylated oil production technology, Chem. Ind. Eng. Prog. (2007) 934–939.

[75] L. Kane, S. Romanow, Hand-held detection system for explosives (Innovations), Hydrocarb. Process. 82 (2003) 30–31.

[76] X.J. Zhang, Z.R. Gao, H.J. Cai, Progress in the application of isobutylbutene alkylation process technology, Appl. Chem. Ind. 49 (2020). 741-743+749.

[77] A. Primo, H. Garcia, Zeolites as catalysts in oil refining, Chem. Soc. Rev. 43 (2014) 7548–7561.

[78] P. Cao, B. Li, W. Sun, L. Zhao, Understanding the zeolites catalyzed isobutane alkylation based on their topology effects on the reactants adsorption, Chem. Eng. Sci. 250 (2022) 117387.

[79] C.H. Christensen, K. Johannsen, I. Schmidt, C.H. Christensen, Catalytic benzene alkylation over mesoporous zeolite single crystals: improving activity and selectivity with a new family of porous materials, J. Am. Chem. Soc. 125 (2003) 13370–13371.

[80] A. Corma, H. Garcia, Zeolite-based photocatalysts, Chem. Commun. (2004) 1443–1459.

[81] C.X.A. da Silva, V.L.C. Gonçalves, C.J.A. Mota, Water-tolerant zeolite catalyst for the acetalisation of glycerol, Green Chem. 11 (2009) 38–41.

[82] R. Krishna, B. Smit, S. Calero, Entropy effects during sorption of alkanes in zeolites, Chem. Soc. Rev. 31 (2002) 185–194.

[83] A. Corma, M.I. Juan-Rajadell, J.M. López-Nieto, A. Martinez, C. Martínez, A comparative study of O_4^{2-}/ZrO_2 and zeolite beta as catalysts for the isomerization of n-butane and the alkylation of isobutane with 2-butene, Appl. Catal. A Gen. 111 (1994) 175–189.

[84] B.O.D. Costa, C.A. Querini, Isobutane alkylation with solid catalysts based on beta zeolite, Appl. Catal. A Gen. 385 (2010) 144–152.

[85] Y. Liu, G. Wu, X. Pang, G. Gao, Isobutane alkylation with 2-butene in novel ionic liquid/solid acid catalysts, Fuel 252 (2019) 316–324.

[86] K. Yoo, E.C. Burckle, P.G. Smirniotis, Isobutane/2-butene alkylation using large-pore zeolites: influence of pore structure on activity and selectivity, J. Catal. 211 (2002) 6–18.

[87] H. Zhang, J. Xu, H. Tang, Z. Yang, R. Liu, S. Zhang, Isobutane/2-butene alkylation reaction catalyzed by cu-modified and rare earth X-type zeolite, Ind. Eng. Chem. Res. 58 (2019) 9690–9700.

[88] S. Sahebdelfar, M. Kazemeini, F. Khorasheh, A. Badakhshan, Deactivation behavior of the catalyst in solid acid catalyzed alkylation: effect of pore mouth plugging, Chem. Eng. Sci. 57 (2002) 3611–3620.

[89] C.A. Querini, E. Roa, Deactivation of solid acid catalysts during isobutane alkylation with C4 olefins, Appl. Catal. A Gen. 163 (1997) 199–215.

[90] A. Feller, J.-O. Barth, A. Guzman, I. Zuazo, J.A. Lercher, Deactivation pathways in zeolite-catalyzed isobutane/butene alkylation, J. Catal. 220 (2003) 192–206.
[91] A. Feller, J.A. Lercher, Chemistry and technology of isobutane/alkene alkylation catalyzed by liquid and solid acids, Adv. Catal. 48 (2004) 229–295.
[92] Y.F. Chu, A.W. Chester, Reactions of isobutane with butene over zeolite catalysts, Zeolites 6 (1986) 195–200.
[93] A.L. Mota Salinas, G. Sapaly, Y. Ben Taarit, J.C. Vedrine, N. Essayem, Continuous supercritical $iC_4/C_4^=$ alkylation over H-Beta and H-USY Influence of the zeolite structure, Appl. Catal. A Gen. 336 (2008) 61–71.
[94] B. Li, P. Huang, P. Cao, W. Gao, W. Zheng, C. Lian, W. Sun, L. Zhao, Understanding the structural properties of zeolites for isobutane alkylation based on adsorption/diffusion behaviors, Microporous Mesoporous Mater. 341 (2022) 112040.
[95] A. Feliczak-Guzik, Hierarchical zeolites: synthesis and catalytic properties, Microporous Mesoporous Mater. 259 (2018) 33–45.
[96] D.M. Ruthven, M.F.M. Post, Chapter 12 Diffusion in zeolite molecular sieves, in: H. van Bekkum, E.M. Flanigen, P.A. Jacobs, J.C. Jansen (Eds.), Studies in Surface Science and Catalysis, Elsevier, 2001, pp. 525–577.
[97] M. Hartmann, A.G. Machoke, W. Schwieger, Catalytic test reactions for the evaluation of hierarchical zeolites, Chem. Soc. Rev. 45 (2016) 3313–3330.
[98] M. Guisnet, F.R. Ribeiro, Deactivation and Regeneration of Zeolite Catalysts, 2011.
[99] Z. Liu, J. Yuan, J.M. van Baten, J. Zhou, X. Tang, C. Zhao, W. Chen, X. Yi, R. Krishna, G. Sastre, A. Zheng, Synergistically enhance confined diffusion by continuum intersecting channels in zeolites, Sci. Adv. 7 (2021). eabf0775.
[100] L.-E. Sandoval-Díaz, J.-A. González-Amaya, C.-A. Trujillo, General aspects of zeolite acidity characterization, Microporous Mesoporous Mater. 215 (2015) 229–243.
[101] C.L. Thomas, Chemistry of cracking catalysts, Ind. Eng. Chem. 41 (1949) 2564–2573.
[102] X. Zhao, J. Xu, F. Deng, Solid-state NMR for metal-containing zeolites: from active sites to reaction mechanism, Front, Chem. Sci. 14 (2020) 159–187.
[103] G.S. Nivarthy, K. Seshan, J.A. Lercher, The influence of acidity on zeolite H-BEA catalyzed isobutane/n-butene alkylation, Microporous Mesoporous Mater. 22 (1998) 379–388.
[104] J.W. Ward, The nature of active sites on zeolites: III. The alkali and alkaline earth ion-exchanged forms, J. Catal. 10 (1968) 34–46.
[105] I.M. Gerzeliev, V.A. Ostroumova, M.N. Baskhanova, Z.A. Saitov, V.A. Temnikova, D.O. Khusaimova, On the choice of ion exchange method for fau type zeolite to synthesize an active and selective catalyst for isobutane alkylation with butylenes, Pet. Chem. 57 (2017) 1182–1185.
[106] I.M. Gerzeliev, V.A. Ostroumova, M.N. Baskhanova, Enhancement of ion exchange in a FAU type zeolite during the synthesis of an active and selective catalyst for isobutane alkylation with butylenes, Pet. Chem. 58 (2018) 676–680.
[107] I.M. Gerzeliev, V.A. Temnikova, O.V. Deniskin, M.N. Baskhanova, D.O. Khusaimova, A.L. Maksimov, Effect of chemical composition of zeolite catalysts on their catalytic properties in isobutane alkylation with butylenes, Pet. Chem. 59 (2019) 706–710.
[108] J.W. Ward, The nature of active sites on zeolites: VI. The influence of calcination temperature on the structural hydroxyl groups and acidity of stabilized hydrogen Y zeolite, J. Catal. 11 (1968) 251–258.
[109] J.W. Ward, The nature of active sites on zeolites: IV. The influence of water on the acidity of X and Y type zeolites, J. Catal. 11 (1968) 238–250.
[110] J.W. Ward, The nature of active sites on zeolites: VIII. Rare earth Y zeolite, J. Catal. 13 (1969) 321–327.
[111] M. Ravi, V.L. Sushkevich, J.A. van Bokhoven, Towards a better understanding of Lewis acidic aluminium in zeolites, Nat. Mater. 19 (2020) 1047–1056.

[112] A. Corma, A. Martinez, C. Martinez, Isobutane/2-butene alkylation on MCM-22 catalyst. Influence of zeolite structure and acidity on activity and selectivity, Catal. Lett. 28 (1994) 187–201.

[113] K.S. Yoo, P.G. Smirniotis, Zeolites-catalyzed alkylation of isobutane with 2-butene: influence of acidic properties, Catal. Lett. 103 (2005) 249–255.

[114] A. Corma, A. Martínez, P.A. Arroyo, J.L.F. Monteiro, E.F. Sousa-Aguiar, Isobutane/2-butene alkylation on zeolite beta: influence of post-synthesis treatments, Appl. Catal. A Gen. 142 (1996) 139–150.

[115] A. Corma, A. Martinez, C. Martinez, The role of extraframework aluminum species in USY catalysts during isobutane/2-butene alkylation, Appl. Catal. A Gen. 134 (1996) 169–182.

[116] O.I. Kuznetsov, G.M. Panchenkov, V.A. Plakhotnik, B. Razali, Y.I. Tolkacheva, Alkylation of isobutane with butenes in the presence of HNaY zeolite modified by cations of nickel, calcium and rare-earth elements, Pet. Chem. USSR 21 (1981) (Engl. Transl.); (United States).

[117] Z. Yang, R. Zhang, H. Zhang, H. Tang, R. Liu, S. Zhang, Comparative catalytic study on butene/isobutane alkylation over LaX and CeX zeolites: the influence of calcination atmosphere, Chin. J. Chem. Eng. 46 (2022) 173–183.

[118] Y.A. Khamzin, R.R. Shiriyazdanov, E.G. Telyashev, A.R. Davletshin, Factors determining the stability of alkylation catalysts based on Y zeolites, Chem. Technol. Fuels Oils 53 (2017) 675–682.

[119] A. Guzman, I. Zuazo, A. Feller, R. Olindo, C. Sievers, J.A. Lercher, On the formation of the acid sites in lanthanum exchanged X zeolites used for isobutane/cis-2-butene alkylation, Microporous Mesoporous Mater. 83 (2005) 309–318.

[120] J.W. Ward, The nature of active sites on zeolites. X. The acidity and catalytic activity of X zeolites, J. Catal. 14 (4) (1969) 365–378.

[121] E.F.T. Lee, L.V.C. Rees, Effect of calcination on location and valency of lanthanum ions in zeolite Y, Zeolites 7 (2) (1987) 143–147.

[122] M. Jin, M. Oh, M. Choi, Hierarchical BEA zeolite with trimodal micro-/meso-/macroporosity as a selective and long-lived catalyst for isobutane/2-butene alkylation, ACS Catal. 12 (2022) 4067–4077.

[123] M. Guisnet, N. Gnep, Mechanism of short-chain alkane transformation over protonic zeolites. Alkylation, disproportionation and aromatization, Appl. Catal. A Gen. 146 (1996) 33–64.

[124] G.S. Nivarthy, A. Feller, K. Seshan, J.A. Lercher, Alkylation of isobutane with light olefins catalyzed by zeolite beta, Microporous Mesoporous Mater. 35 (2000) 75–87.

[125] L.F. Albright, M.A. Spalding, J.A. Nowinski, R.M. Ybarra, R.E. Eckert, Alkylation of isobutane with C4 olefins. 1. First-step reactions using sulfuric acid catalyst, Ind. Eng. Chem. Res. 27 (1988) 381–386.

[126] F.W. Kirsch, J.D. Potts, D.S. Barmby, Isoparaffin-olefin alkylations with crystalline aluminosilicates: I. Early studies—C4-olefins, J. Catal. 27 (1972) 142–150.

[127] Y. Sekine, Y.-I. Tajima, Y.-S. Ichikawa, M. Matsukata, E. Kikuchi, Alkylation of isobutane by 1-butene over H-beta zeolite in CSTR (Part 2) deactivation mechanism of zeolite catalyst and optimization of CSTR conditions, J. Jpn. Pet. Inst. 55 (2012) 308–318.

[128] A. Feller, A. Guzman, I. Zuazo, J.A. Lercher, On the mechanism of catalyzed isobutane/butene alkylation by zeolites, J. Catal. 224 (2004) 80–93.

[129] G.S. Nivarthy, Y. He, K. Seshan, J.A. Lercher, Elementary mechanistic steps and the influence of process variables in isobutane alkylation over H-BEA, J. Catal. 176 (1998) 192–203.

[130] K. Yoo, P.G. Smirniotis, The influence of Si/Al ratios of synthesized beta zeolites for the alkylation of isobutane with 2-butene, Appl. Catal. A Gen. 227 (2002) 171–179.

[131] K. De Jong, C. Mesters, D. Peferoen, P. Van Brugge, C. De Groot, Paraffin alkylation using zeolite catalysts in a slurry reactor: chemical engineering principles to extend catalyst lifetime, Chem. Eng. Sci. 51 (1996) 2053–2060.

[132] A. Corma, A. Martínez, C. Martínez, Isobutane/2-butene alkylation on ultrastable Y zeolites: influence of zeolite unit cell size, J. Catal. 146 (1994) 185–192.
[133] J. Weitkamp, H. Karge, H. Pfeifer, W. Hölderich, Zeolites and related microporous materials: state of the art 1994, in: Proceedings of the 10th International Zeolite Conference, Garmisch-Partenkirchen, Germany, July 17–22, Elsevier, Garmisch-Partenkirchen, Germany, 1994.
[134] C. Guo, S. Yao, J. Cao, Z. Qian, Alkylation of isobutane with butenes over solid superacids, SO_4^{2-}/ZrO_2 and SO_4^{2-}/TiO_2, Appl. Catal. A Gen. 107 (1994) 229–238.
[135] A. Corma, V. Gómez, A. Martínez, Zeolite beta as a catalyst for alkylation of isobutane with 2-butene. Influence of synthesis conditions and process variables, Appl. Catal. A Gen. 119 (1994) 83–96.
[136] A. Corma, A. Martinez, C. Martinez, Influence of process variables on the continuous alkylation of isobutane with 2-butene on superacid sulfated zirconia catalysts, J. Catal. 149 (1994) 52–60.
[137] V.R. Sarsani, Y. Wang, B. Subramaniam, Toward stable solid acid catalysts for 1-butene + isobutane alkylation: investigations of heteropolyacids in dense CO_2 media, Ind. Eng. Chem. Res. 44 (2005) 6491–6495.
[138] O. Toshio, Y. Masayuki, N. Kyutae, M. Makoto, Alkylation of isobutane with butenes catalyzed by a cesium hydrogen salt of 12-tungstophosphoric acid, Chem. Lett. 23 (1994) 1451–1454.
[139] P. Botella, A. Corma, J.M. López-Nieto, The influence of textural and compositional characteristics of nafion/silica composites on isobutane/2-butene alkylation, J. Catal. 185 (1999) 371–377.
[140] C. Lyon, B. Subramaniam, C. Pereira, Enhanced isooctane yields for 1-butene/isobutane alkylation on SiO_2-supported, Stud. Surf. Sci. Catal. (2001) 221–228. Elsevier.
[141] X. Song, A. Sayari, Sulfated zirconia-based strong solid-acid catalysts: recent progress, Catal. Rev. 38 (1996) 329–412.
[142] K. Arata, Preparation of superacids by metal oxides for reactions of butanes and pentanes, Appl. Catal. A Gen. 146 (1996) 3–32.
[143] B.H. Davis, R.A. Keogh, R. Srinivasan, Sulfated zirconia as a hydrocarbon conversion catalyst, Catal. Today 20 (1994) 219–256.
[144] M.Y. Smirnova, G.A. Urguntsev, A.B. Ayupov, A.A. Vedyagin, G.V. Echevsky, Isobutane/butene alkylation on sulfated alumina: influence of sulfation condition on textural, structural and catalytic properties, Appl. Catal. A Gen. 344 (2008) 107–113.
[145] J. Yang, M. Zhang, F. Deng, Q. Luo, D. Yi, C. Ye, Solid state NMR study of acid sites formed by adsorption of SO_3 onto γ-Al_2O_3, Chem. Commun. (2003) 884–885.
[146] M.A.L. Ferreira, E.H. Rueda, Theoretical characterization of alumina and sulfated-alumina catalysts for n-butene isomerization, J. Mol. Catal. A Chem. 178 (2002) 147–160.
[147] A.E. Koklin, V.M.K. Chan, V.I. Bogdan, Conversion of isobutane-butenes mixtures on H-USY and SO_4^{2-}/ZrO_2 catalysts under supercritical conditions: isobutane alkylation and butenes oligomerization, Russ. J. Phys. Chem. B 8 (2014) 991–998.
[148] R.R. Anand, C.D. Devidas, Alkylation of phenol by C9 and C12 olefins, Bull. Chem. Soc. Jpn. 61 (1988) 1379–1381.
[149] G.A. Olah, J. Kaspi, J. Bukala, Heterogeneous catalysis by solid superacids. 3. Alkylation of benzene and transalkylation of alkylbenzenes over graphite-intercalated Lewis acid halide and perfluorinated resin sulfonic acid (Nafion-H) catalysts, J. Org. Chem. 42 (1977) 4187–4191.
[150] H. Hasegawa, T. Higashimura, Selective alkylation of aromatic hydrocarbons with styrene by solid polymeric oxo acids, Polym. J. 12 (1980) 407–409.
[151] M.A. Harmer, W.E. Farneth, Q. Sun, Towards the sulfuric acid of solids, Adv. Mater. 10 (1998) 1255–1257.

[152] W. Shen, D. Dubé, S. Kaliaguine, Alkylation of isobutane/1-butene over periodic mesoporous organosilica functionalized with perfluoroalkylsulfonic acid group, Catal. Commun. 10 (2008) 291–294.
[153] A. Heidekum, M.A. Harmer, W.F. Hoelderich, Highly selective fries rearrangement over zeolites and nafion in silica composite catalysts: a comparison, J. Catal. 176 (1998) 260–263.
[154] M. Misono, T. Okuhara, Solid superacid catalysts, Chemtech (United States) 23 (1993).
[155] W. Shen, Y. Gu, H. Xu, D. Dubé, S. Kaliaguine, Alkylation of isobutane/1-butene on methyl-modified Nafion/SBA-15 materials, Appl. Catal. A Gen. 377 (2010) 1–8.
[156] S. Hommeltoft, L. Jørgensen, Fixed bed alkylation using SLP type catalyst, Erdöl Erdgas Kohle 114 (1998).
[157] F. Cardona, N.S. Gnep, M. Guisnet, G. Szabo, P. Nascimento, Reactions involved in the alkylation of isobutane with 2-butene and with propene on a USHY zeolite, Appl. Catal. A Gen. 128 (1995) 243–257.
[158] A. Feller, A. Guzman, I. Zuazo, J.A. Lercher, 9 A novel process for solid acid catalyzed isobutane/butene alkylation, in: M. Anpo, M. Onaka, H. Yamashita (Eds.), Stud Surf Sci Catal, Elsevier, 2003, pp. 67–72.
[159] S.B. He, Research on Solid Acid Catalyst and Process for Alkylation of Isobutane and Butene, 2005.

CHAPTER

3

Thermodynamics and reaction mechanism

Due to the substantial economic value associated with the alkylate, the investigations into the mechanism of alkylation have persisted for several decades. The earliest investigations reported that the carbonium ion mechanism involves a structured sequence of consecutive and simultaneous reaction steps, providing a comprehensive framework for describing the observed alkylation. Subsequent to these pioneering efforts, Albright emerged as a principal contributor, offering significant information of the alkylation mechanism concerning processes catalyzed by H_2SO_4 [1–5]. In contrast, there exists a notable scarcity of information pertaining to the catalytic role of HF. The difference between zeolites as representatives of solid acid alkylation catalysts, H_2SO_4, HF, and ionic liquids (ILs) as liquid catalysts is emphasized.

3.1 Thermodynamics of C4 alkylation

The C4 alkylation in the petrochemical industry is commonly used to obtain alkylate with multiple components, which is the major component of high-octane gasoline. The simplified thermodynamics reaction network is displayed as shown below (Fig. 3.1).

The thermodynamics reaction pathways of C4 alkylation are presented in Fig. 3.2. A comprehensive understanding of the thermodynamic characteristics of the C4 alkylation process under diverse temperature conditions requires the calculation of the standard-state enthalpy change ($\Delta_r H^\theta$) and entropy change ($\Delta_r S^\theta$) for the reaction. These values facilitate an in-depth analysis of the thermodynamic characteristics inherent to the

Aliphatic Alkylation
https://doi.org/10.1016/B978-0-443-21984-9.00005-2

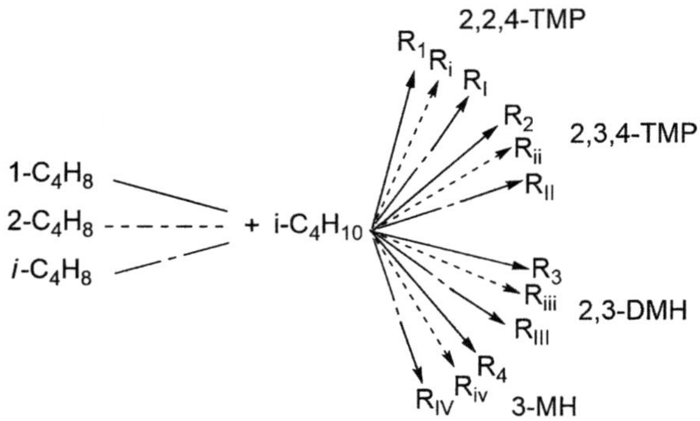

FIG. 3.1 The reaction network of C4 alkylation. *Taken from reference J. Ling, P. Cao, K. Li, L.X. Wen, J.L. Wu, F. Meng, X.H. Li, Effect of changes in C4 olefin composition on the thermodynamic behavior of alkylation reactions, Mod. Chem. Ind. 40 (2020) 150–154.*

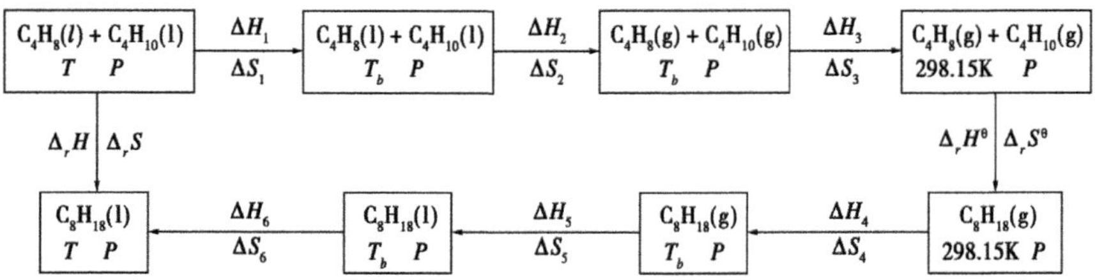

FIG. 3.2 Reaction pathway of the C4 alkylation. *From reference J. Ling, P. Cao, K. Li, L.X. Wen, J.L. Wu, F. Meng, X.H. Li, Effect o+f changes in C4 olefin composition on the thermodynamic behavior of alkylation reactions, Mod. Chem. Ind. 40 (2020) 150–154.*

process at different temperatures, as reported by Huang's work [6]. Here is given a systematic description, as follows:

$$\Delta G_{Tj}^{l} = \Delta H_{Tj}^{l} - T \cdot \Delta S_{Tj}^{l} \tag{3.1}$$

$$K_{Tj} = \exp\left(-\Delta G_{Tj}^{l}/RT\right) \tag{3.2}$$

The Gibbs free energy ΔG_{Tj}^{l} and equilibrium constant K_{Tj} for jth can be obtained using Eq. (3.2). The equilibrium constant can establish a relationship between the material compositions of the reaction system, thus expressing the interdependence of reactants and products, using Eq. (3.3).

$$K = K_f K_x = K_f \cdot \left[x_j/(x_{1-\mathrm{C_4H_8}} \cdot x_{i-\mathrm{C_4H_{10}}})\right] \quad (j = 1,2,3,4) \tag{3.3}$$

Butenes, butanes, and similar hydrocarbon substances with close carbon chain and molecular structures can be regarded as ideal liquid mixtures. Therefore, $K_f = 1$ and $x_{1\text{-}C_4H8}$, $x_{i\text{-}C_4H10}$, and x_j represent the molar fractions of 1-butene, isobutane, and the jth reaction product at equilibrium, respectively.

The equations can be further expressed by conservation of substances and elements in the reaction system:

$$x_{1-C_4H_8} + x_{i-C_4H_{10}} + \sum_{r=1}^{4} X_j = 1 \tag{3.4}$$

$$4x_{1-C_4H_8} + 4x_{i-C_4H_{10}} + 8\sum_{r=1}^{4} X_j = n_C/N \tag{3.5}$$

$$8x_{1-C_4H_8} + 10x_{i-C_4H_{10}} + 18\sum_{r=1}^{4} X_j = n_H/N \tag{3.6}$$

where n_C and n_H represent the quantities of carbon and hydrogen atoms in the reaction system, respectively. N is the total quantity of substances in the reaction system. Assuming initial reactants of 1 mol each for 1-butene and isobutane, i.e., $n_C = 8$, $n_H = 18$. The expression for the equilibrium conversion rate of reactants, X, is given by

$$X = 1 - x_{1-C_4H_8} \cdot N \tag{3.7}$$

The expression for the equilibrium selectivity of the jth reaction product, S_j, is as follows:

$$S_j = x_j/(1 - x_{1-C_4H_8} - x_{i-C_4H_{10}}) \quad (j = 1,2,3,4) \tag{3.8}$$

The enthalpy changes with respect to temperature for the C4 alkylation are illustrated in Table 3.1. It is apparent that the C4 alkylation is an exothermal reaction in the range of 59.15–83.05 kJ/mol, which depends on the olefin used.

Furthermore, the heat release increases as the temperature rises. For the main reactions R_1, R_i, and R_I, the exothermic heat release sequentially decreases during the C4 alkylation with 1-butene, 2-butene, and isobutene. Notably, the heat release during the alkylation with isobutene as feedstock is relatively lower, which is beneficial to control the temperature of the C4 alkylation.

The Gibbs free energy changes with respect to temperature for the C4 alkylation with 1-butene are presented in Table 3.2. The Gibbs free energy changes for C4 alkylation involving different olefins are all negative, indicating spontaneous reaction tendencies. Moreover, with an increase in temperature, the spontaneous nature of the reactions diminishes, suggesting that higher temperatures are not conducive to the reaction. The absolute values of Gibbs free energy changes for side reactions are relatively higher, which indicates a more pronounced thermodynamic preference for byproducts over main products, which becomes more prominent with rising temperatures. By considering kinetic factors, the low reaction temperature can effectively promote the main reaction while suppressing the

TABLE 3.1 The enthalpy-temperature relationship for the alkylation reaction of different butenes with isobutane.

	$\Delta H_{Tj}/(\mathrm{kJ\,mol^{-1}})$						
Reaction	**−10°C**	**0°C**	**10°C**	**20°C**	**30°C**	**40°C**	**50°C**
R_1	−75.78	−76.06	−76.38	−76.73	−77.13	−77.57	−78.06
R_2	−80.71	−81.01	−81.34	−81.71	−82.11	−82.56	−83.05
R_3	−76.29	−76.36	−76.46	−76.60	−76.77	−76.98	−77.25
R_4	−74.75	−74.92	−75.12	−75.36	−75.63	−75.95	−76.32
R_i	−67.69	−67.95	−68.24	−68.56	−68.91	−69.31	−69.75
R_{ii}	−72.63	−72.90	−73.20	−73.53	−73.89	−74.30	−74.74
R_{iii}	−68.20	−68.25	−68.32	−68.42	−68.55	−68.72	−68.94
R_{iv}	−66.67	−66.81	−66.98	−67.18	−67.42	−67.69	−68.01
R_I	−59.15	−59.45	−59.78	−60.14	−60.55	−61.00	−61.50
R_{II}	−64.09	−64.40	−64.74	−65.12	−65.53	−65.99	−66.49
R_{III}	−59.66	−59.74	−59.86	−60.00	−60.19	−60.41	−60.68
R_{IV}	−58.12	−58.31	−58.52	−58.77	−59.05	−59.38	−59.76

Data from reference J. Ling, P. Cao, K. Li, L.X. Wen, J.L. Wu, F. Meng, X.H. Li, Effect of changes in C4 olefin composition on the thermodynamic behavior of alkylation reactions, Mod. Chem. Ind. 40 (2020) 150–154.

TABLE 3.2 The Gibbs free energy-temperature relationship for the C4 alkylation with butene.

	$\Delta G_{Tj}/(\mathrm{kJ\,mol^{-1}})$						
Reaction	**−10°C**	**0°C**	**10°C**	**20°C**	**30°C**	**40°C**	**50°C**
R_1	−47.67	−46.59	−45.51	−44.42	−43.31	−42.19	−41.05
R_2	−51.57	−50.46	−49.33	−48.20	−47.05	−45.89	−44.71
R_3	−51.33	−50.39	−49.43	−48.48	−47.52	−46.55	−45.58
R_4	−54.06	−53.28	−52.48	−51.68	−50.87	−50.05	−49.22
R_i	−41.80	−40.81	−39.82	−38.82	−37.81	−36.78	−35.74
R_{ii}	−45.70	−44.68	−43.64	−42.60	−41.55	−40.48	−39.40
R_{iii}	−45.47	−44.61	−43.74	−42.88	−42.02	−41.15	−40.27
R_{iv}	−48.19	−47.49	−46.79	−46.08	−45.36	−44.64	−43.91
R_I	−34.69	−33.76	−32.82	−31.86	−30.89	−29.91	−28.91
R_{II}	−38.60	−37.62	−36.64	−35.64	−34.63	−33.61	−32.57
R_{III}	−38.36	−37.55	−36.74	−35.92	−35.10	−34.27	−33.43
R_{IV}	−41.09	−40.44	−39.78	−39.12	−38.45	−37.76	−37.07

Data from reference J. Ling, P. Cao, K. Li, L.X. Wen, J.L. Wu, F. Meng, X.H. Li, Effect of changes in C4 olefin composition on the thermodynamic behavior of alkylation reactions, Mod. Chem. Ind. 40 (2020) 150–154.

occurrence of side reactions, thereby enhancing the selectivity toward the desired products.

Among the C4 alkylation with three types of butenes, the absolute values of the Gibbs free energy changes for both the primary and side reactions of isobutene are the lowest. This implies that the spontaneity of the C4 alkylation is relatively weaker. An increase in content of isobutene in feedstock is unfavorable for the efficient conversion of olefins.

The equilibrium constants for different reactions are quite large, as shown in Table 3.3), indicating that alkylation is a typical irreversible reaction. With an increase in temperature, the equilibrium constants for all reactions decrease, with a reduction magnitude spanning between 2 and 3 orders of magnitude.

The thermodynamic characteristics of the main reactions of alkylation were presented in Table 3.4 [7]. At low temperatures, the equilibrium selectivities of the desired product TMPs, the byproduct DMHs, and MH remain relatively constant with variations in the isobutene content. At high temperatures, the equilibrium selectivity of the desired product TMPs decreases with an increase in the volume fraction of isobutene, while the equilibrium selectivity of the byproduct MH increases with

TABLE 3.3 The temperature-dependent equilibrium constant for the C4 alkylation reaction with butenes.

	K_{Tj}						
Reaction	**−10°C**	**0°C**	**10°C**	**20°C**	**30°C**	**40°C**	**50°C**
R_1	2.90×10^9	8.14×10^8	2.49×10^8	8.22×10^7	4.84×10^7	2.90×10^7	1.09×10^7
R_2	1.72×10^{10}	4.46×10^9	1.26×10^9	3.88×10^8	2.21×10^8	1.28×10^8	4.52×10^7
R_3	1.55×10^{10}	4.32×10^9	1.32×10^9	4.35×10^8	2.57×10^8	1.54×10^8	5.83×10^7
R_4	5.39×10^{10}	1.54×10^{10}	4.81×10^9	1.62×10^9	9.62×10^8	5.82×10^8	2.23×10^8
R_i	1.98×10^8	6.38×10^7	2.22×10^7	8.26×10^6	3.27×10^6	1.37×10^6	6.00×10^5
R_{ii}	1.18×10^9	3.50×10^8	1.13×10^8	3.90×10^7	1.44×10^7	5.66×10^6	2.34×10^6
R_{iii}	1.06×10^9	3.39×10^8	1.18×10^8	4.37×10^7	1.74×10^7	7.30×10^6	3.23×10^6
R_{iv}	3.69×10^9	1.21×10^9	4.28×10^8	1.63×10^8	6.56×10^7	2.80×10^7	1.25×10^7
R_I	7.71×10^6	2.86×10^6	1.13×10^6	4.75×10^5	2.10×10^5	9.74×10^4	4.71×10^4
R_{II}	4.59×10^7	1.57×10^7	5.74×10^6	2.24×10^6	9.27×10^5	4.03×10^5	1.84×10^5
R_{III}	4.12×10^7	1.52×10^7	5.99×10^6	2.52×10^6	1.12×10^6	5.21×10^5	2.54×10^5
R_{IV}	1.43×10^8	5.42×10^7	2.19×10^7	9.35×10^6	4.22×10^6	1.99×10^6	9.82×10^5

Data from reference J. Ling, P. Cao, K. Li, L.X. Wen, J.L. Wu, F. Meng, X.H. Li, Effect of changes in C4 olefin composition on the thermodynamic behavior of alkylation reactions, Mod. Chem. Ind. 40 (2020) 150–154.

TABLE 3.4 Thermodynamic characteristics of some main reactions of alkylation.

Reaction	ΔH (kJ mol^{-1})	ΔG (kJ mol^{-1})	ΔS (kJ mol^{-1} K^{-1})
Isobutane + butylene-1 →2,2,4-trimethylpentane	−72.5	−19.3	−190.8
Isobutane + butylene-1 → 2,3, 3- trimethylpentane	−134.1	−78.2	−200.4
Butylene-2 + butylene-2 → 2,2,4-trimethylpentane	−98.2	−41.9	−201.7
Isobutane + butylene-1 → 2,4-dimethylhexane	−79.2	−28.7	−181.5
Isobutane + butylene-1 → 2,3-dimethylhexane	−76.7	−25.4	−184.2
Isobutane + butylene-2 → 2,5-dimethylhexane	−63.6	−11.7	−186.4
Isobutane + propylene → 2,4-dimethylpentane	−60.5	−0.8	−116.4
Butylene-1 + propylene → 2,3-dimethylpentane	−140.1	−91.1	−175.8
Butylene-1 + propane → 2,2-dimethylpentane	−84.1	−33.6	−181.2
Propylene + propylene → isohexene	−140.7	−92.9	−171.4
Isobutylene + isohexene → isodecane	−111.3	−36.5	−170.3
Isobutane +butylene-1 → 2-methyheptane	−77.1	−26.0	−184.3

Data from reference A.E. Nurmakanova, Predicting alkylate yield and its hydrocarbon composition for sulfuric acid catalyzed isobutane alkylation with olefins using the method of mathematical modeling procedia chemistry, Procedia Chemistry 15(2015) 54–64.

an increase in the volume fraction of isobutene. Consequently, at higher temperatures, particular attention should be given to the impact of changes in the isobutene volume fraction within the C4 composition on the yield of the desired product in alkylation reactions.

Huang et al. calculated the adiabatic temperature rise of the C4 alkylation to investigate the relationship between the C4 feedstock composition and the alkylation process [6]. As a typical exothermal reaction, adiabatic temperature rise is an indicator of reaction progress, which can be indirectly used as a basis for adjusting operating parameters.

Assuming that the releasing heat from the alkylation is completely used for material heating, the adiabatic temperature rise can be calculated as

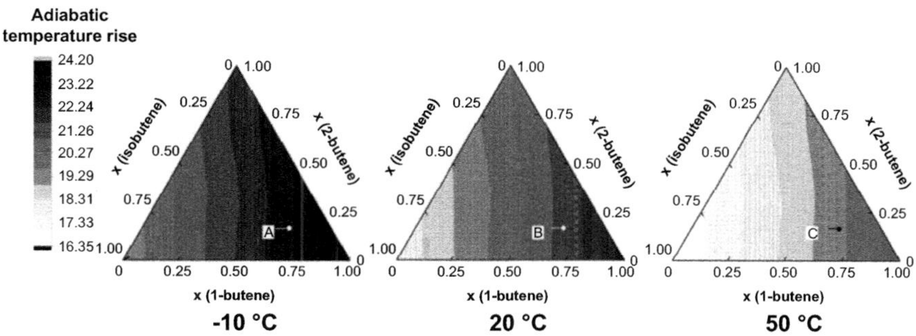

FIG. 3.3 Adiabatic temperature rise-composition of mixed C4 alkylation. *From reference F.L. Huang, P. Cao, E.L. Hou, J. Ling, J.R. Liu, K. Li, X.H. Li, Study of the adaptability of alkylation units to changes in the composition of C4 olefins, Chemical Engineering (China) 49 (2021) 54–59.*

$$\Delta T(T) = \frac{-\Delta_r H(T)}{\sum_i C_{pi}(T) \cdot n_i(\xi_i)} \tag{3.9}$$

where $\Delta T(T)$ is the adiabatic temperature rise of each reaction at certain temperature with units of K. $-\Delta_r H(T)$ is the total reaction heat of each reaction with units of kJ/mol. $C_{pi}(T)$ is the molar heat capacity at a constant pressure of the i component with units of J/mol K. n_i is the amount of substance of the i component with units of mol.

Under the conditions of I/O = 8, acid/hydrocarbon = 1.4, adiabatic temperature rise of C4 alkylation at different compositions and temperatures is shown in Fig. 3.3. Under different temperature and composition conditions, the adiabatic temperature rise in the system of C4 alkylation varies between 16.35°C and 24.11°C. As the molar fraction of 1-butene increases, the adiabatic temperature rise increases, whereas with an increase in the molar fraction of isobutene, the adiabatic temperature rise decreases. For the alkylation with the same composition, the heat effect remains relatively stable within the temperature range of −10°C to 50°C. This stability is due to a significant increase in the heat capacity of liquid isobutene and sulfuric acid as the temperature rises, transitioning from 131.13 kJ/(mol K) at −10°C to 151.14 kJ/(mol K) at 50°C. As shown in Fig. 3.3, the adiabatic temperature rise decreases as the temperature increases, dropping from 22.90°C to 21.28°C and 19.57°C, indicating an enhanced heat absorption and temperature stability capacity of the system.

To promote the efficient conversion of the key component and effectively reduce the adiabatic temperature rise while maintaining a lower reaction temperature, adjusting the I/O ratio and acid-to-hydrocarbon ratio becomes an effective means of controlling the reaction temperature stability. Taking the industrial C4 composition as an example, where the

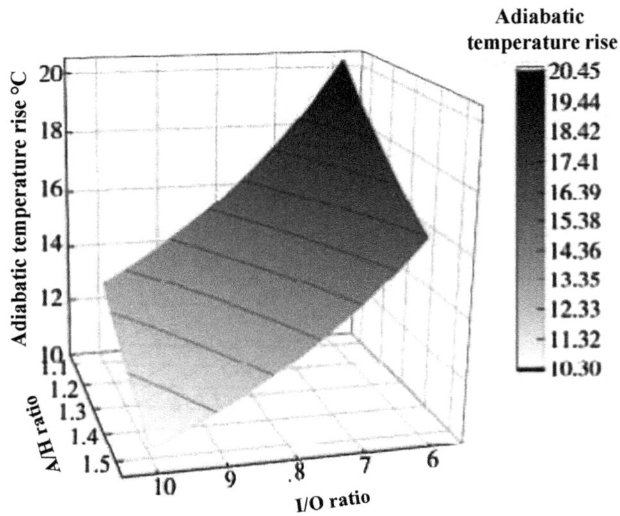

FIG. 3.4 Relationship of I/O-A/O-adiabatic temperature rise. *From reference F.L. Huang, P. Cao, E.L. Hou, J. Ling, J.R. Liu, K. Li, X.H. Li, Study of the adaptability of alkylation units to changes in the composition of C4 olefins, Chemical Engineering (China) 49 (2021) 54–59.*

composition of 1-butene, 2-butene, and isobutene is 1:2:1, the adiabatic temperature rise of the mixed system at 10°C is shown in Fig. 3.4.

The I/O ratio increases from 6.0 to 10.0, and the acid-to-hydrocarbon ratio increases from 1.1 to 1.5, leading to a decrease in the adiabatic temperature rise of the C4 alkylation from 20.42°C to 10.32°C. Increasing the I/O ratio and acid-to-hydrocarbon ratio proves to be an effective way to mitigate the adiabatic temperature rise and maintain the reaction temperature. Under conditions of lower I/O ratios, increasing these ratios significantly suppresses the adiabatic temperature rise. Considering the actual effects of factors such as sulfuric acid diffusion, viscosity, and the extent of acid-hydrocarbon mixing in the reaction, it is advisable to choose appropriate conditions above the minimum I/O ratio.

3.2 Carbonium ion

For the C4 alkylation, Schmerling et al. [8] proposed the carbonium ions leading to the generation of the carbonium ion. The formation step of the carbonium ion is presented below:

$$H_2C::\underset{\underset{CH_3}{\cdot\cdot}}{C}:CH_3 + H^+OSO_3H^- \rightleftharpoons H_3C:\overset{+}{\underset{\underset{CH_3}{\cdot\cdot}}{C}}:CH_3 + OSO_3H^- \tag{3.10}$$

Carbonium ion is a charged carbon atom with only six outer shell electrons, resulting in a positive charge. Its general formula can be expressed as shown below (Scheme 1):

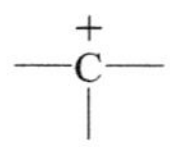

SCHEME 1 Carbonium ions.

Such a representation of carbonium ions is very useful as it permits an explanation of all the hydrocarbon reactions catalyzed by acid-type catalysts. Substituents around the carbonium ion can be either hydrogen atoms or methyl groups, with four forms (Scheme 2):

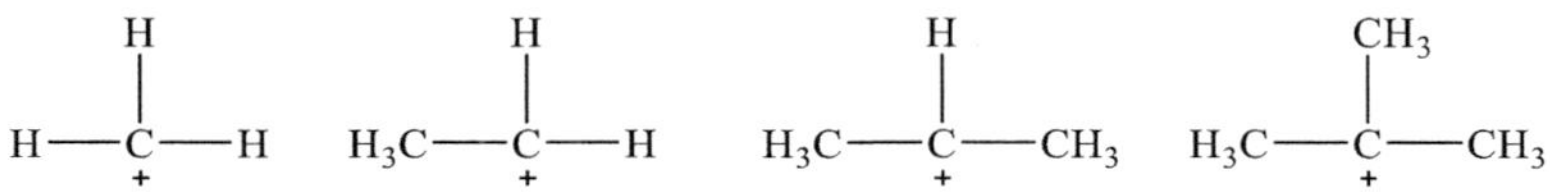

SCHEME 2 Four forms of carbonium ions.

The stability increases from left to right, meaning that the carbonium ion of the tertiary carbon atom is the most stable. In alkylation reactions catalyzed by strong acids, the addition reaction between the hydrogen protons provided by the acid and the olefin is the primary reaction leading to the formation of carbonium ions.

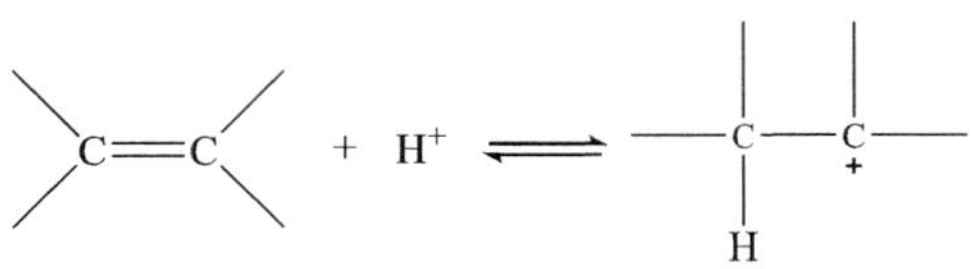

(3.11)

When the unsaturated hydrocarbons react with hydrogen protons, carbonium ions can be formed on primary or secondary carbon atoms. These ions may undergo rearrangement or isomerization to yield relatively stable carbonium ions, such as:

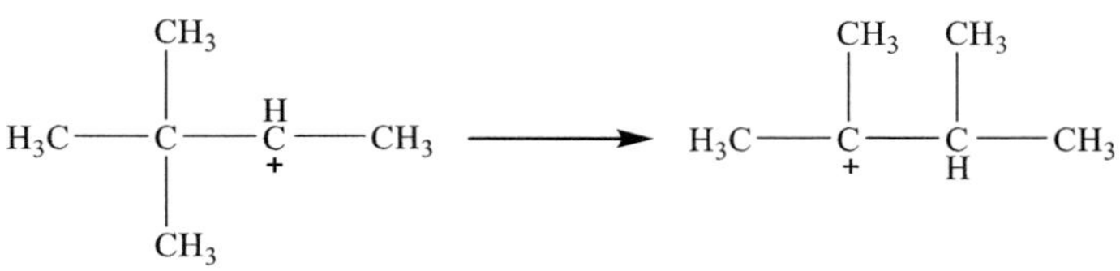

(3.12)

The addition reaction between olefins and protons can generate carbonium ions. Conversely, carbonium ions can lose protons to regenerate olefins. Moreover, not only is it possible to regenerate the original olefin, but there is also a greater possibility to produce more stable olefins, as listed in Eq. (3.13).

$$CH_3CH_2CH{=}CH_2 + H^+ \longrightarrow CH_3CH_2\overset{+}{C}HCH_3 \begin{cases} \nearrow CH_3CH_2CH{=}CH_2 \ \text{Seldom} \\ \searrow CH_3CH{=}CHCH_3 \ \text{Main product} \end{cases} \tag{3.13}$$

The carbonium ions can further undergo electrophilic addition reactions with carbon-carbon double bonds. This constitutes a primary step in the elongation of the alkyl chain in alkylation reactions.

$$(CH_3)_3CH^+ + (CH_3)_2C{=}CH_2 \longrightarrow (CH_3)_3C{-}CH_2{-}\underset{+}{\overset{CH_3}{\overset{|}{C}}}{-}CH_3 \tag{3.14}$$

Given the molecular composition of isobutane, wherein the tertiary carbon atom is linked to three methyl groups, the possibility arises for the hydrogen atom on this tertiary carbon to undergo detachment in the form of a hydride ion, thereby converting isobutane into a tertiary butyl carbonium ion. Consequently, C_8^+ carbonium ions accepts the hydride ion, forming the alkylate, namely isooctane.

$$(CH_3)_3C{-}CH_2{-}\underset{+}{\overset{CH_3}{\overset{|}{C}}}{-}CH_3 + (CH_3)_3CH \longrightarrow (CH_3)_3CCH_2\overset{CH_3}{\overset{|}{C}}HCH_3 + (CH_3)_3C^+ \tag{3.15}$$

Overall, the reactions involving carbonium ions can be categorized into three types [8]:

(1) The addition of the carbonium ion to an unsaturated molecule, utilizing a pair of π electrons, resulting in the formation of the higher molecular weight carbonium ion.

(2) The decomposition of the carbonium ion into another ion, such as the smaller carbonium ion or the proton, along with the generation of an unsaturated compound.

(3) The isomerization of the carbonium ion through the migration of hydrogen, alkyl, or aryl groups, accompanied by a pair of electrons.

(4) The ultimate conversion of the carbonium ion to the reaction product occurs through the elimination of a proton or another cation, or by the abstraction of a hydride ion or another anion from another molecule.

3.3 Overall product distributions

The product distributions depend on various factors, such as acidity, pressure, reaction time, temperature, and the molar ratio of I/O, among which the acidity of the catalyst is the most crucial. For instance, in the H_2SO_4 process, the Hammett acidity parameter (H_0) must be in the range of −10 to −11 to produce alkylate (higher acidity catalyzes cracking reactions to produce LEs, while lower acidity yields significant HEs). This reaction typically occurs at temperatures 6–8°C and pressures 15–30 bar, with a molar ratio of I/O of approximately 10/1 and reaction time of about 30 min [9].

For C4 alkylation, the preliminary judgments and examinations of the reaction mechanism can be made by analyzing the reaction product distributions. Using three different olefin raw materials, namely isobutene, 1-butene, and 2-butene, with sulfuric acid and hydrofluoric acid as catalysts, respectively, the obtained compositions of alkylate are shown in Table 3.5.

TABLE 3.5 The distribution of alkylate.

Feedstock Olefin	**Isobutene**		**1-Butene**		**2-Butene**	
Catalyst	**H_2SO_4**	**HF**	**H_2SO_4**	**HF**	**H_2SO_4**	**HF**
I/O ratio	3.0	5.2	3.0	4.8	3.0	5.6
Composition (%)						
2,3-Dimethylbutane	5.69	1.5	4.9	0.6	4.5	2.5
2-Methylpentane	1.8	0.9	1.6	–	1.5	0.5
3-Methylpentane	0.8	0.0	0.5	–	0.5	–
C_6 total	8.2	2.4	7.0	0.6	6.5	3.0
2,2,3-Trimethylbutane	0.3	–	0.4	–	0.3	–
2,3-Dimethylpentane	2.0	2.7	1.7	1.7	1.6	1.4
2,4-Dimethylpentane	4.1	2.3	3.2	1.3	3.3	2.4
2-Methylhexane	0.2	–	0.2	–	0.2	–
3-Methylhexane	0.1	–	0.2	–	0.3	–
C_7 total	6.7	5.0	5.7	3.0	5.7	3.8
2,2,4-TMP	24.7	49.0	28.8	29.5	32.6	37.9

Continued

TABLE 3.5 The distribution of alkylate—cont'd

Feedstock Olefin	Isobutene		1-Butene		2-Butene	
Catalyst	H_2SO_4	HF	H_2SO_4	HF	H_2SO_4	HF
2,2,3-TMP	1.7	1.5	2.3	0.9	2.4	2.4
2,3,4-TMP	6.3	9.4	11.8	14.1	11.8	19.4
2,3,3-TMP	9.5	6.8	15.2	8.2	17.0	10.1
2,4-DMH	2.5	3.3	3.5	4.9	2.6	2.6
2,5-DMH	3.5	2.9	3.4	1.9	3.1	2.8
2,3-DMH	1.6	2.4	3.3	25.2	2.2	3.4
3,4-DMH	–	–	0.2	–	0.4	–
C_8 total	49.8	75.3	68.5	84.7	72.1	78.6
C_{9+}	35.5	17.3	18.8	11.5	15.7	14.7

Data from reference B.L. Gao, High Octane Gasoline Component Production Technology, China Petrochemical Press, Beijing, 2006.

From Table 3.5, it can be observed that.

(1) The C4 alkylation not only produces C_8 components but also generates C_6, C_7, and C_9^+ components, indicating that the reaction mechanism is complex.
(2) Regardless of H_2SO_4 or HF as catalysts, the distribution of C4 alkylation with different C4 olefins is generally similar. The majority of the generated C_8 components are dominated by 2,2,4-TMP, followed by 2,3,4-TMP and 2,3,3-TMP.
(3) The variety of products in H_2SO_4-catalyzed alkylation is greater than in HF-catalyzed alkylation, possibly due to H_2SO_4-catalyzed alkylation involving more reaction processes.

3.4 General alkylation mechanism

The mechanism of the alkylation reaction is generally believed to follow the carbonium ion reaction mechanism proposed by Schmerling [10,11]. This mechanism comprises two key steps: the formation of carbonium ions and hydride transfer. The general reaction mechanism is shown below. The alkylation reaction is initiated by the protonation of olefins promoted by strong acids to produce tertiary carbonium ions.

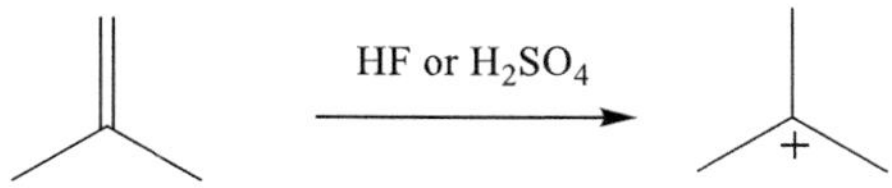

(3.16)

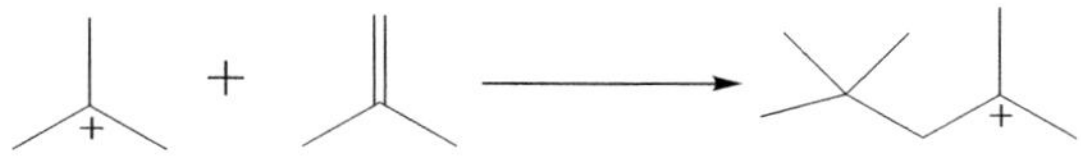

(3.17)

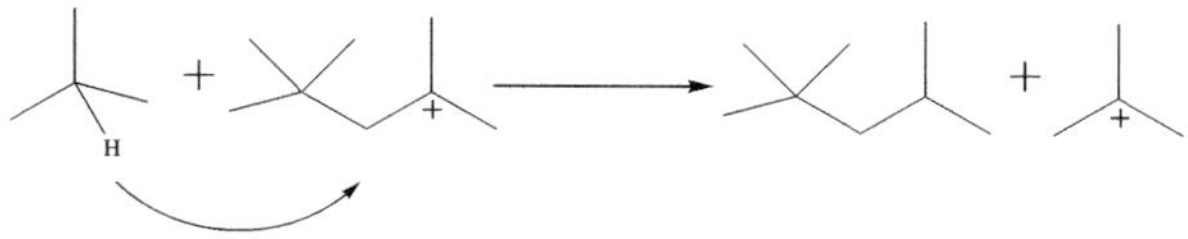

(3.18)

Subsequently, the olefin is added to the carbonium ion, resulting in the formation of trimethylpentane cations ($TMPs^+$). This process follows Markovnikov's rule, in which carbonium ion undergoes isomerization rearrangement to produce the thermodynamically favored isomers. During the isomerization reaction, intramolecular hydrogen transfers or methyl transfers occur, primarily resulting in the formation of different types of $TMPs^+$. The principal termination mechanism of the alkylation is the extraction of a hydrogen ion from isobutane by $TMPs^+$, as elucidated in the Reaction (3.18).

Moreover, the alkylation process also produces several other byproducts. These byproducts are generated through complex reactions such as polymerization, cracking, cyclization, and hydrogen transfer. The byproducts include C_5–C_7 cracking products or other LEs, and C_9^+ polymerization products or other HEs, and some oligomers. The oligomers are primarily composed of butene oligomers, known as ASO, leading to deactivation and reduced recycle time of liquid catalysts such as H_2SO_4 [9].

3.5 Alkylation mechanism of H_2SO_4 as catalyst

For C4 alkylation with concentrated H_2SO_4 as the catalyst, the chain reactions typically comprise three fundamental stages: initiation stage, propagation stage, and termination stage. Within the initiation stage, the formation of carbonium ions into the tertiary butyl moiety, resulting from the olefin protonation, plays a pivotal role in catalyzing the subsequent alkylation reactions. This observation underscores the significance of the carbonium ion into the tertiary butyl group as a key participant in the alkylation mechanism.

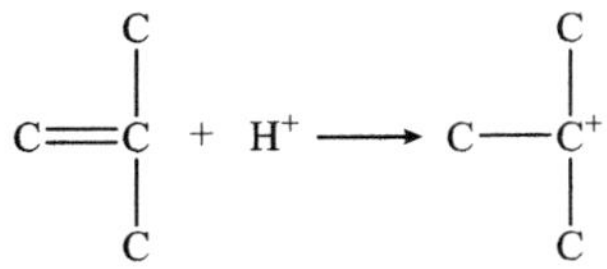

(3.19)

The principal source of tertiary butyl carbonium ions is isobutene, where the protonation of isobutene results in the generation of these ions, as shown in Eqs. (3.20) and (3.21). Carbonium ions arising from the protonation of normal butene can also undergo hydride transfer, giving rise to tertiary butyl carbonium ions.

$$C{-}C{=}C{-}C + H^+ \longrightarrow C{-}\underset{+}{C}{-}C{-}C \quad (3.20)$$

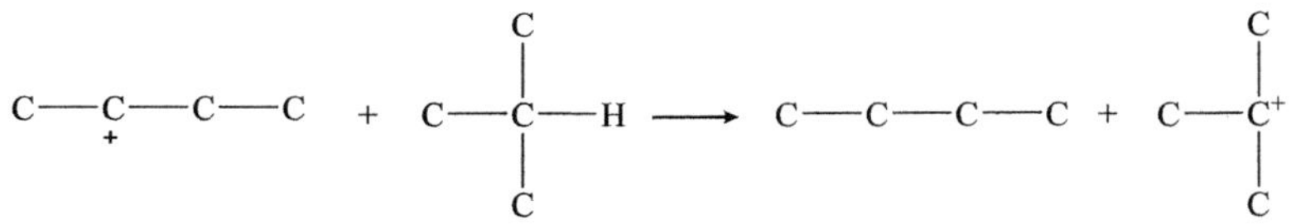

(3.21)

The processes outlined in Eqs. (3.22)–(3.24) elucidate the prolongation of the chain.

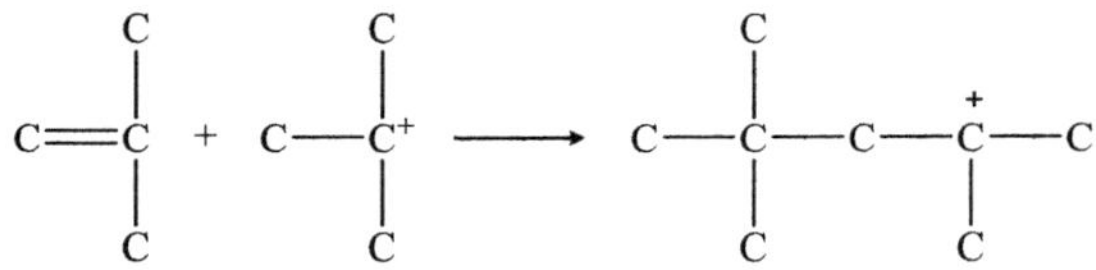

(3.22)

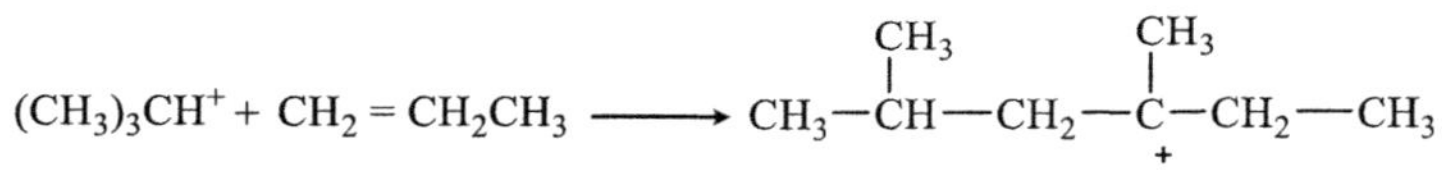

(3.23)

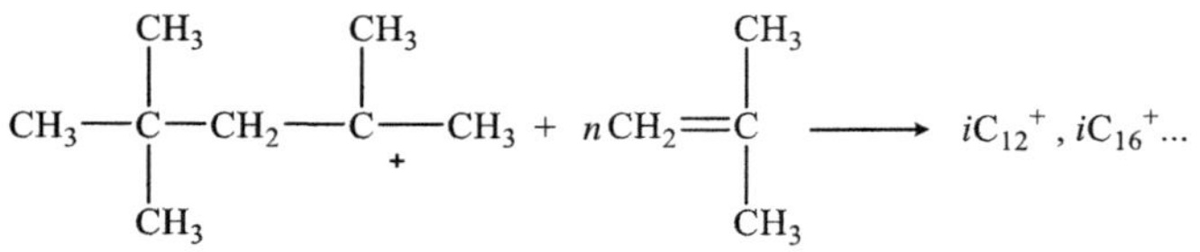

(3.24)

The principal chain termination is the extraction of hydride ion from isobutane by carbonium ions, as elucidated in Eqs. (3.25) and (3.26). Although an alternative pathway involves the loss of H^+ by carbonium ions (Eq. 3.27), leading to chain termination, there are infrequent occurrences of olefin generation in such instances.

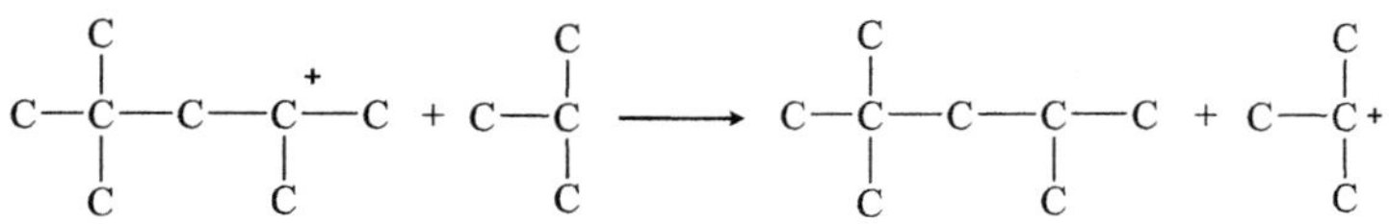

(3.25)

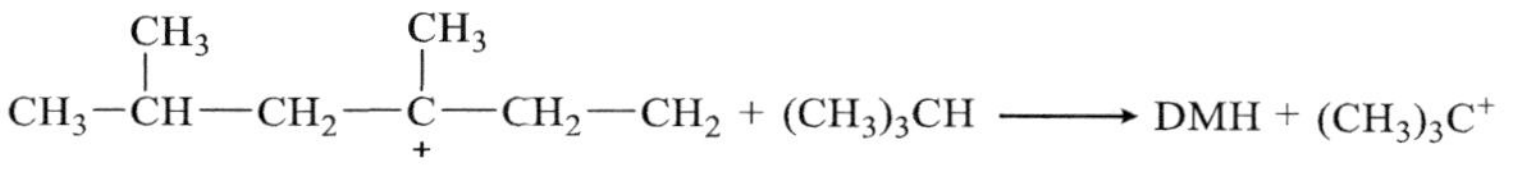

(3.26)

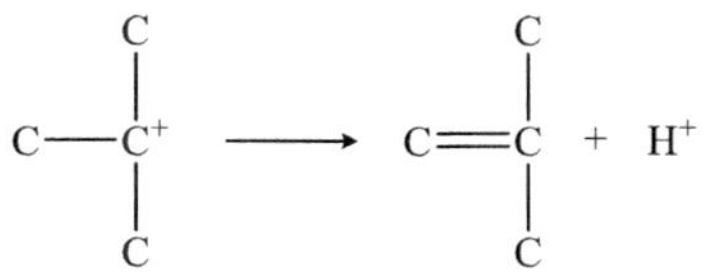

(3.27)

Generally, the alkylation process catalyzed by concentrated H_2SO_4 involves the following complex reactions:

(1) Dimerization and oligomerization of hydrocarbons

The formation of DMH components is attributed to two primary factors. Firstly, as delineated in Eqs. (3.25) and (3.26), the C_8^+ derived from the reaction between the tert-butyl carbonium cations and normal butene undergoes hydrogen transfer to yield DMHs. Secondly, the occurrence of dimerization plays a crucial role, encompassing co-dimerization reactions of 1-butene and isobutene, dimerization of 1-butene, and dimerization of isobutene. Various isomeric forms of C_7 and C_8 components undergo isomerization reactions at the stage of positive carbocation, subsequent to dimerization, and generate by the extraction of hydride ions. Given the existence of dimerization reactions, it unavoidably gives rise to trimerization and oligomerization, particularly in the case of isobutene oligomerization, contributing to the presence of HEs components in the alkylate.

(2) Cracking reactions

The olefins produced through oligomerization reactions, facilitated by the strong catalyst, undergo protonation to yield carbonium ions. These carbonium ions possess the capacity to undergo cracking reactions autonomously prior to the extraction of hydrogen ions. The resultant carbonium ions, characterized by reduced molecular weights, subsequently engage in the capture of hydride ions, culminating in the generation of alkanes. This process constitutes the principal mechanism about the production of C_5, C_7, and analogous alkanes.

$$iC_{12}H_{25}{}^{+} \rightarrow iC_7H_{15}{}^{+} + iC_{15}H_{15} \tag{3.28}$$

$$iC_7H_{15}{}^{+} + iC_4H_{10} \rightarrow iC_7H_{16} + (CH_3)_3C^{+} \tag{3.29}$$

$$iC_5H_{10} + H^{+} \rightarrow iC_5H_{11}{}^{+} \tag{3.30}$$

$$iC_5H_{11}{}^{+} + iC_4H_{10} \rightarrow iC_5H_{12} + (CH_3)_3C^{+} \tag{3.31}$$

(3) Isomerization of carbonium cations

The difference in the content of TMPs and DMHs within the resultant products can be readily elucidated through the isomerization of carbonium ions.

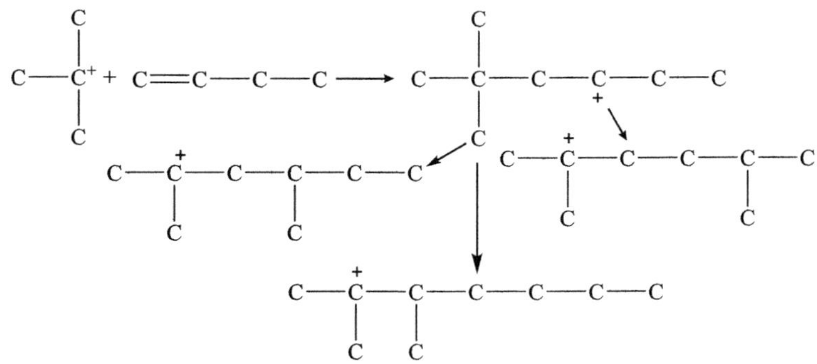

(3.32)

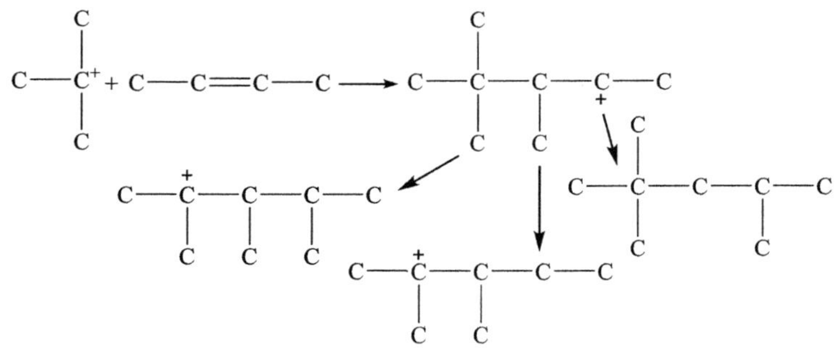

(3.33)

(4) Self-alkylation

The self-alkylation reaction results in the formation of undesirable light, normal alkanes, negatively impacting the overall product quality. For C4 reaction with 2-butene, the tert-butyl carbonium cations can undergo a proton transfer and generate isobutylene. Subsequently, isobutylene reacts with another tert-butyl carbonium cation to form C_8^+. A feed of butene initiates the process in parallel, leading to the formation of C_8^+. Simultaneously, the C_8^+ undergoes hydride transfer with isobutane, producing normal alkane and tert-butyl carbonium cation. Fig. 3.5 illustrates the self-alkylation mechanism.

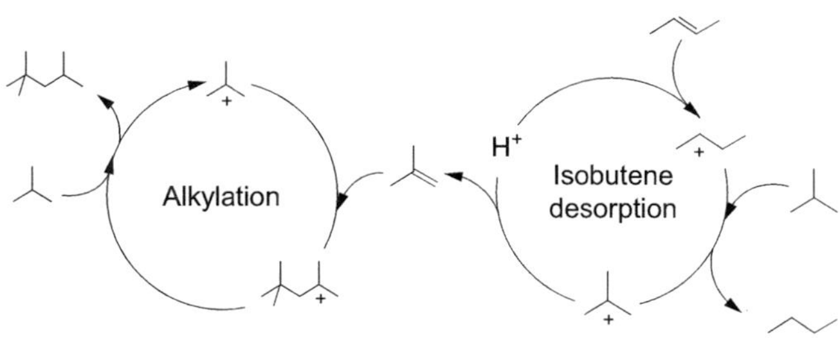

FIG. 3.5 Self-alkylation mechanism. *From reference A. Feller, A. Guzman, I. Zuazo, J.A. Lercher, On the mechanism of catalyzed isobutane/butene alkylation by zeolites, J. Catal. 224 (2004) 80–93.*

(5) Formation of ASO

For the investigation into the mechanism of alkylation reactions utilizing C^{14}-labeled normal butene, a fraction of the labeled normal butene was observed to undergo conversion into isobutane [12]. Product analyses and C^{14}-distribution determinations were obtained for the initial products and for the products formed at steady-state conditions. The initial products primarily come from olefin polymerization reactions, while at steady state the predominant reaction was direct alkylation. In addition, a considerable difference in reactivity between 1-butene and isobutylene was observed with the isobutylene having a greater tendency to polymerize.

In the generation of TMPs, $t\text{-}C_4H_9^+$ plays a crucial role in the H_2SO_4 alkylation process. However, distinct from HF alkylation, the primary source of $t\text{-}C_4H_9^+$ is isobutene, in which isobutene reacts with H_2SO_4 to form acid-soluble hydrocarbons and $t\text{-}C_4H_9^+$. This observation is employed to explain the phenomenon of greater variety of products in H_2SO_4 alkylation compared to HF alkylation.

$$R^+ + i - C_4H_{10} \rightarrow RH + t - C_4H_9^+ \tag{3.34}$$

Despite variations in the depiction of the mechanism of alkylation across different literature, alkylation reactions conform to specific reaction principles: (1) The formation of carbonium ions from olefins constitutes the fundamental driver for the expeditious advancement of alkylation reactions, with tert-butyl carbonium ions typically exhibiting the highest reactivity in the majority of alkylation processes. (2) Excessive acidity is not conducive to the progression of alkylation reactions. The stronger acid sites are more prone to facilitating olefin polymerization reactions on the catalyst surface. For the H_2SO_4 system, the catalytic performance of 99 wt% H_2SO_4 is shown to be less effective than that of the 95 wt% one. (3) Hydride transfer reactions emerge as the rate-determining steps in alkylation, a principle consistently observed even in the absence of catalysts for alkylation reactions. Under acidic conditions, the dimerization and oligomerization of olefins proceed rapidly. Consequently, the selection of novel catalysts for alkylation should involve a careful consideration of balancing the inhibition of olefin polymerization with the control of the hydride transfer rate.

3.6 Alkylation mechanism of HF as catalyst

The available data on the HF-catalyzed alkylation is quite limited. Albright [2] presented some composition data for alkylate produced by using HF with 1-butene, 2-butene, or isobutylene as feedstock. When isobutylene is employed as the olefin in the HF process, the alkylate contain a significantly higher content of TMP components, especially 2,2,4-TMPs, but exhibit a lower content of LE components compared to the H_2SO_4 process. On the other hand, the alkylate produced from 1-butene in the HF process has considerably higher amounts of DMHs, fewer TMPs, and a much lower RON than that produced from 2-butene as a feedstock. The alkylate produced with the HF catalyst and 2-butene as olefin shows some similarities to those produced using the H_2SO_4 catalyst, but the compositions of the TMPs are notably different. The composition and quality of alkylate produced from 1-butene and 2-butene are nearly identical when H_2SO_4 is used, but exhibit significant differences when HF is the catalyst.

Differences in the alkylate produced using either H_2SO_4 or HF are undoubtedly caused by their significantly different physical properties. Physical properties of the acid have important effects on the character of the emulsion, including viscosity, interfacial tensions between the acid and hydrocarbon phases, and density. For H_2SO_4 alkylation, the dissolved ratio of isobutane to olefin in the acid phase is extremely important to the quality of the alkylate [1,2,4,5]. Presumably, a similar conclusion can be obtained for HF alkylation, but factors such as dissolved ratio vary significantly depending on the acid used.

Since ASO is also produced in HF, these reaction steps involving the ASO ions, hydride transfer from the ASO, etc., likely occur to some extent when HF is used. However, extensive data, similar to those obtained by Li et al. [3,4] for H_2SO_4 alkylation, are needed for HF alkylation.

Esteves et al. [13] performed a theoretical study using the Car-Parrinello molecular dynamics as well as DFT simulation for the isobutylene-isobutane alkylation process catalyzed by HF. The computational results suggest that when the system contains few HF molecules, the π-complex forms and the olefin is not protonated. However, an increase in the number of HF molecules promotes the formation of the carbonium ion, which acts as a long-lived species. Selected geometries for the complex isobutylene $(HF)_n$ ($n=5$) were calculated (Fig. 3.6). Solvation effects play a crucial role in the formation of the carbocation ion. The strong solvation of fluoride is the driving force for the protonation of the olefin, leading to the formation of carbocation.

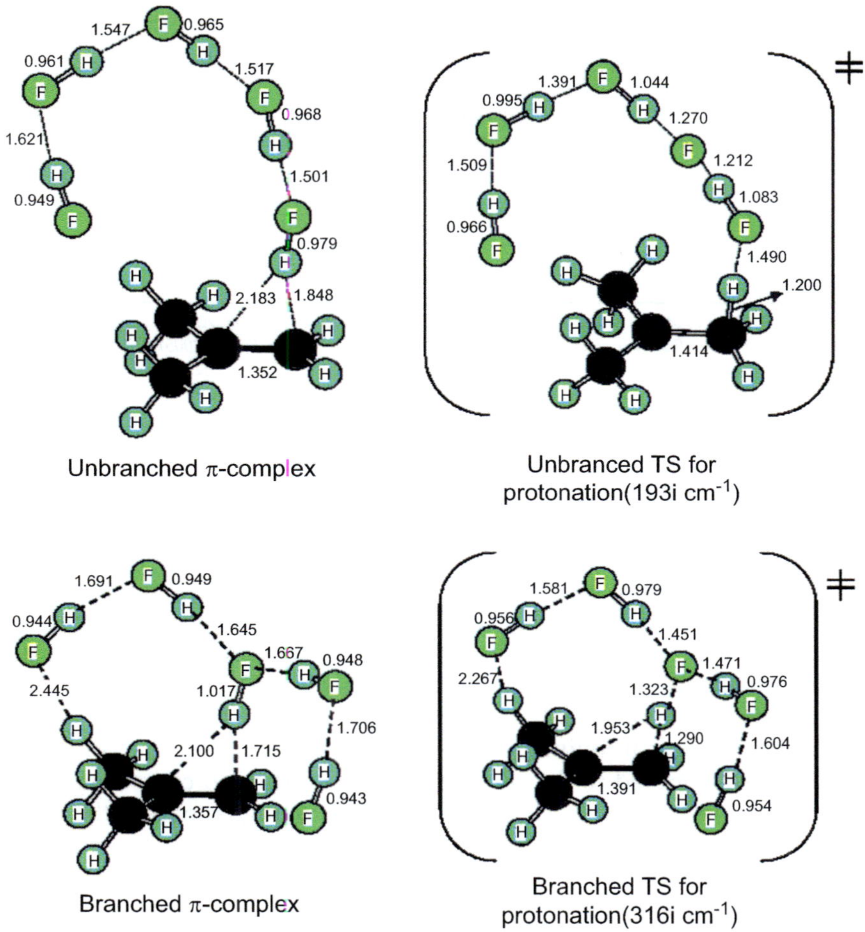

FIG. 3.6 Selected geometries for the complex isobutylene $(HF)_n$ ($n=5$) calculated at the B3LYP/6-31++G** level, and the respective transition states for double bond protonation with (A) an unbranched HF chain (top) and (B) a branched HF chain (bottom). *From reference P. Esteves, C. Araújo, B. Horta, L. Alvarez, C. Zicovich-Wilson, A. Ramírez-Solís, The isobutylene--isobutane alkylation process in liquid HF revisited, J. Phys. Chem. B 109 (2005) 12946–12955.*

The acidity of the HF promotes and stabilizes the tert-butyl cation as a long-lived intermediate. Alkylation and hydride transfer are proposed to take place at the interface between the organic and HF phases, where the contact between the carbocation and olefin molecules is possible. Thermodynamic quantities for the protonation of isobutylene with unbranched and branched HF chains were calculated as listed in Table 3.6. Liquid HF displaces the isobutylene/t-$C_4H_9^+$ equilibrium toward the carbocation side and facilitates phase separation, contributing to the improvement of alkylate quality.

TABLE 3.6 Thermodynamic quantities for the protonation of isobutylene with unbranched and branched HF chains.

	$\Delta H^{\ddagger}$ (298 K) (kcal/mol) $\varepsilon=1$	$\Delta H^{\ddagger}$ (298 K) (kcal/mol) $\varepsilon=78.39$ (PCM)	$\Delta S^{\ddagger}$ (298 K) (cal/mol K) $\varepsilon=1$	$\Delta G^{\ddagger}$ (298 K) (kcal/mol) $\varepsilon=1$	$\Delta G^{\ddagger}$ (298 K) (kcal/mol) $\varepsilon=78.39$ (PCM)
Unbranched	9.8	0.6	−10.12	12.9	3.6
Branched	1.7	−2.4	−9.54	4.5	0.5

Data from reference P. Esteves, C. Araújo, B. Horta, L. Alvarez, C. Zicovich-Wilson, A. Ramírez-Solís, The isobutylene–isobutane alkylation process in liquid HF revisited, J. Phys. Chem. B 109 (2005) 12946–12955.

3.7 Alkylation mechanism of ILs as catalyst

The immobilized chloroaluminate ILs (CILs), specifically [C_4mim]Cl-$AlCl_3$ ($x=0.67$) and 1-(3-triethoxysilyl)-propyl-3-methyl imidazolium chloroaluminate ([SiC_3mim]Cl-$AlCl_3$, $x=0.67$), onto siliceous materials were used as the catalyst for C4 alkylation. The concentrations of olefins remain relatively constant throughout the reactions. The shift in selectivity toward the formation of TMPs and DMHs is elucidated based on the generation of distinct carbonium ions, as illustrated in Fig. 3.7. The formation of TMPs is attributed to the reaction between tert-butyl carbonium ions and butene. In contrast, DMHs are suggested to originate from the reaction of either tert-butyl carbonium ions with butene or the dimerization of sec-butyl carbonium ions with butene.

Li et al. [14] conducted an in-depth investigation into the alkylation mechanism of 2-butene with deuterated isobutane in continuous-flow equipment, utilizing [BMIM]Cl-$AlCl_3$ and [BMIM]-$AlCl_3$-CuCl as catalysts. It is observed that there is a significant shorter induction period in IL-catalyzed alkylation compared to H_2SO_4. A notable disparity in isobutane solubility between IL and H_2SO_4 was identified, with IL displaying a strong capability to dissolve isobutane at the initiation of alkylation. Analysis of deuterated product distributions revealed that TMP fractions predominantly arise from the self-alkylation of isobutane, the direct alkylation reaction of C4 hydrocarbons, and the cleavage of C_{12}^{+} intermediates, as shown in Fig. 3.8.

DFT calculations and MD simulations were employed to investigate the complex reaction mechanism of C4 alkylation catalyzed by neat chloroaluminate ionic liquid (NIL) and Cu-containing CIL [14]. Transition states (TS) for key reactions were identified to elucidate alkylate formation selectivity in different ILs catalytic systems, which revealed that Cu species

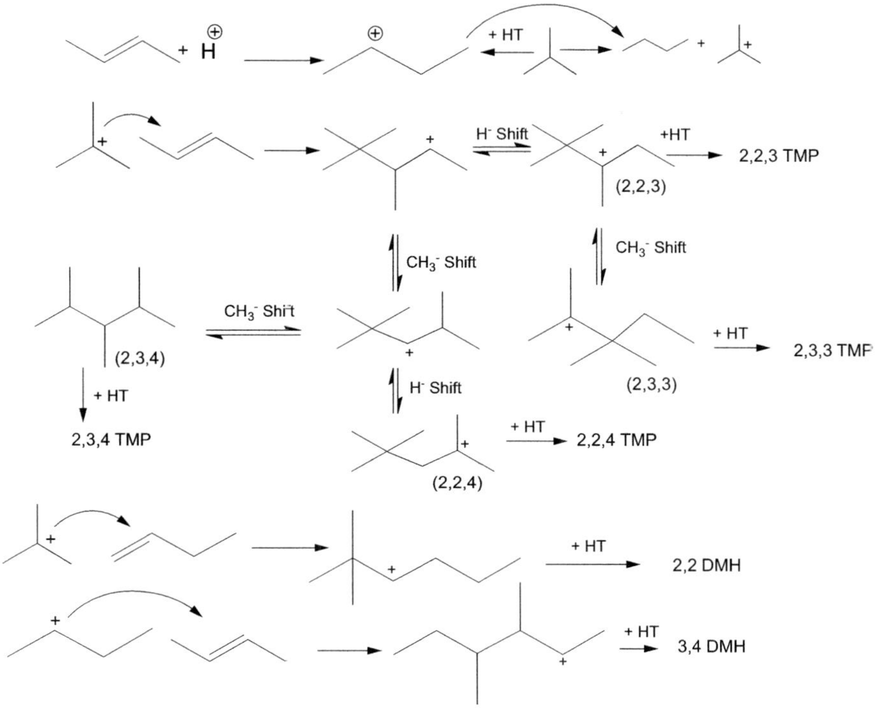

FIG. 3.7 Possible carbonium ion mechanism for the formation of different TMPs and DMHs. *From reference P. Kumar, W. Vermeiren, J.-P. Dath, W.F. Hoelderich, Production of alkylated gasoline using ionic liquids and immobilized ionic liquids, Appl. Catal. A Gen. 304 (2006) 131–141.*

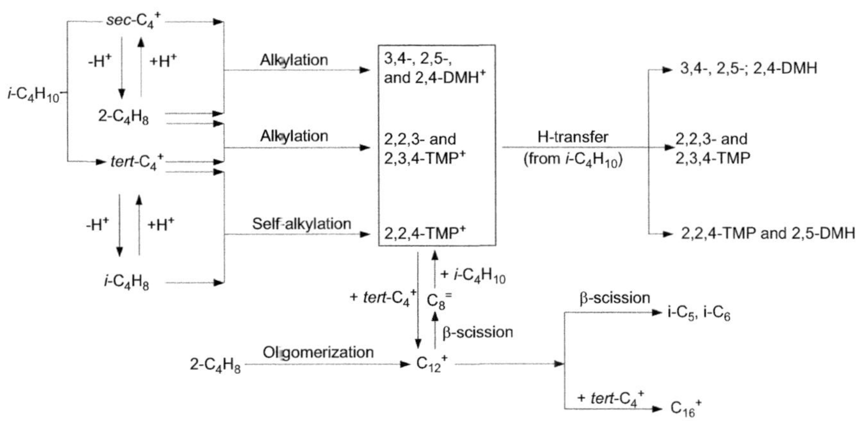

FIG. 3.8 Scheme of reaction mechanism for the CIL alkylation. *From reference Y. Liu, L. Wang, R. Li, R. Hu, Reaction mechanism of ionic liquid catalyzed alkylation: alkylation of 2-butene with deuterated isobutene, J. Mol. Catal. A – Chem. 421 (2016) 29–36.*

effectively hinder the polymerization of C4⁼ olefins and promote C_8^+ carbonium ions. In comparison to the competitive hydride transfer reaction, the NIL exhibits rapid polymerization reaction rates, leading to poor product selectivity. The Mayer bond order and electron localization function maps illustrate that H-transfer from isobutane to the C_8^+ carbonium ion occurs through a concerted H—Cu bond formation, generating the desired isooctane. The Cu complex participates in hydride transfer without compromising catalytic activity compared to neat chloroaluminate anions. MD simulations demonstrated that the effects of the Cu complex, combined with high isobutane concentration and low isobutylene concentration at the CIL-hydrocarbon interface, contribute to the improvement of the alkylation selectivity. The Cu species in CIL play a crucial role in impeding the formation of C_{12}^+ carbonium ions and promoting hydride transfer between isobutane and C_8^+ carbonium ions.

In addition, Li et al. [15] focused on delineating the pathways for the formation of carbocation, LEs, C_7 isoparaffins, C_8 isoparaffins, and HEs in the context of alkylation, as shown in Fig. 3.9. The significance of understanding carbocation formation is emphasized as it serves as a crucial intermediate species in the alkylation process.

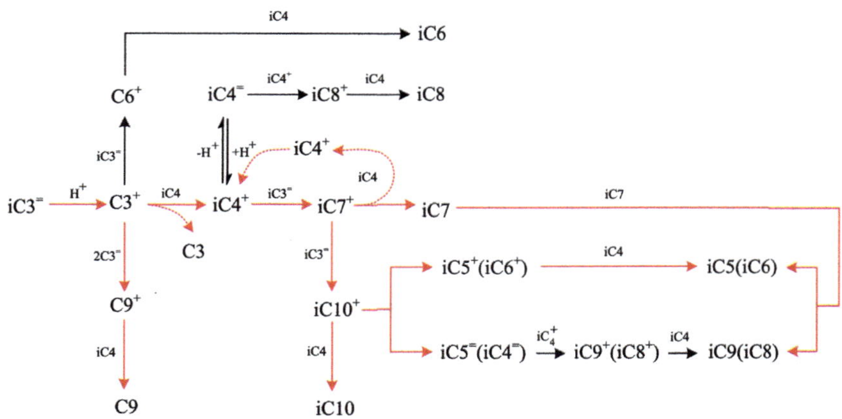

FIG. 3.9 Reaction pathway of isobutane alkylation with propene using the CIL catalyst. *From reference Q. Li, R. Zhang, Y. Li, X. Meng, H. Liu, X. Wang, C. Xu, H. Dong, Q. Su, X. Zhang, Reaction behaviors and mechanism of isobutane/propene alkylation catalyzed by composite ionic liquid, Ind. Eng. Chem. Res. 61 (2022) 8624–8633.*

(1) Carbocation formation

Carbocation emerges as a pivotal intermediate in the alkylation mechanism. The origin of Brønsted acid, derived from HCl in the presence of CIL, is considered essential for comprehending carbocation formation.

Cui et al. confirmed the presence of HCl in ILs through ^{27}Al NMR, illustrating two possible pathways, as depicted in Eqs. (3.46) and (3.47).

$$[Al_2Cl_7]^- + H_2O \rightarrow [Al_2Cl_6OH]^- + HCl \uparrow \tag{3.35}$$

$$[Al_2Cl_6OH]^- \rightleftharpoons [AlCH_5O]^- + HCl \uparrow \tag{3.36}$$

The proton, supplied by Brønsted acid, initiates the chain reaction by reacting with alkene. For C4 alkylation with propene, the addition reaction between propene and HCl yields 2-chloropropane. Subsequently, 2-chloropropane and isobutane undergo hydride transfer, leading to the formation of propane, tert-butyl carbocation, and chloride ion.

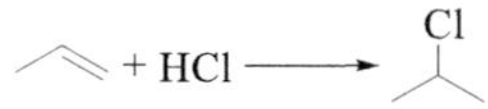

(3.37)

(3.38)

(2) Formation of C_5 and C_6 isoparaffins

The presence of C_5 and C_6 isoparaffins in the product is detected, and their formation involves two primary mechanisms. Firstly, cracking occurs, where large carbocations undergo β-scission, resulting in the cleavage of an olefin and the formation of a smaller carbocation (e.g., C_5^+ or C_6^+). Subsequently, C_5^+ or C_6^+ abstracts hydride ions from isobutane, yielding C_5 or C_6 isoparaffins. Using C_5 formation as an example, the processes are detailed in Eqs. (3.39) and (3.40).

$$C_{10}^+ \rightarrow C_5^= + C_5^+ \tag{3.39}$$

$$C_5^+ \rightarrow iC_4 \rightarrow C_5 + iC_4^+ \tag{3.40}$$

Another pathway for LEs component formation is the disproportionation reaction, as illustrated in Eq. (3.41).

$$2C_7 \rightarrow C_5 + C_9 \tag{3.41}$$

Additionally, the formation of C_6 isoparaffins, exemplified in Eqs. (3.42) and (3.43), is attributed to propene dimerization. A notable step involves the addition reaction, where sec-propyl carbocation formed by the reaction of 2-chloropropane (Eq. 3.37) with propene leads to the generation of hexyl carbocation. This intermediate then extracts a hydride ion from isobutane, transforming into hexanes.

$$C_3^+ + C_3^= \rightarrow iC_6^+ \tag{3.42}$$

$$C_6^+ + iC_4 \rightarrow C_6 + iC_4^+ \tag{3.43}$$

(3) Formation of C_7 Isoparaffins

C_7 isoparaffins play a crucial role in isobutane/propene alkylation. The sequence involves tert-butyl carbocation reacting with propene to form sec-heptyl carbocation, followed by intramolecular methyl transfer, generating various heptyl carbocations. The C_7 isoparaffins are produced as heptyl carbocations abstract hydride ions from isobutane. Simultaneously, the hydride transfer reaction between heptyl carbocation and isobutane leads to the formation of tert-butyl carbocation. This formation pathway, termed direct alkylation, involves a significant production of tert-butyl carbocations through hydride transfer (Eqs. 3.44–3.46).

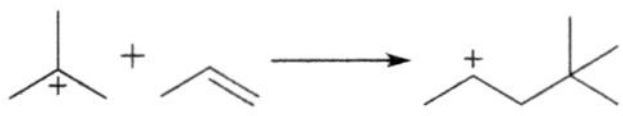
(3.44)

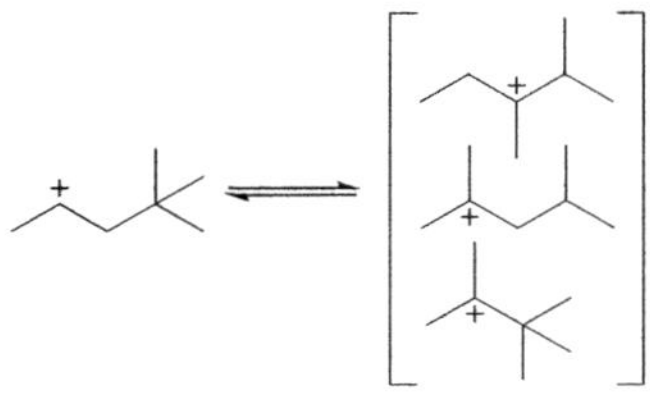
(3.45)

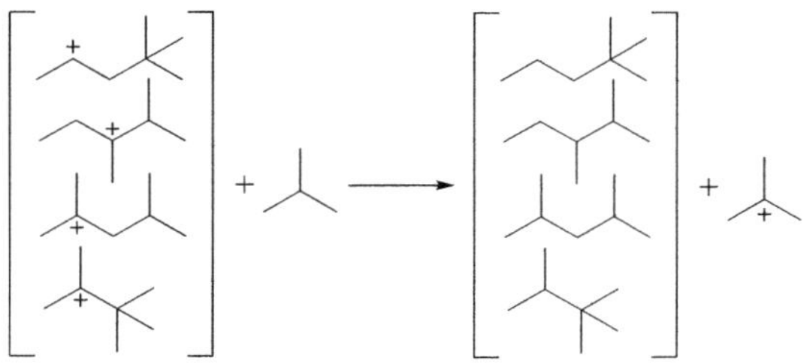
(3.46)

(4) Formation of C_8 isoparaffins

The generation of C_8 carbocations through cracking, followed by their subsequent splitting to form smaller carbocations and olefins, is considered less likely. The primary pathways for C_8 component formation involves isobutene generation by deprotonation of tert-butyl carbocation (Eq. 3.47). Isobutane then alkylates with tert-butyl carbocation, resulting in TMP^+ (Eq. 3.48), which undergoes intramolecular methyl transfer to form DMH^+ (Eq. 3.49). Hydride ions are abstracted from isobutane by $TMPs^+$ ($DMHs^+$), leading to the formation of TMPs (DMHs), such as 2,2,4-TMP, 2,3,4-TMP, 2,3,3-TMP, 2,5-DMH, and 2,4-DMH. A self-alkylation reaction (Eq. 3.51) further contributes to C_8 isoparaffin formation.

$$iC_4^+ \rightleftharpoons iC_4^= + H^+ \tag{3.47}$$

$$iC_4^+ \rightleftharpoons iC_4^= \rightarrow TMP^+ \tag{3.48}$$

$$TMP^{+} \rightleftharpoons DMH^{+} \tag{3.49}$$

$$TMP^{+}(DMH^{+}) + iC_4 \rightarrow TMP(DMH) + iC_4{}^{+} \tag{3.50}$$

$$C_3{}^{=} + 2C_4 \rightarrow C_3 + C_8 \tag{3.51}$$

The energy barrier comparison suggests that tert-butyl carbocation deprotonation (Eq. 3.52) has a higher barrier than its reaction with olefin. An alternative pathway involves the alkylation of products through large carbocation cracking, as exemplified by the reaction between butene and tert-butyl carbocation to form $C_8{}^{+}$ (Eq. 3.52), which then extracts a hydride ion from isobutane to generate C_8.

$$iC_4{}^{+} + iC_4{}^{=} \rightarrow C_8{}^{+} \tag{3.52}$$

$$C_8{}^{+} + iC_4 \rightarrow C_8 + iC_4{}^{+} \tag{3.53}$$

A third pathway is the disproportionation reaction, where two moles of C_7 isoparaffins react to form C_8 (and C_6), as detailed in Eq. (3.54).

$$2C_7 \rightarrow C_8 + C_6 \tag{3.54}$$

(5) Formation of $C_9{}^{+}$

Components with carbon numbers equal to or exceeding nine are classified as heavy ends. The first formation pathway involves oligomerization, where more than 2 moles of propene polymerized to form $C_9{}^{+}$ olefins. Protonation of these olefins leads to the formation of $C_9{}^{+}$ carbocations, which abstracts hydride ions from isobutane to produce $C_9{}^{+}$ isoparaffins.

$$3C_3{}^{=} + iC_4 + H^{+} \rightarrow C_9 + iC_4{}^{+} \tag{3.55}$$

The second formation pathway was secondary alkylation between propene and carbocations generated from direct alkylation (Eq. 3.56) or the alkylation of products through large carbocation cracking (Eq. 3.57). For instance, the reaction between propene and heptyl carbocations resulted in the formation of decyl carbocations, which then abstracted hydride ions from isobutane to generate decanes.

$$C_3{}^{=} + C_7{}^{+} \rightarrow C_{10}{}^{+} \tag{3.56}$$

$$C_{10}{}^{+} + iC_4 \rightarrow C_{10} + iC_4{}^{+} \tag{3.57}$$

The third pathway involved disproportionation (Eq. 3.41), analogous to the formation of C_8 (Eq. 3.54).

3.8 Alkylation mechanism of solid acids

The reaction mechanism of solid acid catalyzed C4 alkylation is also based on the carbocation mechanism. Similarly, solid acid catalyzed alkylation can also be divided into steps such as carbocation generation, olefin addition, hydride transfer, olefin oligomerization, β-scission, and self-alkylation. The difference between the liquid acid and solid acid catalysts lies in the activated method. In the liquid acid-catalyzed systems, the protonation of alkene follows Markovnikov's rule, where the proton provided by the acid attacks the most easily substituted carbon atom, forming the carbonium ion. The proton can be provided by H_2SO_4, HF, CF_3SO_2OH, etc. For solid acids, the alkene is adsorbed by the basic surface oxygen atoms of the zeolite, then the proton transfer to adsorbed alkenes leads to the formation of alkoxide. However, there is no unified carbocation generation with solid acids because of the lower acidity than H_2SO_4, especially aluminosilicate zeolites. The total reaction occurs through the synergistic effect of the confined space of the pore channels and acidity. Moreover, the effect of solid acids on the formation of carbocation ions differs from that of liquid acids. The reaction mechanism of zeolite-based solid acid-catalyzed C4 alkylation is described below.

Some efforts have been made to study the mechanism of zeolites. The mechanism of aluminosilicate zeolite-catalyzed alkylation on zeolites involves the protonation of reactants.

It is generally believed that in the catalytic conversion of alkenes, carbonium ions are formed through the adsorption of protons onto double bonds [16].

$$>C=C< + H^+ \longrightarrow R^+ \tag{3.58}$$

The detailed conversion of alkenes involves three states: π-complex, carbonium ion-like species, and alkoxy group. Senchenya and Kazansky established the formation of surface ethoxy groups from ethylene on high silica H-zeolites with AIMD by the simplest cluster $HO(H)Al(OH)_2OH$ [17]. It was found that ethylene is first adsorbed onto OH-groups as a π-complex, as shown in Fig. 3.10A, and then turned into the most stable ethoxy groups, as shown in Fig. 3.10C, with proton transfer rather than the carbonium ions. Moreover, the TS appears, as shown in Fig. 3.10B, which is similar to the free ethyl carbonium ion. The same TS in the reverse reaction of surface ethoxy group decomposition was also obtained. The energy of carbonium ion-like species has the highest potential barrier, which indicates that it is the most free species among the three species. For larger molecules (such as isobutylene), the proton transfer to the alkene can generate a tert-butyl carbonium intermediate, which forms a stable ion pair species with the zeolite conjugate base [18,19]. This ion can pass through a TS to an alkoxy group, as shown in Fig. 3.10.

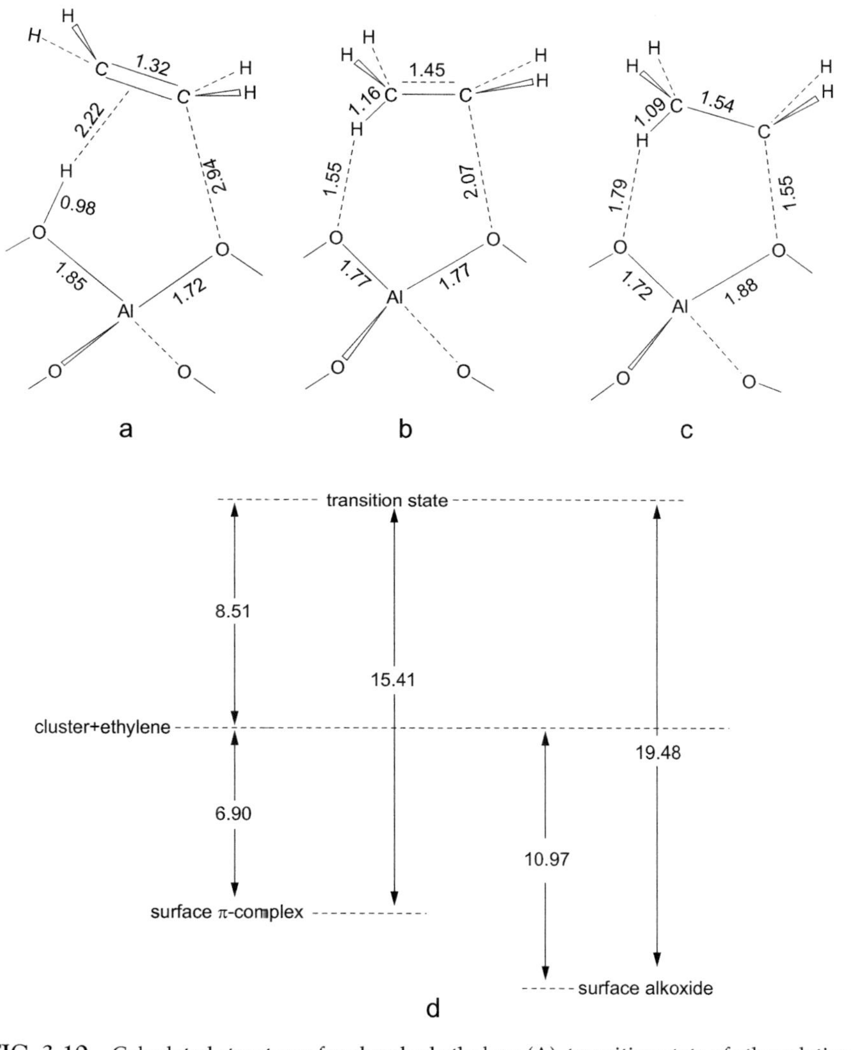

FIG. 3.10 Calculated structures for absorbed ethylene (A), transition state of ethoxylation reaction (B), final ethoxide (C), and energy diagram (kcal/mol) for ethoxy group formation (D). *Modified from reference I.N. Senchenya, V.B. Kazansky, Quantum chemical studies of ethylene interaction with zeolite OH-groups, Catal. Lett. 8 (1991) 317–325.*

Furthermore, the stability of carbonium ion-like species determines the rate of protonation of olefins to a certain extent. Essentially, the protonation reaction of olefins on zeolite is an acid–base interaction process. Therefore, when studying the protonation reaction, in addition to the basicity of the hydrocarbon and the pore structure of the zeolite acid, the acid strength of the zeolite is also an important influencing factor.

Engelhardt and Hall et al. [20] regarded carbonium ion as a metastable reaction intermediate and believed that its lifetime depends on the acid strength of the initial OH group. They found that the stronger the hydroxyl groups, the longer the lifetime and chemical activity of the carbonium are. Since olefins are easy to polymerize on zeolites, this brings difficulties to data analysis. It is difficult to study each elementary reaction clearly through experimental means alone. As a complement to experimental methods, theoretical calculation methods are well suited to study each step of the zeolite-catalyzed alkylation reaction and can depict the reaction mechanism at the atomic and molecular levels. Rigby et al. used 3T cluster models with different acid strengths to study the effect of acid strength on the ethylene protonation reaction [21,22]. They found that as the acid strength increases, the reaction activation energy of protonation slowly decreases, but the reaction mechanism remains unchanged.

Zheng et al. calculated the protonation reactions of propylene, isobutylene, styrene, and α-methylstyrene on 8T zeolite models with different acid strengths (closer to the structure of molecular sieves), and focused on examining the effect of acid strength on reaction mechanisms and intermediates. The influence of stability and reactivity is shown in Fig. 3.11 [23]. The authors found that propylene is eventually converted into a stable

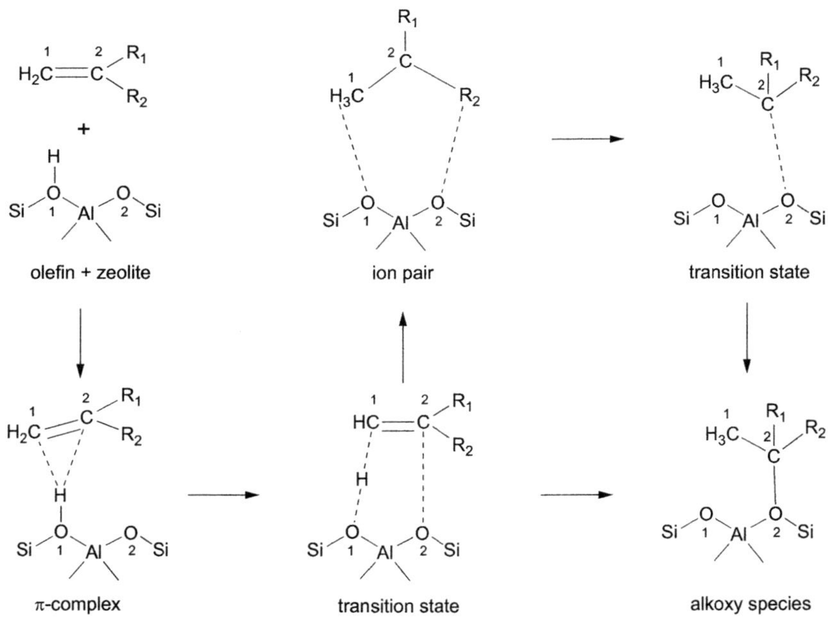

FIG. 3.11 Reaction mechanism of olefin protonation on solid acids. *From reference H. Fang, A. Zheng, S. Li, J. Xu, L. Chen, F. Deng, New insights into the effects of acid strength on the solid acid-catalyzed reaction: theoretical calculation study of olefinic hydrocarbon protonation reaction, J Phy Chem C 114 (2010) 10254–10264.*

reaction intermediate, isopropyl alkoxide, while isopropyl carbocation is not stable on the weak acid model and exists only in the TS. In the medium-strong acid model, propylene undergoes the same process, except that the energies of the Π complex, TS, and alkoxylate gradually decrease. Surprisingly, when the acid strength of the zeolite is increased to a superacid, the isopropyl carbocation can exist as a stable ion pair. In these two acid models, the isopropyl carbocation pair can be converted to isopropyl alkoxide through the TS. Furthermore, as the acid strength increases, ionic species are more sensitive to acid strength than covalent species. That is, the energy of ionic species decreases more with the increase of acid strength than that of covalent species, as shown in Fig. 3.12 [23].

The direct protonation of alkanes to generate adsorbed carbonium ions is also studied [16]. Then the carbonium ions are transferred into adsorbed carbonium ions, which often involves protolytic cracking and protolytic dehydrogenation, as shown in Eq. (3.59). Patrilyak et al. also revealed a similar mechanism with the protonation of isobutane on the superacid surface [24,25]. The superacid mechanism differs from the formation of the carbonium ion, which forms tert-butyl carbonium ion, as listed in Eq. (3.60). The starting and deprotonated B-sites of the active site of the zeolite catalyst are indicated in Eqs. (3.61) and (3.62) by H-Z and Z^-, respectively.

$$RH + H^+ \rightarrow RH_2{}^+ \rightarrow R_1H + R_2{}^+ \quad (3.59)$$

$$RH + H^+ \rightarrow RH_2{}^+ \rightarrow R^+ + H_2 \quad (3.60)$$

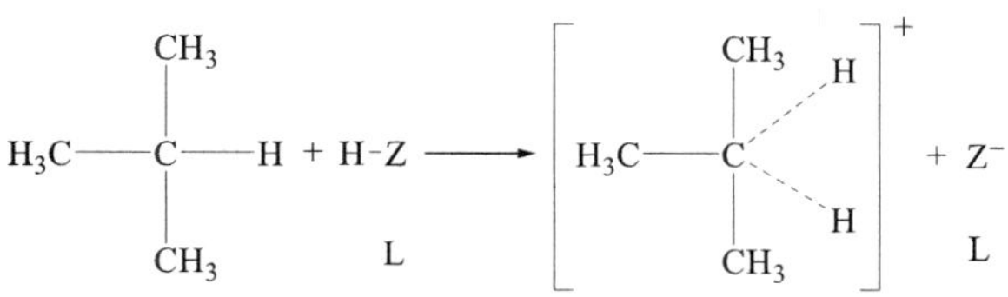

(3.61)

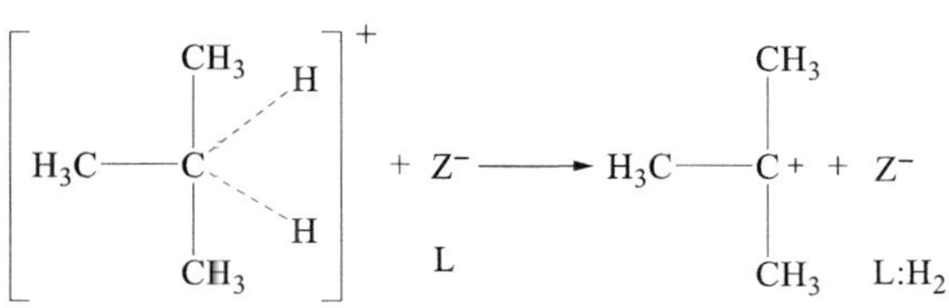

(3.62)

Corma and Martinez summarized the mechanism of C4 alkylation [26]. The first reaction mechanism, related to Olah et al. [27,28], is divided into parts: primary reactions and other reactions occurring during alkylation. The details of the primary reaction can be written as Eqs. (3.63) and (3.64). It is proposed that alkane protolysis involves formation of triangular TSs in the case of zeolites, which is different from the reactions involved in the mechanism of alkylation catalyzed by liquid acids. A pentacoordinated carbon atom is formed by the isobutane and proton in the first stage.

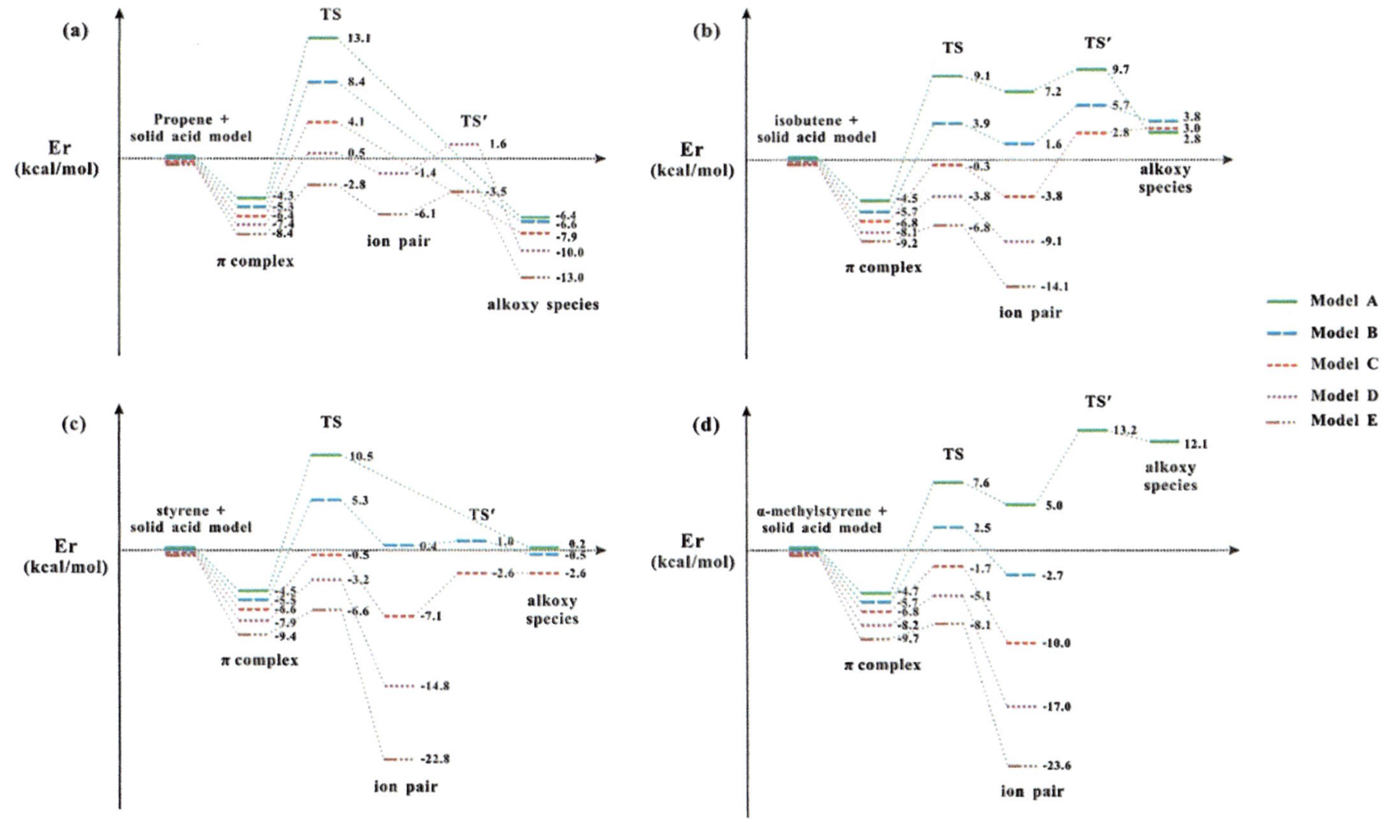

FIG. 3.12 Energy profiles of the protonation reactions on solid acid models A, B, C, D, and E for (a) propene, (b) isobutene, (c) styrene, and (d) α-methylstyrene. Energies (Er) were relative to the olefinic hydrocarbon and the solid acid model at infinite separation. *From reference H. Fang, A. Zheng, S. Li, J. Xu, L. Chen, F. Deng, New insights into the effects of acid strength on the solid acid-catalyzed reaction: theoretical calculation study of olefinic hydrocarbon protonation reaction, J Phy Chem C 114 (2010) 10254–10264.*

Then the olefin is added with a pentacoordinated carbon atom to generate a new cyclohydrocarbon with a positive charge.

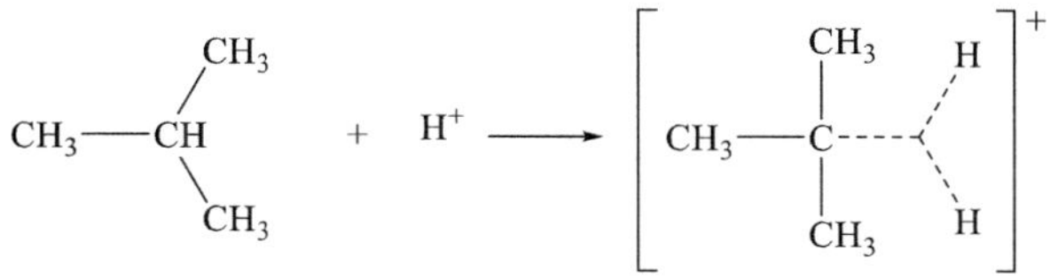

(3.63)

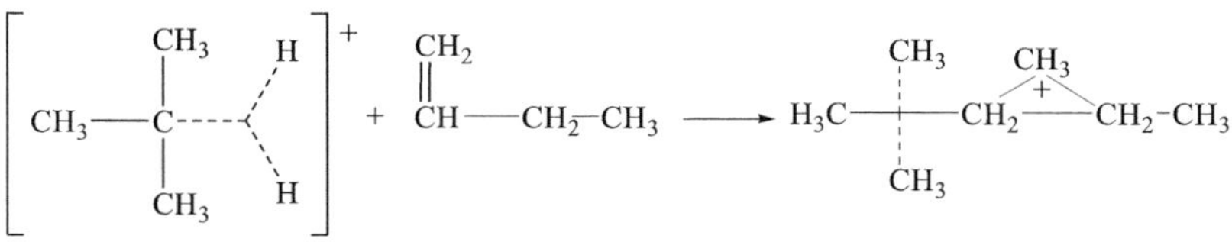

(3.64)

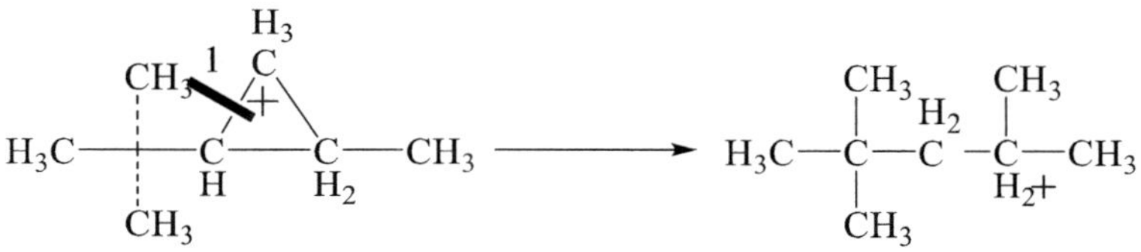

(3.65)

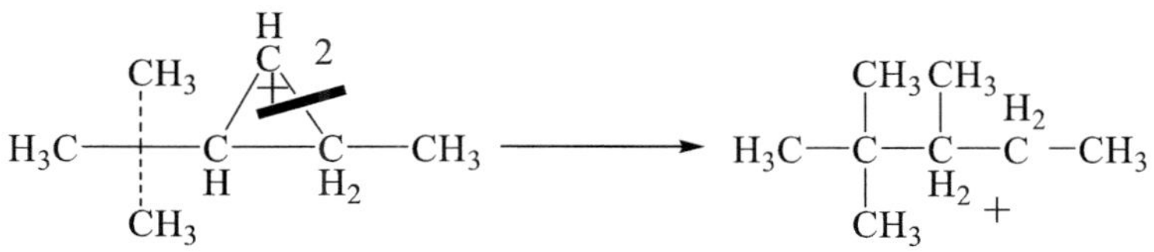

(3.66)

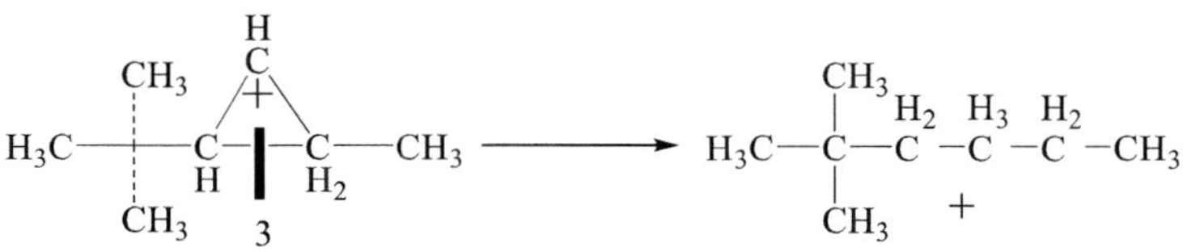

(3.67)

Alkenes are added to alkylate with carbocations, producing an octyl carbon chain followed by a cyclopropane ring. Three C—C bonds in the cyclopropane ring can be broken to generate three C_8^+, including 2,2,4-TMP^+, 2,2,3-TMP^+, and 2,2-DMH^+, as seen in Eqs. (3.65)–(3.67). Then C_8 paraffin is generated by transferring the hydride from the isobutane into C_8 carbonium intermediates and a new carbocation is generated.

Here 2,2-DMH and 2,2,3-TMP are predicted as the two primary alkylation products. However, Chu et al. and Weitkamp et al. [29,30] found that no 2,2-DMH and very small amounts of 2,2,3-TMP occur in the reaction products, as listed in Tables 3.7–3.9. According to Tables 3.7–3.9, the

TABLE 3.7 Comparison of dimethylhexane distribution on solid and liquid acid catalysts.

	Distribution (mol%)			
Catalyst	**REHY**	**Equilibrium (127°C)**	**HF (20°C)**	**H_2SO_4 (20°C)**
2,5-DMH	23.9	30.4	21.4	64.7
2,4-DMH	49.2	25.5	34.8	
2,3-DMH	6.0	4.0	38.4	29.4
2,2-DMH	–	22.2	–	2.0
3,3-DMH	–	12.7	–	–
3,4-DMH	20.9	5.1	5.4	4.0
Molar ratio: 2,5-DMH+2,4-DMH+2,3-DMH/3,4-DMH	3.8	11.7	17.5	23.5

From reference A. Corma, A. Martínez, Chemistry, catalysts, and processes for isoparaffin–olefin alkylation: actual situation and future trends, Catal. Rev. 35 (1993) 483–570.

TABLE 3.8 C_8 product distribution obtained during the alkylation of isobutane with 1-butene at 90°C and 21 bar, for a level of conversion (C4 wt%) of 77, on an REHY zeolite catalyst.

C_8 product	**Mol% distribution**	**Equilibrium composition**
TMP	90.5	
Selectivity		
2,2,4	29.8	61.4
2,2,3	5.6	16.5
2,3,4	25.8	10.8
2,3,3	38.8	11.3
DMH		
Selectivity		
2,2	–	
3,3	–	
2,3	6.0	
2,4	49.2	
2,5	23.9	
3,4	20.9	
HMp	0.6	

From reference A. Corma, A. Martínez, Chemistry, catalysts, and processes for isoparaffin–olefin alkylation: actual situation and future trends, Catal. Rev. 35 (1993) 483–570.

TABLE 3.9 Distribution of individual C_8 isomers (mol%) at 80°C on a CeY-98 zeolite catalyst, at a time on stream of 1 min.

C_8 product	Mol%
2-MHp	0.2
3-MHp + 2-EH	0.4
4-MHp	0.1
2,3-DMH	4.8
2,4-DMH	6.2
2,5-DMH	3.0
3,4-DMH	6.9
2-M,3-EP	0.6
2,2,3-TMP	4.4
2,2,4-TMP	22.3
2,3,3-TMP	27.9
2,3,4-TMP	23.2

From reference A. Corma, A. Martínez, Chemistry, catalysts, and processes for isoparaffin–olefin alkylation: actual situation and future trends, Catal. Rev. 35 (1993) 483–570.

amounts of 2,2-DMH and 2,2,3-TMP are much lower than the thermodynamic equilibrium concentration, and it is difficult to explain that 2,2-DMH and 2,2,3-TMP are the only major products. For the alkylation process, during the reaction of isobutane and propylene marked with ^{13}C at 80°C and 1 MPa, Daage and Fajula [26] observed the migration of secondary carbocations through hydride or methylene to give tertiary carbocations at a faster rate than desorption while the desorption rate is faster than the isomerization of tertiary carbocation. Nonetheless, Chu et al. [29] found that the 2,3,4-TMP and 2,2,3-TMP decrease while the 2,3,3-TMP and 2,2,4-TMP increase with decreasing olefin WHSV. In addition, the molar ratios of 2,2,3/2,2,4-TMP and 2,3,4/2,3,3-TMP increase significantly with increasing olefin WHSV, but 2,2,4 + 2,2,3/2,3,4 + 2,3,3-TMP remains relatively low, which suggest that the latter products are not predominantly formed by isomerization from 2,2,4- and 2,2,3-TMP. Consequently, other reactions occurring during alkylation reaction are also the reason for generating the C8 products.

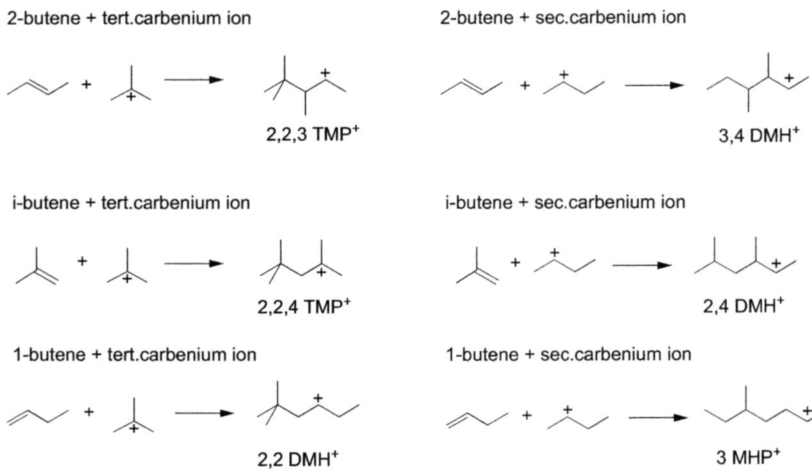

FIG. 3.13 Primary products of the addition of the different butene isomers to sec- and tert-butyl ions. *From reference A. Feller, J.A. Lercher, Chemistry and technology of isobutane/alkene alkylation catalyzed by liquid and solid acids, Adv. Catal. 48 (2004) 229–225.*

Feller et al. further organized the reaction mechanism of C4 alkylation with liquid acid or solid acid catalysts [31]. The alkylation mechanism is divided into seven parts, including initiation steps, alkene addition and isomerization, hydride transfer, oligomerization and cracking, self-alkylation, product and acid degradation, as well as pathways to allylic and cyclic compounds. After the formation of carbonium ion-like species with olefin, the olefin addition leads to different isooctyl cations depending on the isomer and the type, as shown in Fig. 3.13 [31].

Under alkylation reaction conditions, olefins can easily be protonated at the Brønsted acid sites of solid acid catalysts. During protonation, the hydride transfer of the isobutane competes with the polymerization of the carbon cation with another olefin. Since the former can produce alkylation products, a relatively high proportion of isobutane is usually fed to the reactor along with butenes. However, the fact that polymerization is faster than alkylation leads to the expectation that olefin polymerization occurs to a large extent [32,33]. Therefore, the products initially formed from olefin self-alkylation can be expressed as Eqs. (3.68) and (3.69). The olefin can be 1-butene or 2-butene. Considering that isobutene can be formed from the isomerization of n-butene and the dehydrogenation of isobutane, n-butene can co-dimerize with the isobutene to form the product, as shown in Eq. (3.70), and the isobutane can also dimerize by itself to generate the product as listed in Eq. (3.71), explaining the formation of 2,5-DMH, 2,4-DMH, and 2,6-DMH products. Nonetheless, Corma and Martinez thought that TMPs cannot be formed by the dimerization reactions from the structural perspective [32,33]. They assumed that the self-alkylation of isobutane can give 2,2,4-TMP as the main product, as listed in Eqs. (3.72), (3.73), and (3.74).

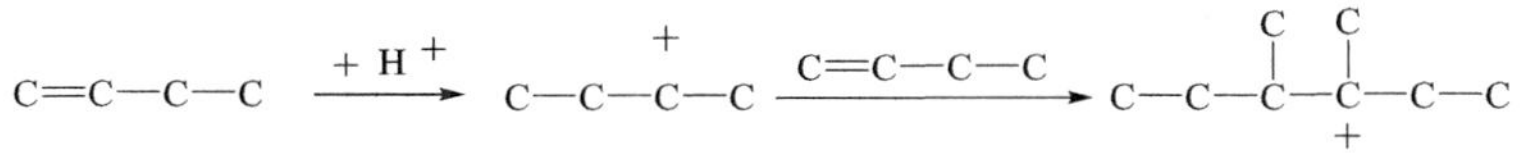
(3.68)

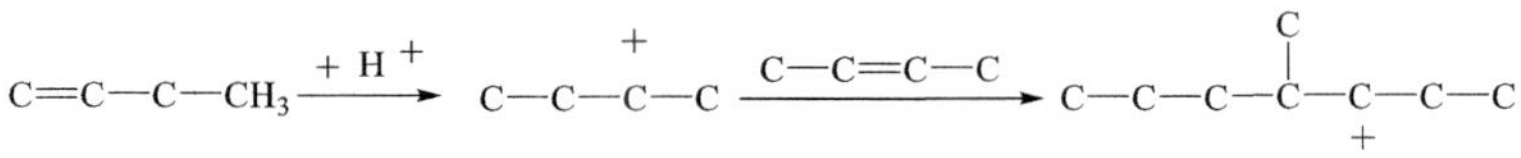
(3.69)

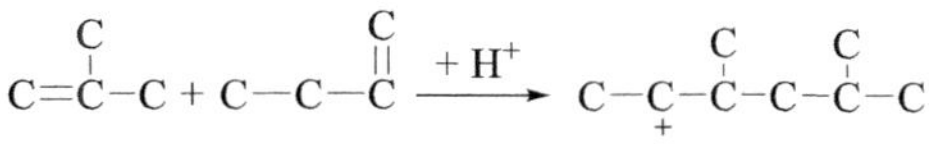
(3.70)

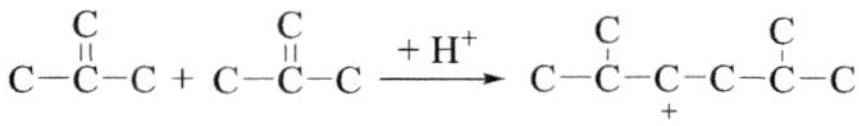
(3.71)

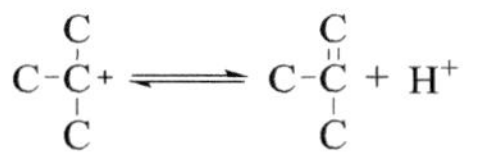
(3.72)

(3.73)

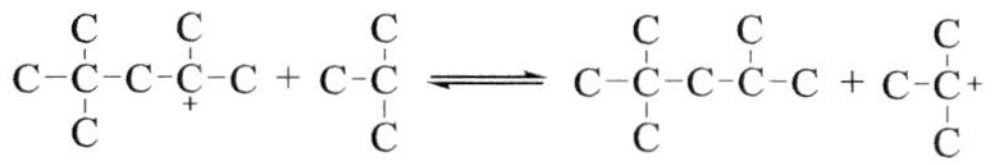
(3.74)

The relative significance of isobutane self-alkylation has been demonstrated through the alkylation of isobutene and isobutane [34]. When compared to using normal butene as the alkylating agent, the obtained products exhibit a higher TMP/DMH ratio. Within TMPs, 2,2,4-TMP is the most prominent isomer (60%), followed by the 2,3,3- and 2,3,4-isomers at 20% and 18%, respectively. Lastly, a small amount of 2,2,3-TMP (2%) is formed. However, more direct evidence supporting the contribution of isobutane self-alkylation to the formation of 2,2,4-TMP is obtained by alkylating isobutane with 1-butene-^{14}C [34]. As per the molecular weight range, the relative enrichment of butenes in the liquid product is shown in Table 3.10. In the case of TMP, the 2,2,4-TMP isomer (which, as mentioned earlier, dominates the self-alkylation process) contains only about half of the butene content as compared to 1-butene-^{14}C relative to other TMP isomers. For 1-butene-^{14}C, only half of the 2,2,4-TMP results from the isomerization of 1-butene, while the other half is produced through isobutane self-alkylation.

However, the polymerization rate of isobutene is higher than that of normal butene, which needs serious consideration [12]. Corma and Martinze proposed the third pathway for forming isooctane, which involves the decomposition of polymeric C_{12} and C_{16} products formed

TABLE 3.10 Butene-1-^{14}C and isobutane alkylation with zeolite catalyst.

Name	Weight distribution	^{14}C distribution	Relative % ^{14}C weight
i-C_4 to *i*-C_5	4.1	1.4	34.1
2,3-DMB to 3-MP	3.4	2.6	76.5
2,4-DMP + 2,2,3-TMB	4.0	2.4	60.0
2-MH to 3-MH	2.0	1.5	75.0
2,2,4-TMP	16.8	10.4	61.9
2,2,3-TMP	4.1	4.7	114.6
(2,3,4- + 2,3,3-)TMP + DMH + MHP	51.4	64.5	125.4
2,2,5-TMH plus	10.2	12.5	122.5

From reference A. Corma, A. Martínez, Chemistry, catalysts, and processes for isoparaffin–olefin alkylation: actual situation and future trends, Catal. Rev. 35 (1993) 483–570.

through consecutive alkylation and/or oligomerization, as listed in Eqs. (3.75)–(3.80) [32]:

$$C_8^+(\mathrm{TMPs\,or\,DMHs}) + 2 - C_4^= \rightleftharpoons i - C_{12}^+ \tag{3.75}$$

$$i - C_{12}^+ + 2 - C_4^= \rightleftharpoons i - C_{16}^+ \tag{3.76}$$

$$i\text{-}\overset{+}{C}_{12} \longrightarrow i\text{-}\overset{+}{C}_6 + i\text{-}C_6^= \tag{3.77}$$

$$i\text{-}\overset{+}{C}_{16} \longrightarrow i\text{-}\overset{+}{C}_8 + i\text{-}C_8^= \tag{3.78}$$

$$i\text{-}\overset{+}{C}_{16} \longrightarrow i\text{-}\overset{+}{C}_5 + i\text{-}C_5^= + i\text{-}C_6^= \tag{3.79}$$

$$i\text{-}\overset{+}{C}_{16} \longrightarrow i\text{-}\overset{+}{C}_5 + i\text{-}C_5^= + i\text{-}C_6^= \tag{3.80}$$

Self-alkylation affects the quality of alkylate. The self-alkylation mechanism, based on 2-butene and isobutane, is listed in Fig. 3.14 [31].

When utilizing solid acid as a catalyst for alkylation reactions, catalytic activity is generally observed only at elevated temperatures. The interaction between olefins and the H^+ species of solid acid readily facilitates the generation of carbonium ions, serving as reactive intermediates adsorbed on the surface of the solid acid. These carbonium ions exhibit a pronounced proclivity to undergo reactions with the olefins themselves, leading to the formation of olefinic oligomers. This propensity is attributed to the enhanced adsorption of olefins on the solid acid surface compared to isomeric alkanes, thereby influencing the reaction pathways of

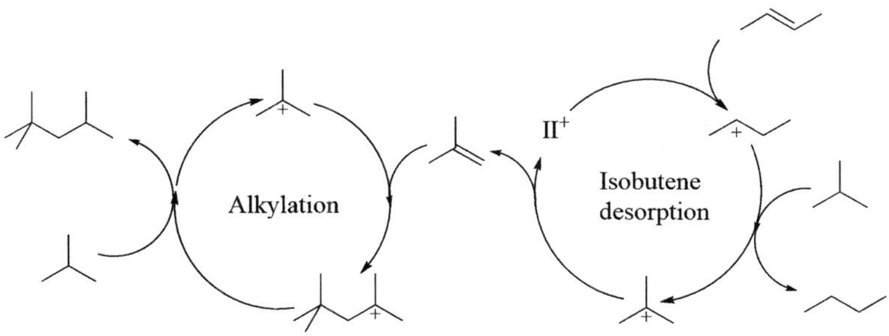

FIG. 3.14 Self-alkylation mechanism, depicted with 2-butene as the feed alkene. *From reference A. Feller, J.A. Lercher, Chemistry and technology of isobutane/alkene alkylation catalyzed by liquid and solid acids, Adv. Catal. 48 (2004) 229–225.*

carbonium ions away from predominant reactions with isomeric alkanes.

Through a comparative analysis of the mechanisms of the C4 alkylation on liquid acids and solid acids, significant differences were found in the generation of isomeric carbonium ions. When liquid acid serves as the catalyst, the facile generation of isomeric carbonium ions during the reaction process is observed. These isomeric carbonium ions subsequently engage in reactions with olefins, resulting in the production of alkylate. In contrast, when solid acid is utilized as the catalyst, the inherent characteristics of active centers on the solid acid surface and the influence of steric hindrance impede the generation and stabilization of carbocations. As a consequence, the occurrence of oligomerization reactions is facilitated, predisposing the catalyst to coke formation and deactivation.

References

[1] L.F. Albright, K. Li, Alkylation of isobutane with light olefins using sulfuric acid. Reaction mechanism and comparison with HF alkylation, Ind. Eng. Chem. Pro. Des. Dev. 9 (1970) 447–454.

[2] L.F. Albright, Comparison of alkylation processes, Chem. Eng. J. 73 (1966) 209.

[3] K. Li, R.E. Eckert, L.F. Albright, Alkylation of isobutane with light olefins using sulfuric acid. Operating variables affecting both chemical and physical phenomena, Ind. Eng. Chem. Pro. Des. Dev. 9 (1970) 441–446.

[4] K. Li, R.E. Eckert, L.F. Albright, Alkylation of isobutane with light olefins using sulfuric acid. Operating variables affecting physical phenomena only, Ind. Eng. Chem. Pro. Des. Dev. 9 (1970) 434–440.

[5] L.F. Albright, Alkylation processes using sulfuric acid as catalyst, Chem. Eng. J. 73 (14) (1966) 119.

[6] J. Ling, P. Cao, K. Li, L.X. Wen, J.L. Wu, F. Meng, X.H. Li, Effect of changes in C4 olefin composition on the thermodynamic behavior of alkylation reactions, Mod. Chem. Ind. 40 (2020) 150–154.

[7] A.E. Nurmakanova, Predicting Alkylate Yield and its Hydrocarbon Composition for Sulfuric Acid Catalyzed Isobutane Alkylation with Olefins Using the Method of

Mathematical Modeling, Procedia Chemistry (2015), https://doi.org/10.1016/j.proche.2015.10.009.
[8] L. Schmerling, Reactions of hydrocarbons, Ind. Eng. Chem. 45 (1953) 1447–1455.
[9] G.A. Olah, Á. Molnár, Hydrocarbon Chemistry, John Wiley & Sons, 2003.
[10] L. Schmerling, The mechanism of the alkylation of paraffins, J. Am. Chem. Soc. 67 (1945) 1778–1783.
[11] P.D. Bartlett, F.E. Condon, A. Schneider, Exchanges of halogen and hydrogen between organic halides and isoparaffins in the presence of aluminum halides, J. Am. Chem. Soc. 66 (1944) 1531–1539.
[12] J. Hofmann, A. Schriesheim, Ionic reactions occurring during sulfuric acid catalyzed alkylation. I. Alkylation of isobutane with butenes, J. Am. Chem. Soc. 84 (1962) 953–957.
[13] P. Esteves, C. Araújo, B. Horta, L. Alvarez, C. Zicovich-Wilson, A. Ramírez-Solís, The isobutylene–isobutane alkylation process in liquid HF revisited, J. Phys. Chem. B 109 (2005) 12946–12955.
[14] Y. Liu, L. Wang, R. Li, R. Hu, Reaction mechanism of ionic liquid catalyzed alkylation: alkylation of 2-butene with deuterated isobutene, J. Mol. Catal. A – Chem. 421 (2016) 29–36.
[15] Q. Li, R. Zhang, Y. Li, X. Meng, H. Liu, X. Wang, C. Xu, H. Dong, Q. Su, X. Zhang, Reaction behaviors and mechanism of isobutane/propene alkylation catalyzed by composite ionic liquid, Ind. Eng. Chem. Res. 61 (2022) 8624–8633.
[16] V.B. Kazansky, Adsorbed carbocations as transition states in heterogeneous acid catalyzed transformations of hydrocarbons, Catal. Today 51 (1999) 419–434.
[17] I.N. Senchenya, V.B. Kazansky, Quantum chemical studies of ethylene interaction with zeolite OH-groups, Catal. Lett. 8 (1991) 317–325.
[18] P. Viruela-Martin, C.M. Zicovich-Wilson, A. Corma, Ab initio molecular orbital calculations of the protonation of propylene and isobutene by acidic hydroxyl groups of isomorphously substituted zeolites, J. Phys. Chem. 97 (1993) 13713–13719.
[19] V.B. Kazansky, M.V. Frash, R.A. van Santen, Quantumchemical study of the isobutane cracking on zeolites, Appl. Catal. A Gen. 146 (1996) 225–247.
[20] J. Engelhardt, W.K. Hall, Peculiarities observed in H-D exchange between perdeuterioisobutane and H-zeolites, J. Catal. 151 (1995) 1–9.
[21] A. Rigby, G. Kramer, R. Van Santen, Mechanisms of hydrocarbon conversion in zeolites: a quantum mechanical study, J. Catal. 170 (1997) 1–10.
[22] A. Rigby, M. Frash, Ab initio calculations on the mechanisms of hydrocarbon conversion in zeolites: skeletal isomerisation and olefin chemisorption, J. Mol. Catal. A – Chem. 126 (1997) 61–72.
[23] H. Fang, A. Zheng, S. Li, J. Xu, L. Chen, F. Deng, New insights into the effects of acid strength on the solid acid-catalyzed reaction: theoretical calculation study of olefinic hydrocarbon protonation reaction, J. Phys. Chem. C 114 (2010) 10254–10264.
[24] K. Patrilyak, Y.N. Sidorenko, V. Bortyshevskii, Alkylation on zeolites (in Russian), Naukova Dumka, Kiev, 1991, pp. 130–144.
[25] V. Bortyshevskii, K. Patriliak, V. Gutyria, V. Baiburskii, Some peculiarities of the reaction of isoparaffin alkylation by olefins, Dokl. Akad. Nauk SSSR 282 (1985) 1382–1385.
[26] A. Corma, A. Martínez, Chemistry, catalysts, and processes for isoparaffin–olefin alkylation: actual situation and future trends, Catal. Rev. 35 (1993) 483–570.
[27] G.A. Olah, Stable carbocations. CXVIII. General concept and structure of carbocations based on differentiation of trivalent (classical) carbenium ions from three-center bound penta- of tetracoordinated (nonclassical) carbonium ions. Role of carbocations in electrophilic reactions, J. Am. Chem. Soc. 94 (1972) 808–820.
[28] G.A. Olah, G. Liang, G.D. Mateescu, J.L. Riemenschneider, Stable carbocations. CL. Fourier transform carbon-13 nuclear magnetic resonance and x-ray photoelectron spectroscopic study of the 2-norbornyl cation, J. Am. Chem. Soc. 95 (1973) 8698–8702.

[29] Y.F. Chu, A.W. Chester, Reactions of isobutane with butene over zeolite catalysts, Zeolites 6 (1986) 195–200.
[30] J. Weitkamp, Isobutane/butene alkylation on cerium exchanged X and Y zeolites, in: B. Imelik, C. Naccache, Y.B. Taarit, J.C. Vedrine, G. Coudurier, H. Praliaud (Eds.), Studies in Surface Science and Catalysis, Elsevier, 1980, pp. 65–75.
[31] A. Feller, J.A. Lercher, Chemistry and technology of isobutane/alkene alkylation catalyzed by liquid and solid acids, Adv. Catal. 48 (2004) 229–295.
[32] A. Corma, M. Faraldos, A. Mifsud, Influence of the level of dealumination on the selective adsorption of olefins and paraffins and its implication on hydrogen transfer reactions during catalytic cracking on USY zeolites, Appl. Catal. 47 (1989) 125–133.
[33] A. Corma, M. Faraldos, A. Martínez, A. Mifsud, Hydrogen transfer on USY zeolites during gas oil cracking: influence of the adsorption characteristics of the zeolite catalysts, J. Catal. 122 (1990) 230–239.
[34] F. Kirsch, J. Lauer, J. Potts, Reactivity and functionality of a zeolite alkylation catalyst, in: ACS Los Angeles Meeting, March, 1971.

CHAPTER

4

Reaction kinetics

As is well known, the fundamental understanding of the reaction kinetics is not only helpful to provide deeper insights into the reaction mechanism but also plays an important role in the design and optimization of the novel reactor and reaction process. The alkylation reaction is a typical process controlled by both chemical reactions and diffusion, which is very complex and contains simultaneous considerable reactions, including alkylation, polymerization, cracking, disproportionation, and self-alkylation, based on carbonium ions as intermediates. It is generally believed that the mass transfer of isobutane is the controlling step to determine the reaction rate. In the alkylation process, the mixing degree of acid and hydrocarbon has a great influence on the selectivity and yield of the alkylate because of the large difference in density and viscosity of the two phases. There exist more than three dozen isoparaffins during the reaction, and it is hard to detect each of them through the existing analytical methods. There is no doubt that the strong acid reaction system simultaneously increases the difficulty of analytical detection. Considerable efforts have been devoted to establishing the alkylation kinetics to explain the formation pathway of several key components, such as trimethylpentanes (TMPs), dimethylhexanes (DMHs), heavy ends (HEs), and light ends (LEs) in alkylate. This chapter introduces the alkylation kinetics with H_2SO_4, ionic liquids, and H_2SO_4 with additives as catalysts and summarizes the network pathway, kinetics parameters, and prediction results for each catalytic system.

4.1 Kinetics using sulfuric acid as catalyst

C4 alkylation mechanism has been extensively investigated for decades. Schmerling et al. first proposed the chain reaction mechanism based on the classic carbonium ion mechanism, which could well explain the formation of LEs, DMHs, and HEs [1]. However, several features of alkylation were not adequately explained, i.e., the secondary reactions

Aliphatic Alkylation
https://doi.org/10.1016/B978-0-443-21984-9.00007-6

and the formation of acid soluable oil (ASO) in the acid phase. Kramer et al. proposed that the hydride transfer reaction is a limiting step influencing the alkylation reaction rate [2]. The indication of the heterogeneous reaction is observed through the influence of the stirring rate and the correlation between the reaction rate and the concentration of methyl cyclopentane in the emulsion. Albright et al. proposed a reaction mechanism related to the catalyst used, which explains the startup phenomena, secondary reactions, and role of ASO in the acid phase [3].

For the C4 alkylation kinetics with sulfuric acid as the catalyst, Lee et al. built a simple kinetic model for the H_2SO_4-catalyzed C4 alkylation based on uniform hydrocarbon drops and short contact times [4]. However, this simplified kinetic model only considered two key reactions of isobutane with 1-butene and the oligomerization of 1-butene without several important isoparaffins in alkylate. When using 98 wt% H_2SO_4, the rate constants for alkylation and oligomerization at 298 K were calculated to be 2.25×10^8 and $9.5 \times 10^8\,cm^3/(mol\ s)$, respectively. The simplified model is displayed as Eqs. (4.1) and (4.2).

$$A + A \xrightarrow{k_A} \text{oligomers} \tag{4.1}$$

$$A + B \xrightarrow{k_B} \text{octanes} \tag{4.2}$$

where A represents 1-butene and B represents isobutane.

The rate constants for the alkylation of isobutane with 1-butene (k_A) and the oligomerization of 1-butene (k_B) are shown in Table 4.1.

However, the understanding of other key reaction steps remains incomplete, especially considering the presence of numerous isoparaffins in the alkylate. Albright et al. examined the two-step process for the C4 alkylation, but the rate constants were not provided [3].

The progress in alkylation kinetics has been demonstrated in the work of Sun et al., who proposed an alkylation kinetic model of C4 alkylation using H_2SO_4 as the catalyst [5]. The authors found that the generation of undesired DMHs most likely results from the addition of 1-butene to

TABLE 4.1 Rate constants for the alkylation of isobutane with 1-butene (k_A) and the oligomerization of 1-butene (k_B) at 298.15 K.

Sulfuric acid (wt%)	k_B (cm^3/g mol s)	k_A (cm^3/g mol s)	k_A/k_B
98	2.25×10^8	9.5×10^8	4.2
94	1.32×10^8	5.3×10^8	4.0
90	6.64×10^7	2.3×10^8	3.5

Data from reference L.-m. Lee, P. Harriott, The kinetics of isobutane alkylation in sulfuric acid, Ind. Eng. Chem. Process. Des. Dev. 16 (1977) 282–287.

isobutyl carbocation (iC_4^+) with their subsequent degradation due to the scission reactions, leading to the formation of LEs. Three key components of products are involved in the kinetic model, i.e., TMPs, DMHs, and HEs. An anti-Arrhenius kinetic behavior is observed from experimental data, where the rate constants are shown to change over a temperature interval ranging from 276.2 to 285.2 K. The kinetic model is based on the classic carbonium ion mechanism. Initially, the *tert*-butyl cation is produced by the protonation of butene,

$$iC_4^{=} + H^+ \xrightarrow{k_1} iC_4^+ \tag{4.3}$$

where $iC_4^{=}$ is the isobutene and iC_4^+ represents the *tert*-butyl cation. Regardless of whether the feedstock butene is 1-butene or 2-butene, they undergo isomerization for generating $iC_4^{=}$. This phenomenon is notably pronounced, particularly in the presence of H_2SO_4 [6]. The transformation between different butenes proceeds by the following equilibrium reaction,

$$1-C_4H_8 \overset{K_1}{\rightleftarrows} 2-C_4H_8 \overset{K_2}{\rightleftarrows} iC_4^{=} \tag{4.4}$$

Notably, it has been observed that when isobutane undergoes alkylation with olefins, the production of TMPs persists through a mechanism referred to as self-alkylation. The crucial step in self-alkylation involves the formation of isobutene, and the process is characterized by reversibility as outlined in Eq. (4.3),

$$iC_4^+ \xrightarrow{k_2} iC_4^{=} + H^+ \tag{4.5}$$

The *tert*-butyl cation adds to an olefin to produce the corresponding C_8^+. For example, when the *tert*-butyl cation reacts with $iC_4^{=}$ or 2-butene, the TMPs components are formed

$$iC_4^+ + 2-\text{butene}(\text{or } iC_4^{=}) \xrightarrow{k_3} \text{TMPs}^+ \tag{4.6}$$

where TMPs is the abbreviation of trimethylpentanes. These C_8^+ tend to isomerize via methyl shifts and hydride transfer reactions to form several isomer cations [7]. For simplification, all these C_8^+ are regarded as one pseudo-component TMP^+. By rapid hydride transfer with isobutane, each TMP^+ changes to the corresponding TMP components, producing the iC_4^+ to keep on the chain propagation.

$$\text{TMP}^+ + iC_4 \xrightarrow{k_4} \text{TMP} + iC_4^+ \tag{4.7}$$

where iC_4 stands for isobutane. The origin of undesirable DMHs with low RON entails several potential pathways. Among them, the addition of $iC_4^{=}$ and allylic carbocation is the result of a hydride ion abstraction from the allylic position of an olefin molecule.

$$TMP^{+} + iC_4^{=} \rightarrow TMP + C{=}C(CH_3) - C^{+} \quad (4.8)$$

$$C{=}C(CH_3){-}C^{+} + iC_4^{=} \rightarrow C{=}C(CH_3){-}C{-}C{-}(CH_3)C^{+}{-}C \quad (4.9)$$

Through appropriate hydride transfer reactions, the unsaturated cation transforms into a DMH molecule. Another possible pathway to give DMHs is through the addition of 1-butene to iC_4^{+} [8],

$$iC_4^{+} + 1 - \text{butene} \xrightarrow{k_5} DMHs^{+} \quad (4.10)$$

Similar to TMPs^{+}, the DMHs^{+} are saturated through the hydride abstraction from isobutane, giving an iC_4^{+}; meanwhile,

$$DMHs^{+} + iC_4 \xrightarrow{k_6} DMHs + iC_4^{+} \quad (4.11)$$

Olefins participate in reactions leading to the formation of HEs components, primarily involving polymerization reactions, as shown in Eq. (4.12). The polymerization of C4 olefins occurs through the addition of $iC_4^{=}$ or 2-butene to TMPs^{+} or DMHs^{+}, respectively.

$$TMPs^{+}(\text{or } DMHs^{+}) + iC_4^{=}(\text{or } 2 - \text{butene}) \rightarrow iC_{12}^{+} \quad (4.12)$$

The addition of $iC_4^{=}$ or 2-butene to iC_{12}^{+} results in the formation of iC_{16}^{+}. For the sake of simplicity, this study considers iC_{12}^{+} and iC_{16}^{+} as pseudo-components, collectively denoted as polymer cations (iC_m^{+}). When these iC_m^{+} interact with isobutane, the formation of HEs occurs, accompanied by the generation of iC_4^{+}.

$$iC_m^{+} + iC_4 \xrightarrow{k_7} iC_m + iC_4^{+} \quad (4.13)$$

As part of the five groups present in alkylate, the LEs encompass C_5–C_7 isoparaffins. The formation of LEs is attributed to the fragmentation of substantial isoalkyl cations. Due to the oxidizing capacity of concentrated H_2SO_4, the isoalkyl cation generates from HEs through the following reaction:

$$iC_m + H^{+} \rightarrow iC_m^{+} \quad (4.14)$$

These heavy isoalkyl cations undergo fragmentation in accordance with the β-scission rule, resulting in the formation of smaller isoalkyl cations and olefins [9].

$$iC_m^{+} \underset{k_9}{\overset{k_8}{\rightleftharpoons}} iC_x^{+} + iC_y^{=} \quad (4.15)$$

Iso-alkyl cations resulting from fragmentation (iC_x^{+}) and olefin protonation ($iC_y^{=}$) are considered the primary precursors of LEs. Through the transfer of hydride ions, these isoalkyl cations transform into the corresponding LEs. The degradation of DMHs is attributed to the generation of DMH^{+} through the oxidation of DMH, subsequently undergoing

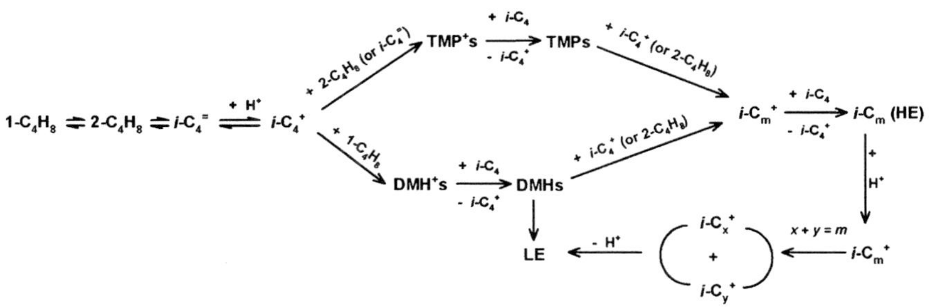

FIG. 4.1 Diagram of the reaction pathways network. *FFrom reference W. Sun, Y. Shi, J. Chen, Z. Xi, L. Zhao, Alkylation kinetics of isobutane by C4 olefins using sulfuric acid as catalyst, Ind. Eng. Chem. Res. 52 (2013) 15262–15269.*

fragmentation into smaller LEs through a scission reaction. For simplification, one single reaction step is used.

$$\mathrm{DMH} \xrightarrow{k_{10}} \mathrm{LE} \tag{4.16}$$

The diagram of the whole reaction pathway network is presentedpresented in Fig. 4.1.

There exist more than two dozen branched-chain species, including the corresponding carbonium components in C4 alkylation. Owing to the limit of the analysis method and the faster transformation between various species, only three or four groups of components were measured. To avoid the overfit or over-parametrization, Sun et al. simplified species with similar properties and close molecular weight as one pseudo-component, such as TMPs, DMHs, HEs, and LEs, respectively, being regarded as one single component. Furthermore, steps (4.4) and (4.5) were considered as the instantaneous reaction and step (4.7) is assumed to be governed by chemical equilibrium. Based on the total concentration of butene, the calculation of the concentration distribution for these three butenes is shown below.

$$c_{1-\mathrm{C_4H_8}} = \alpha^* c_{\mathrm{C_4H_8}} \tag{4.17}$$

$$c_{2-\mathrm{C_4H_8}} = \beta^* c_{\mathrm{C_4H_8}} \tag{4.18}$$

$$c_{i\mathrm{C_4^=}} = (1-\alpha-\beta)^* c_{\mathrm{C_4H_8}} \tag{4.19}$$

where α and β represent $1/(K_1 + K_1{}^*K_2 + 1), K_1/(K_1 + K_1{}^*K_2 + 1)$, respectively.

According to the summarized reaction steps and the simplified assumptions, the formulated kinetic model is displayed below:

$$\begin{aligned}\frac{dc_1}{dt} &= -(1-\alpha-\beta)k_1c_1 - (1-\alpha)k_3c_1c_3 - \alpha k_5c_1c_3 \\ &\quad -(1-\alpha)k_7c_1c_2c_4 + k_2c_3\end{aligned} \tag{4.20}$$

$$\frac{dc_2}{dt} = -k_4c_2c_4 - k_6c_2c_5 - (1-\alpha)k_7c_1c_2c_4 \quad (4.21)$$

$$\frac{dc_3}{dt} = (1-\alpha-\beta)k_1c_1 + k_4c_2c_4 + k_6c_2c_5 - (1-\alpha)k_3c_1c_3 - \alpha k_5c_1c_3 - k_2c_3 + (1-\alpha)k_7c_1c_2c_4 \quad (4.22)$$

$$\frac{dc_4}{dt} = (1-\alpha)k_3c_1c_3 - k_4c_2c_4 - (1-\alpha)k_7c_1c_2c_4 \quad (4.23)$$

$$\frac{dc_5}{dt} = \alpha k_5c_1c_3 - k_6c_2c_5 \quad (4.24)$$

$$\frac{dc_6}{dt} = k_4c_2c_4 \quad (4.25)$$

$$\frac{dc_7}{dt} = k_6c_2c_5 - k_{10}c_7 \quad (4.26)$$

$$\frac{dc_8}{dt} = (1-\alpha)k_7c_1c_2c_4 + k_9c_9c_{10} - k_8c_8 \quad (4.27)$$

$$\frac{dc_9}{dt} = k_8c_8 - k_9c_9c_{10} \quad (4.28)$$

$$\frac{dc_{10}}{dt} = k_8c_8 - k_9c_9c_{10} \quad (4.29)$$

The initial conditions are $t=0$, $c_1=c_1^0$; $c_2=c_2^0$; $c_3=0$; $c_4=0$; $c_5=0$; $c_6=0$; $c_7=0$; $c_8=0$; $c_9=0$; $c_{10}=0$. The corresponding species in (4.20)–(4.29) are 1, butene; 2, iC_4; 3, iC_4^+; 4, TMP^+; 5, DMH^+; 6, TMP; 7, DMH; 8, HE; 9, iC_x^+; 10, iC_y^+.

The concentration profiles of three groups of key components in the C4 alkylation were plotted in Fig. 4.2. All three species have dramatic changes within 2 min. Initially, the HE components go up sharply and then come down shortly. This demonstrates that the C4 alkylation is a fast reaction. After 5 min, each of the species reaches the platform individually, although the DMHs continue to go down slightly.

The estimated k_1–k_{10} are listed in Table 4.2 with 95% confidence intervals. However, an anti-Arrhenius behavior can be found for k_1; that is, the value of k_1 falls as the temperature increases. This phenomenon is not unusual or abnormal for hydrocarbon reactions with the presence of free radical molecules [10]. The observations of anti-Arrhenius kinetics are usually ascribed to an entropic contribution to the free energy of activation. This anti-Arrhenius behavior is believed to be caused by a decrease in the equilibrium constant of a multistep reaction [11].

The activation energies of k_1 and k_2 are determined to be −58.3 and 173.2 kJ/mol, respectively, with the corresponding preexponential factors

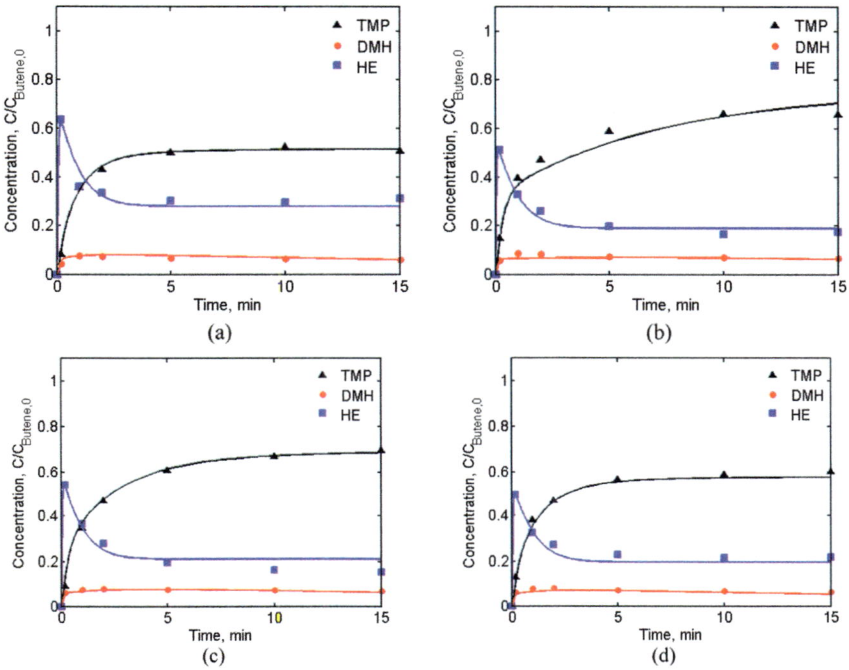

FIG. 4.2 Concentration profile of key components in alkylation of isobutane with butene. Temperature: (A) 276.2K, (B) 279.2K, (C) 282.2K, and (D) 285.2K. Symbols: experimental data; Line: calculated values by kinetic model. *From reference W. Sun, Y. Shi, J. Chen, Z. Xi, L. Zhao, Alkylation kinetics of isobutane by C4 olefins using sulfuric acid as catalyst, Ind. Eng. Chem. Res. 52 (2013) 15262–15269.*

of 5.23×10^{-11} and 6.87×10^{31} min^{-1}. The heat of the reaction is equal to the difference of activation energy between the forward and backward reaction for a reversible reaction based on the fundamental principle of physical chemistry. Consequently, the heat of reaction for the addition of H^+ to $iC_4^=$ can be computed as −231.5 kJ/mol, indicating a highly exothermic nature for alkylation. To further prove the reliability of the kinetic model [5], Sun et al. simulated the alkylation reactor operated at industrial conditions. The material balance equation of the industrial alkylation reactor is applied as follows:

$$x_i^0 - x_i = \tau^*(-r_i) \tag{4.30}$$

The conversion of olefin and alkylate yield versus mean residence time is shown in Fig. 4.3. The olefin conversion surpasses 99.5% when the mean residence time (τ) exceeds 10 min. However, even when τ reaches 20 min, the alkylate yield remains below 200%. Typically, an alkylate yield of over 200% is desired. Based on the results in Fig. 4.3, τ should be maintained at

TABLE 4.2 Estimated rate constants with 95% confidence intervals.

Temperature (K)	k_1 (min^{-1})	k_2 (min^{-1})	k_3, 10^2 (kg mol^{-1} min^{-1})	k_4 (kg mol^{-1} min^{-1})	k_5, 10^2 (kg mol^{-1} min^{-1})	k_6, 10^2 (kg mol^{-1} min^{-1})	k_7 (kg^2 mol^{-2} min^{-1})	k_8 (min^{-1})	k_9 (kg mol^{-1} min^{-1})	k_{10}, 10^2 (min^{-1})
276.2	6.37 ± 0.72	0.15 ± 0.06	8.33 ± 1.86	1.03 ± 0.22	1.26 ± 0.42	4.36 ± 2.00	88.46 ± 6.78	0.63 ± 0.06	5.48 ± 0.82	2.61 ± 0.85
279.2	5.06 ± 0.57	0.45 ± 0.20								
282.2	4.15 ± 0.50	1.00 ± 0.38								
285.2	2.79 ± 0.71	1.59 ± 0.56								

Data from reference W. Sun, Y. Shi, J. Chen, Z. Xi, L. Zhao, Alkylation kinetics of isobutane by C4 olefins using sulfuric acid as catalyst, Ind. Eng. Chem. Res. 52 (2013) 15262–15269.

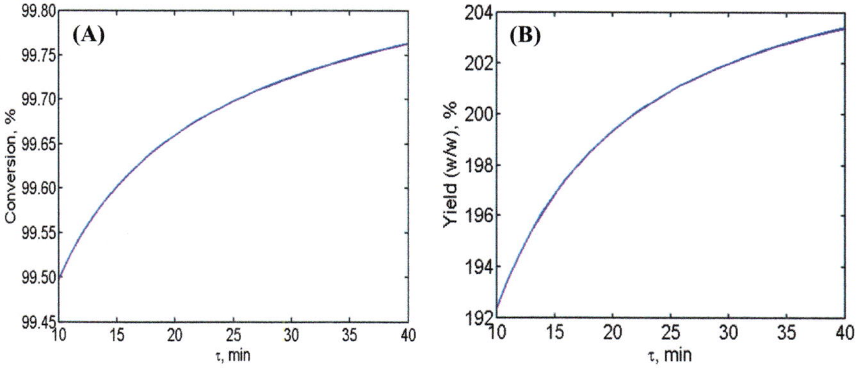

FIG. 4.3 Conversion of olefin versus mean residence time (A), and Alkylate yield versus mean residence time (B). *FFrom reference W. Sun, Y. Shi, J. Chen, Z. Xi, L. Zhao, Alkylation kinetics of isobutane by C4 olefins using sulfuric acid as catalyst, Ind. Eng. Chem. Res. 52 (2013) 15262–15269.*

approximately 21 min or more to meet this criterion. Moreover, the data suggests that as τ is more than 30 min, the yield of alkylate approaches its theoretical value of 203.6%. Notably, in industrial units, the reported τ typically ranges from 20 to 30 min [9,12].

However, there are still some unresolved issues. For instance, the formation pathways of the HEs remain unclear, and the fast isomerization reaction between butenes is not fully considered in the previous model [5]. To avoid the overestimated prediction of the HEs and estimate the fast isomerization reaction between butenes, the complex kinetic model was improved [13]. The DFT calculations were employed to determine the transition state and activation energy of the isomerization reaction involving three C4 olefins. Meanwhile, the reaction pathways were adjusted for the polymerization of C4 olefins to generate HEs. The difference lies in the model proposed by previous reports as follows:

The *tert*-butyl cation reacts with an olefin, leading to the generation of the corresponding C_8 carbonium cation:

$$iC_4^+ + 2-\text{butene}(\text{or}\, iC_4^{=}) \xrightarrow{k_3} \text{TMPs}^+ \tag{4.31}$$

The *tert*-butyl cation experiences isomerization, resulting in various isomers due to methyl shifts and hydride transfer reactions. Subsequently, the $TMPs^+$ react to isobutane to produce the TMP components via the hydride transfer and concurrently generates the iC_4^+ to continue the chain propagation

$$\text{TMP}^+ + iC_4 \xrightarrow{k_4} \text{TMP} + iC_4^+ \tag{4.32}$$

where iC_4 represents isobutane. As an undesired component with low RON, DMH components are formed through various pathways. One such pathway involves the addition of iC_4^+ to 1-butene:

$$iC_4^+ + 1 - \text{butene} \xrightarrow{k_5} \text{DMHs}^+ \tag{4.33}$$

The resulting DMHs^+ abstract a hydride ion from isobutane, yielding the DMHs and regenerating iC_4^+

$$\text{DMHs}^+ + iC_4 \xrightarrow{k_6} \text{DMHs} + iC_4^+ \tag{4.34}$$

HEs are produced by the addition of $iC_4^=$ or 2-butene to TMPs^+ or DMHs^+, resulting in the corresponding HEs cation ($i\text{-}C_m^+$). Through rapid hydride transfer with isobutane, $i\text{-}C_m^+$ transforms into the HEs, accompanied by the formation of iC_4^+.

$$\text{TMPs}^+ + iC_4^= \xrightarrow{k_7} iC_m^+ \tag{4.35}$$

$$\text{DMHs}^+ + 2 - \text{butene} \xrightarrow{k_{15}} iC_m^+ \tag{4.36}$$

$$iC_m^+ + iC_4 \rightarrow i - C_m + iC_4^+ \tag{4.37}$$

In summary, the comprehensive reaction pathway network, considering the isomerization among C4 olefins and the modified formation pathways of HEs components, is depicted in Fig. 4.4.

To prevent overfitting, the same strategies are adapted from the previous model simplification [5]. However, different from the previous model, the current model includes a detailed calculation of the isomerization rates between 1-butene, 2-butene, and isobutylene. Cao et al. calculated the energies and geometries of the reactants, transition states, and products for the isomerization between 1-butene, 2-butene, and isobutylene by using the quantum chemical based on the DFT method, as shown in Fig. 4.5.

The activation energies for these isomerization processes were 63.65, 63.43, and 64.42 kcal/mol, respectively. Notably, it is reasonable to infer that the rapid isomerization favors the transformation of either 1-butene or 2-butene into isobutene. The established model with fast isomerization is as follows:

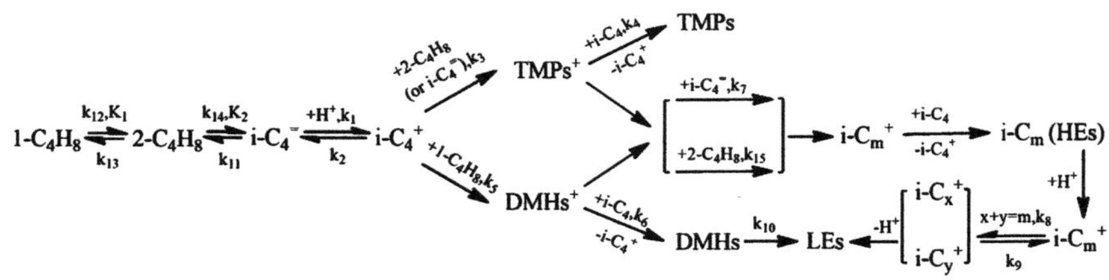

FIG. 4.4 Diagram of the reaction pathway network. *From reference P. Cao, L. Zheng, W. Sun, L. Zhao, Multiscale modeling of isobutane alkylation with mixed C4 olefins using sulfuric acid as catalyst, Ind. Eng. Chem. Res. 58 (2019) 6340–6349.*

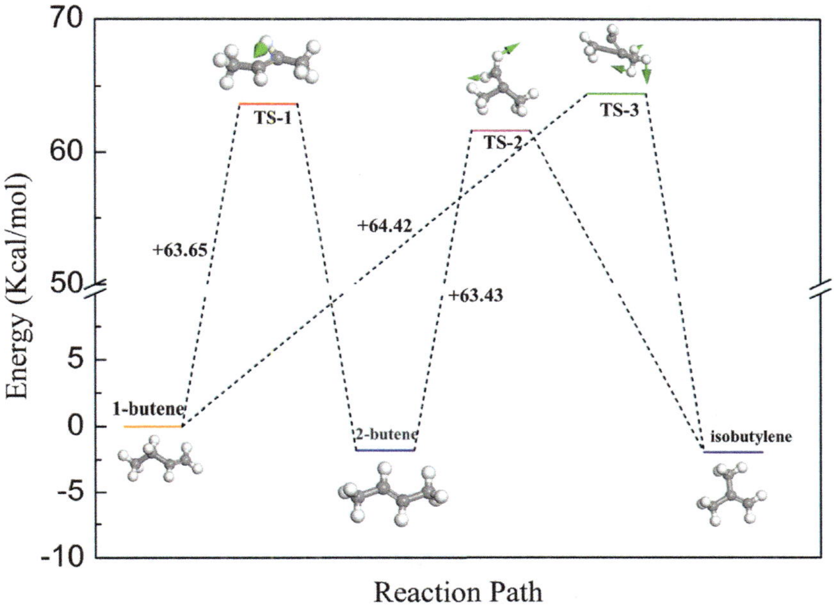

FIG. 4.5 Optimized geometries and potential-energy profile of the reactants, transition states, and products for the isomerization in butenes. The energy of 1-butene is set to zero. *From reference P. Cao, L. Zheng, W. Sun, L. Zhao, Multiscale modeling of isobutane alkylation with mixed C4 olefins using sulfuric acid as catalyst, Ind. Eng. Chem. Res. 58 (2019) 6340–6349.*

$$\frac{dc_1}{dt} = -k_1c_1 + k_2c_3 - k_3c_1c_3 - k_7c_1c_2c_4 - k_{11}c_1 + k_{14}c_{11} \tag{4.38}$$

$$\frac{dc_2}{dt} = -k_4c_2c_4 - k_6c_2c_5 - k_7c_1c_2c_4 - k_{15}c_2c_4c_{11} \tag{4.39}$$

$$\begin{aligned}\frac{dc_3}{dt} &= k_1c_1 + k_4c_2c_5 + k_6c_2c_5 - k_3(c_1 + c_{11})c_3 - k_5c_{12}c_3 - k_2c_3 \\ &\quad + k_7c_1c_2c_4 + k_{15}c_{11}c_2c_4\end{aligned} \tag{4.40}$$

$$\frac{dc_4}{dt} = k_3(c_1 + c_{11})c_3 - k_4c_2c_4 - k_7c_1c_2c_4 - k_{15}c_2c_4c_{11} \tag{4.41}$$

$$\frac{dc_5}{dt} = k_5c_{12}c_3 - k_6c_2c_5 \tag{4.42}$$

$$\frac{dc_6}{dt} = k_4c_2c_4 \tag{4.43}$$

$$\frac{dc_7}{dt} = k_6c_2c_5 - k_{10}c_7 \tag{4.44}$$

$$\frac{dc_8}{dt} = k_7c_1c_2c_4 + k_{15}c_{11}c_2c_4 + k_9c_9c_{10} - k_8c_8 \tag{4.45}$$

$$\frac{dc_9}{dt} = k_8c_8 - k_9c_9c_{10} \tag{4.46}$$

$$\frac{dc_{10}}{dt} = -k_9c_9c_{10} + k_8c_8 + k_{10}c_7 \tag{4.47}$$

$$\frac{dc_{11}}{dt} = -k_3c_{11}c_3 - k_{15}c_{11}c_2c_4 + k_{11}c_1 + k_{12}c_{12} + k_{13}c_{11} - k_{14}c_{11} \tag{4.48}$$

$$\frac{dc_{12}}{dt} = -k_5c_{12}c_3 + k_{13}c_{11} - k_{12}c_{12} \tag{4.49}$$

The initial conditions are $t=0$, $c_1=c_1^0$; $c_2=c_2^0$; $c_3=0$; $c_4=0$; $c_5=0$; $c_6=0$; $c_7=0$; $c_8=0$; $c_9=0$; $c_{10}=0$; $c_{11}=c_{11}^0$; $c_{12}=c_{12}^0$. The corresponding species in Eqs. (4.38)–(4.49) are 1, iC_4H_8; 2, iC_4; 3, iC_4^+; 4, $TMPs^+$; 5, $DMHs^+$; 6, TMPs; 7, DMHs; 8, HEs; 9, iC_x^+; 10, iC_y^+; 11, 2-C_4H_8; 12, 1-C_4H_8.

Similar to the previous work [5], the chemical equilibrium constants, denoted as K_1 and K_2, were determined utilizing the RGibbs module within the ASPEN Plus software. This calculation was based on the Gibbs free energy variation of the isomerization of butene. The temperature-dependent equilibrium constants are presented in Table 4.3.

To prevent overfitting or over-parametrization, the rate constants k_4, k_6, and k_8–k_{10} are independent of temperature, which implies minimal or very low activation energy associated with these reaction steps. Therefore, these rate constants remain constant within the experimental temperature range to minimize the number of adjustable parameters. In addition, the parametrized rate constants of k_1, k_2, k_3, and k_5 were directly applied from the previous model fitting [5]. This approach reduced the number of parameters that need estimation and enhanced the precision of the rate constants.

Consequently, the remaining adjustable parameters consist of k_7 and k_{11}–k_{15}, which correspond to the modified reaction pathways of the HEs and the transformation rates between butenes, respectively. Utilizing experimental data, the fitting results within the temperature range of

TABLE 4.3 Equilibrium distribution and constants of butenes.

Temperature (K)	1-$C_4^=$ (%)	*cis*-2-$C_4^=$ (%)	*trans*-2-$C_4^=$ (%)	*i*-$C_4^=$ (%)	K_1	K_2
276.2	0.003529	0.03410	0.09163	0.8707	35.63	6.925
279.2	0.003798	0.03554	0.09383	0.8668	34.06	6.700
282.2	0.004080	0.03700	0.09602	0.8629	32.60	6.487
285.2	0.004376	0.03847	0.09820	0.8589	31.23	6.285

Adapted from reference P. Cao, L. Zheng, W. Sun, L. Zhao, Multiscale modeling of isobutane alkylation with mixed C4 olefins using sulfuric acid as catalyst, Ind. Eng. Chem. Res. 58 (2019) 6340–6349.

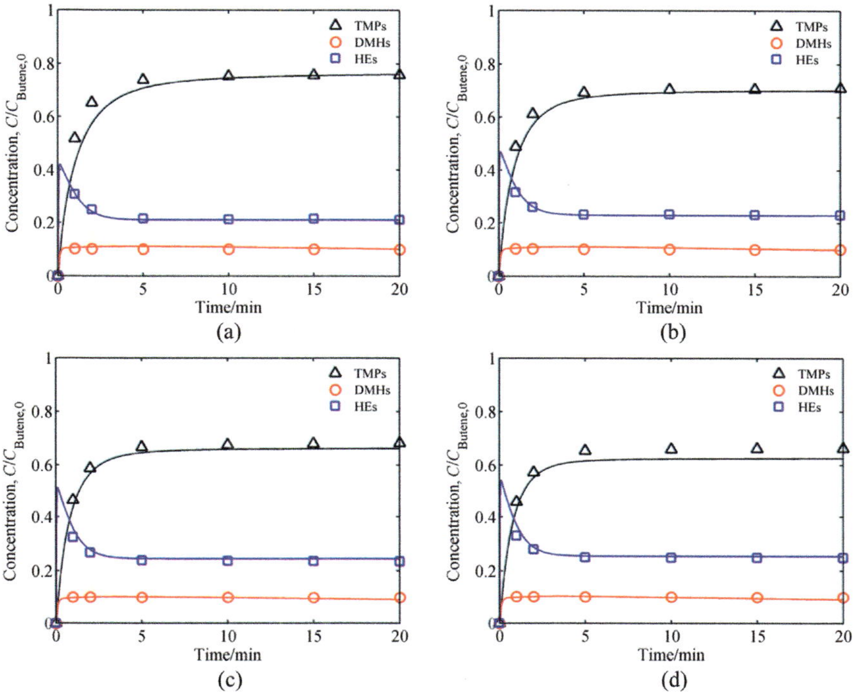

FIG. 4.6 Concentration profiles of key components in the alkylation of isobutane with mixed C4. Temperature: (A) 276.2, (B) 279.2, (C) 282.2, and (D) 285.2 K. Symbols, experimental data; Line, calculated values by kinetic model. *From reference P. Cao, L. Zheng, W. Sun, L. Zhao, Multiscale modeling of isobutane alkylation with mixed C4 olefins using sulfuric acid as catalyst, Ind. Eng. Chem. Res. 58 (2019) 6340–6349.*

276.2–285.2 K are illustrated in Fig. 4.6. The fitting data associated with the TMPs, DMHs, and HEs agree well with the experimental results. Furthermore, the production of HEs components experiences a rapid increase followed by a rapid decline due to the polymerization of butenes.

The estimated parameters of k_7 and k_{11}–k_{15} are listed in Table 4.4 with 95% confidence intervals.

The rate constants k_{12} and k_{14} are one and two orders of magnitude larger than their corresponding reversible reactions, k_{13} and k_{11}, respectively. This discrepancy suggests that the isomerization rates of 1-butene into 2-butene and 2-butene into isobutylene occur much more rapidly than their reversible reactions. This observation agrees with the phenomenon where either 1-butene or 2-butene tends to transform into isobutene through fast isomerization, particularly in the presence of H_2SO_4.

Furthermore, the Arrhenius relationships for k_7 and k_{11}–k_{14} were derived by examining their variation with temperature. The calculated activation energies and preexponential factors are presented in Table 4.5.

TABLE 4.4 Estimated rate constants with 95% confidence intervals.

Temperature (K)	k_7, 10 ($kg\ mol^{-2}\ min^{-2}$)	k_{11} (min^{-1})	k_{12}, 10^2 (min^{-1})	k_{13} (min^{-1})	k_{14}, 10 (min^{-1})	k_{15} ($kg\ mol^{-2}\ min^{-2}$)
276.2	9.99 ± 0.72	7.95 ± 1.89	2.54 ± 0.01	7.13 ± 0.02	5.50 ± 1.31	32.62 ± 19.80
279.2	10.54 ± 1.22	9.59 ± 0.41	3.17 ± 0.01	9.31 ± 0.03	6.42 ± 0.27	
282.2	11.04 ± 0.46	11.41 ± 1.84	3.84 ± 0.10	11.76 ± 0.31	7.40 ± 1.19	
285.2	11.59 ± 0.36	13.30 ± 0.22	4.69 ± 1.54	15.02 ± 4.92	8.36 ± 0.14	

Data from reference P. Cao, L. Zheng, W. Sun, L. Zhao, Multiscale modeling of isobutane alkylation with mixed C4 olefins using sulfuric acid as catalyst, Ind. Eng. Chem. Res. 58 (2019) 6340–6349.

TABLE 4.5 Calculated activation energies (E_a) and preexponential factors (k_0) for each reaction.

	k_7	k_{11}	k_{12}	k_{13}	k_{14}
E_a (kJ/mol)	10.69	37.54	44.33	53.92	30.48
k_0	1.05×10^4 (kg mol^{-2} min^{-2})	1.01×10^8 (min^{-1})	6.18×10^{10} (min^{-1})	1.13×10^{11} (min^{-1})	3.22×10^7 (min^{-1})

Data from reference P. Cao, L. Zheng, W. Sun, L. Zhao, Multiscale modeling of isobutane alkylation with mixed C4 olefins using sulfuric acid as catalyst, Ind. Eng. Chem. Res. 58 (2019) 6340–6349.

The activation energies corresponding to k_{11}–k_{14} are 37.54, 44.33, 53.92, and 30.48 kJ/mol , respectively. In comparison to k_{13} and k_{11}, k_{12} and k_{14} exhibit lower activation energies, indicating that the isomerization of 1-butene to 2-butene and 2-butene to isobutylene is easier than their reversible reactions.

Aiming to determine the influence of operating variables and feed composition on the yield and hydrocarbon composition of alkylate, Ivashkina et al. used the unsteady mathematical model to optimize the operating conditions based on industrial data [14]. The developed formalized reaction network of the process is as follows (Fig. 4.7).

The thermodynamic and kinetic regularities of the H_2SO_4-catalyzed C4 alkylation are determined, as listed in Table 4.6. The entropy and enthalpy of intermediate formation can be employed to compute the preexponential factor of the reaction rate constant and the activation energy.

The calculations reveal that the presence of n-pentane in the raw materials is undesirable, as it adversely affects the quality of the alkylate. To mitigate the adverse impact of impurities on alkylate quality, the following conditions are recommended and deemed necessary to achieve satisfactory results during the isobutane alkylation with a mixture of propylene and butenes [14].

(1) Propylene content in the mixture should not be more than 55%.
(2) High dispersity of the acid-hydrocarbon emulsion, which is achieved by intensive mixing.
(3) Increased acid content in the emulsion (up to 60–65%vol).

Furthermore, Tian et al. adapted the novel method of artificial neural network (ANN) to achieve and optimize for evaluating product distribution and the research octane number (RON) of alkylate in the stirred tank (STR) and rotating packed bed (RPB) [15]. The input parameters for the ANN model included feedstock compositions, operating conditions, and reactor types. The algorithm, transfer function, and framework were examined to choose the optimal ANN model.

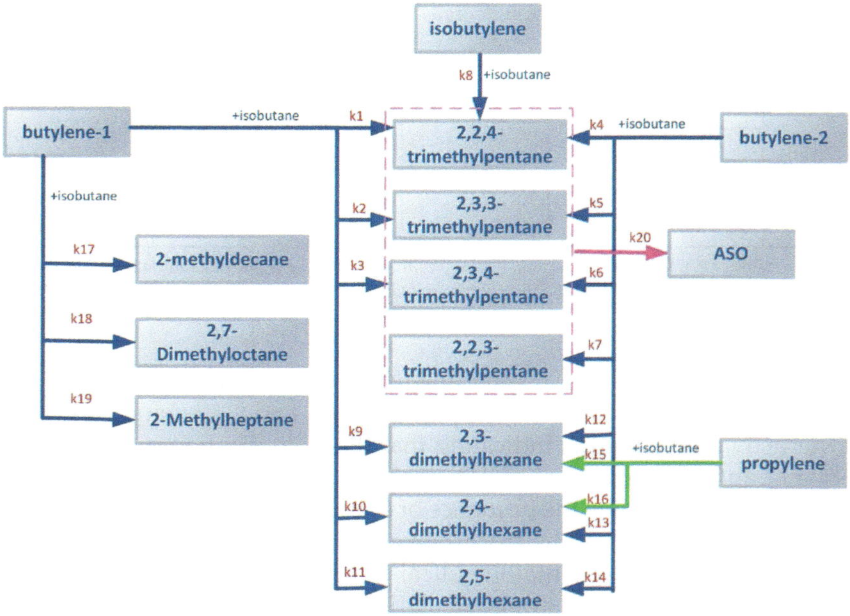

FIG. 4.7 Formalized reaction network of the H_2SO_4-catalyzed isobutane alkylation with alkenes. *From reference E. Ivashkina, I. Dolganova, I. Dolganov, E. Ivanchina, A. Nurmakanova, A. Bekker, Modeling the H_2SO_4-catalyzed isobutane alkylation with alkenes considering the process unsteadiness, Catal. Today 329 (2019) 206–213.*

TABLE 4.6 Kinetic parameters of the reactions of the process of isobutane alkylation with olefins at 10°C and 0.1 MPa.

No.	Reaction	E_a (kJ/mol)	k_0 (s^{-1})
1	Isobutane + butylene-1 → 2,2,4-trimethylpentane	67.03	2.09×10^{12}
2	Isobutane + butylene-1 → 2,3,3-trimethylpentane	64.50	1.25×10^{12}
3	Isobutane + butylene-1 → 2,3,4-trimethylpentane	63.28	1.38×10^{12}
4	Isobutane + butylene-2 → 2,2,4-trimethylpentane	63.92	2.09×10^{12}
5	Isobutane + butylene-2 → 2,3,3-trimethylpentane	61.39	1.25×10^{12}
6	Isobutane + butylene-2 → 2,3,4-trimethylpentane	60.17	1.38×10^{12}
7	Isobutane + butylene-2 → 2,2,3-trimethylpentane	62.78	4.56×10^{4}
8	Isobutane + isobutylene → 2,2,4-trimethylpentane	63.38	1.10×10^{12}
9	Isobutane + butylene-1 → 2,3-dimethylgexane	66.17	7.74×10^{11}
10	Isobutane + butylene-1 → 2,4-dimethylgexane	67.35	8.04×10^{11}
11	Isobutane + butylene-1 → 2,5-dimethylgexane	68.36	8.04×10^{11}

TABLE 4.6 Kinetic parameters of the reactions of the process of isobutane alkylation with olefins at 10°C and 0.1 MPa—cont'd

No.	Reaction	E_a (kJ/mol)	k_0 (s^{-1})
12	Isobutane + butylene-2 → 2,3-dimethylgexane	63.06	7.74×10^{11}
13	Isobutane + butylene-2 → 2,4-dimethylgexane	64.24	8.04×10^{11}
14	Isobutane + butylene-2 → 2,5-dimethylgexane	65.25	8.04×10^{11}
15	Isobutane + propylene → 2,3-dimethylgexane	66.91	6.02×10^{11}
16	Isobutane + propylene → 2,4-dimethylgexane	69.41	8.29×10^{11}
17	2Isobutane + 2butylene-1 → 2-methyldecane	87.13	1.46×10^{3}
18	2Isobutane + 2butylene-1 → 2,7-dimethyloctane	87.11	2.05×10^{3}
19	Isobutane + butylene-1 → 2-methylheptane	67.62	6.11×10^{4}

Data from reference E. Ivashkina, I. Dolganova, I. Dolganov, E. Ivanchina, A. Nurmakanova, A. Bekker, Modeling the H_2SO_4-catalyzed isobutane alkylation with alkenes considering the process unsteadiness, Catal. Today, 329 (2019) 206–213.

An optimal ANN model was established with a network topology of 10–20–30–5 using Bayesian Regularization backpropagation and tan-sigmoid transfer function.

Tian et al. suggested that m_{STR} is advantageous for enhancing the high-RON components and hydrocarbons should be introduced with liquid acid during the process. The inputs of P and N_{STR} were able to control the liquid phase of hydrocarbon and the liquid-liquid mixing, respectively. When $N_{RPB} > N_{STR}$, it indicated that RPB was more effective than STR in enhancing the liquid-liquid two-phase mass transfer and micromixing, suggesting that the rotational speed deeply affected the alkylation process [16].

Recently, Xin et al. studied the reactor model based on the structural pattern and flow state of the reactor [17]. The alkylation unit operates under the following conditions: a feed flow rate of approximately 30–70 tons/h for each reactor, a reactor inlet temperature ranging from 2.0°C to 4.0°C, a reactor outlet temperature between 6.0°C and 6.5°C, a molar ratio of isobutane to olefins at the reactor inlet ranging from 7:1 to 12:1, an acid-to-hydrocarbon ratio (volume ratio) inside the reactor of about 1.1, and an acid concentration (wt%) in the three reactors around 92%–96%.

Based on the work of Albright and Doshi et al. [18] and Sun et al. [13], the reaction network of alkylation reaction from Xin et al. is proposed. Furthermore, to simplify the reactor model, Xin et al. made the detailed assumptions in the report [17].

Additionally, many researchers have investigated the kinetics of H_2SO_4-catalyzed C4 alkylation in novel reactor systems [16,19]. Tian et al. provided a study of employing the novel RPB reactor to enhance C4 alkylation [16]. A correlation model is applied to predict product distribution and the RON of alkylate, which shows reliable relevance to experimental results. The correlation model is related to the factors, including reaction time, reaction temperature, A/HC, and rotational speed of RPB. This was based on Liu's study on the performance of C4 alkylation catalyzed by composite ionic liquid [20]. An empirical correlation was developed to predict product distributions and the RON of alkylate as follows:

$$y = m_1 + m_2t^2 + m_3t + m_4T^2 + m_5T + m_6\left(\frac{A}{\mathrm{HC}}\right)^2 + m_7\left(\frac{A}{\mathrm{HC}}\right) + m_8\left(\frac{N_{\mathrm{RPB}}}{150}\right)^2 + m_9\left(\frac{N_{\mathrm{RPB}}}{150}\right) + m_{10}P^2 + m_{11}P + m_{12}\left(\frac{\mathrm{I}}{\mathrm{O}}\right)^2 + m_{13}\left(\frac{\mathrm{I}}{\mathrm{O}}\right) \quad (4.50)$$

where y refers to the yield of alkylate components (wt%) and m_1–m_{13} are the coefficients of the respective parameters in Eq. (4.50). Optimal coefficients for m_1–m_{13} are obtained through curve fitting and listed in Table 4.7.

Li et al. first investigated the kinetics of H_2SO_4-catalyzed C4 alkylation using a micro-structured chemical system (see Fig. 4.8), which is suitable for the fast reaction. The kinetic model developed by Li et al. encompasses both main and side reactions within the micro-structured chemical system. The model exhibits excellent agreement with the experimental data and is subsequently validated by simulated results using COMSOL software.

Based on the kinetic model established by Sun et al. [5], Li et al. simplified the model for the fitting of kinetic parameters [19]. The simplified formula is shown as follows:

$$C_4^{=} + iC_4 \xrightarrow{k_1} \mathrm{TMP} \quad (4.51)$$

$$C_4^{=} + iC_4 \xrightarrow{k_2} \mathrm{DMH} \quad (4.52)$$

$$2C_4^{=} + iC_4 \xrightarrow{k_3} \mathrm{HE} \quad (4.53)$$

$$2C_4^{=} + iC_4 \xrightarrow{k_4} 2\mathrm{LE} \quad (4.54)$$

According to the simplified formula, the kinetic models are formulated as follows:

$$\frac{dc_1}{dt} = -k_1c_1c_2 - k_2c_1c_2 - k_3c_1c_2^2 - k_4c_1c_2^2 \quad (4.55)$$

TABLE 4.7 Coefficients of the correlation model.

Correlation coefficients	C_8	TMPs	DMHs	C_9^+	RON
m_1	−128.73	185.24	29.57	−144.42	14.69
m_2	−0.24	0.15	−0.07	−0.28	0.03
m_3	5.13	−3.30	1.56	5.89	−0.70
m_4	−0.01	0.00	0.00	−0.01	0.00
m_5	−0.51	0.69	−0.27	−0.66	0.13
m_6	−54.17	34.25	−14.98	−60.29	5.00
m_7	136.31	−88.34	38.06	150.88	−12.31
m_8	−0.44	0.43	−0.15	−0.42	0.00
m_9	5.86	−5.34	2.03	5.98	−0.21
m_{10}	−76.48	69.17	−37.18	−79.40	1.99
m_{11}	90.76	−86.31	41.65	93.56	−2.55
m_{12}	0.06	−0.02	0.03	0.08	−0.01
m_{13}	0.47	−1.79	−0.02	−0.18	0.53

From reference Y. Tian, Z. Li, S. Mei, M. Sheng, J.-f. Chen, G. Chu, L. Zhang, A.C. Fisher, H. Zou, Alkylation of isobutane and 2-butene by concentrated sulfuric acid in a rotating packed bed reactor, Ind. Eng. Chem. Res. 57 (2018) 13362–13371.

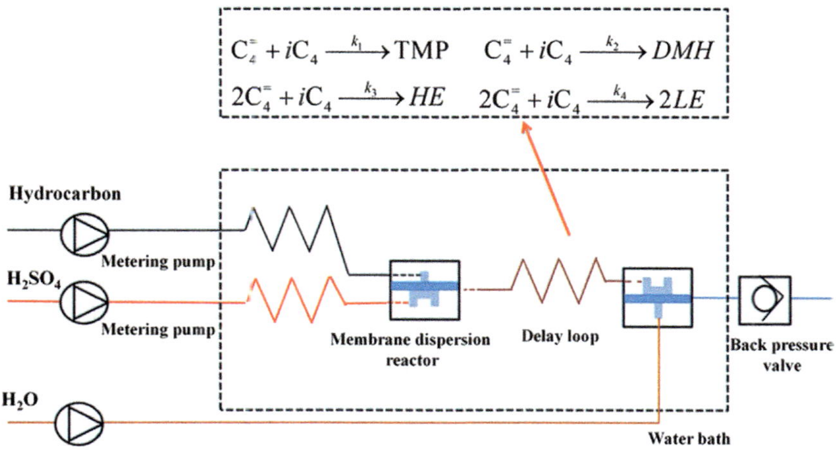

FIG. 4.8 The graphical abstract of Li's work. *From reference L. Li, J. Zhang, C. Du, K. Wang, G. Luo, Kinetics study of sulfuric acid alkylation of isobutane and butene using a microstructured chemical system, Ind. Eng. Chem. Res. 58 (2018) 1150–1158.*

$$\frac{dc_2}{dt} = -k_1c_1c_2 - k_2c_1c_2 - 2k_3c_1{c_2}^2 - 2k_4c_1{c_2}^2 \tag{4.56}$$

$$\frac{dc_3}{dt} = k_1c_1c_2 \tag{4.57}$$

$$\frac{dc_4}{dt} = k_2c_1c_2 \tag{4.58}$$

$$\frac{dc_5}{dt} = k_3c_1{c_2}^2 \tag{4.59}$$

$$\frac{dc_6}{dt} = 2k_4c_1{c_2}^2 \tag{4.60}$$

The species are represented with the following numbers: 1, isobutane; 2, butene; 3, TMP; 4, DMH; 5, HE; 6, LE.

According to the simplified formula, the experimental data are fitted in Matlab with the nonlinear function. The value of k_1 rises from 0.0033 mol mol^{-1} s^{-1} at 0°C to 0.0145 mol mol^{-1} s^{-1} at 15°C. When comparing the values of k_1–k_4 at the same reaction temperature, k_3 stands out as the largest. Compared with the kinetic parameters published by Sun et al., it can be seen that the kinetics parameter for TMP formation in Li's experiments is approximately 0.0051 mol mol^{-1} s^{-1} at 3°C, around 100 times higher than the published value of approximately 0.0001 mol mol^{-1} s^{-1}.

To further verify the reliability of the results, a simulation combining the kinetics parameters and mass transfer was performed using COMSOL software [19]. The mass transfer equations used were as follows:

$$\frac{\partial c_i}{\partial t} + \nabla(-D_i \nabla c_i) + u \cdot \nabla c_i = R_i \tag{4.61}$$

$$N_i = -D_i \nabla c_i + uc_i \tag{4.62}$$

where i presents the components of the system, including isobutane, butene, TMP, DMH, HE, and LE; c is the concentration of components, D is the diffusion coefficient, u is the mass mean velocity, R is the reaction rate, N is the mass transfer flux. As the droplet size increased from 0.05 to 1.6 mm and the diffusion increased to 10 and 100 times, the simulation results revealed that a small droplet was necessary to eliminate the effect of mass transfer and improve the reaction rate constant determined in the micro-structured chemical system.

The above kinetic studies are based on the fresh H_2SO_4 with 98.30 wt%; however, the concentrated H_2SO_4 applied in the typical industrial reactor generally contains 6.0–7.0 wt% organics, 1.0–3.0 wt% water, and 90.0–92.0 wt% H_2SO_4, which is called the used H_2SO_4. Ma et al. [21] used the kinetic model to accurately predict the concentration changes of three key components in the catalytic system of the used H_2SO_4 (see Fig. 4.9). It

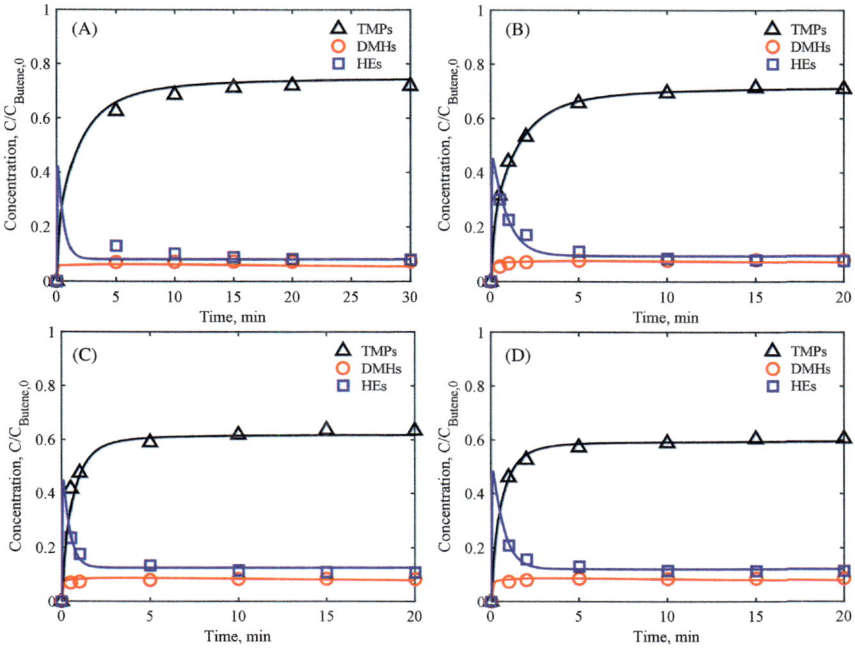

FIG. 4.9 Concentration profiles of key components in the pure used H_2SO_4-catalyzed alkylation. Temperature: (A) 277.15 K. (B) 281.15 K. (C) 285.15 K. (D) 289.15 K. Symbols: experimental data. *From reference Z. Ma, W. Zheng, W. Sun, L. Zhao, Enhanced catalytic performance of H_2SO_4-catalyzed C4 alkylation by formyl functional $[N_{1,1,1,1}][C_{10}SO_4]$ additive, AICHE J. 69 (2023) e18179.*

is assumed that the impurities (H_2O, esters, and ASO) existing in the used H_2SO_4 have no impact on the isomerization of three types of butenes. Hence, the values of rate constants k_{11}–k_{14} were directly acquired from the Arrhenius relationship in the previous work [13]. In addition, the rate constants k_4, k_6, and k_8–k_{10} are independent of the reaction temperature according to the previous work [13]. The remaining rate constants of k_1–k_{10} and k_{15} for C4 alkylation catalyzed by the used H_2SO_4 were estimated, as listed in Table 4.8.

Comparing the estimated rate constants of the used H_2SO_4 system with those in the fresh concentrated H_2SO_4 system [13], the rate constant k_4 for hydride transfer from isobutane to TMP cation is higher than that of the fresh concentrated H_2SO_4 system ($k_4 = 1.03\,\text{kg/(mol min)}$). Considering the experimental results, the high selectivity of TMPs components in the used H_2SO_4 alkylation largely relies on the increased key rate constant k_4 about the generation of TMPs in the kinetic model. Specifically, the fitted activation energies for the rate constants k_3, k_5, and k_7 in the used H_2SO_4 alkylation are 49.79, 67.95, and 76.49 kJ/mol, respectively (Table 4.9).

TABLE 4.8 The estimated rate constants of the used H_2SO_4-catalyzed alkylation with 95% confidence intervals.

T (K)	k_1 (min^{-1})	k_2 (min^{-1})	k_3, 10^3 ($kg\ mol^{-1}\ min^{-1}$)	k_4 ($kg\ mol^{-1}\ min^{-1}$)	k_5, 10^4 ($kg\ mol^{-1}\ min^{-1}$)	k_6 ($kg\ mol^{-1}\ min^{-1}$)
277.15	12.70 ± 4.57	0.70 ± 0.07	0.50 ± 0.02	1.67 ± 0.10	0.66 ± 0.03	4.71 ± 0.77
281.15	11.75 ± 1.40	0.84 ± 0.21	0.71 ± 0.09		1.19 ± 0.44	
285.15	10.27 ± 0.33	2.10 ± 0.63	1.01 ± 0.47		1.98 ± 0.27	
289.15	9.32 ± 0.87	2.12 ± 0.24	1.20 ± 0.18		2.15 ± 0.49	

T (K)	k_7, 10^2 ($kg\ mol^{-2}\ min^{-2}$)	k_8 (min^{-1})	k_9 ($kg\ mol^{-1}\ min^{-1}$)	k_{10} ($10^{-3}\ min^{-1}$)	k_{15} ($kg\ mol^{-2}\ min^{-2}$)
277.15	1.33 ± 0.26	1.62 ± 0.34	4.18 ± 0.11	7.20 ± 0.04	1.46 ± 0.86
281.15	1.37 ± 0.10				
285.15	1.44 ± 0.19				
289.15	1.52 ± 0.08				

Data from reference Z. Ma, W. Zheng, W. Sun, L. Zhao, Enhanced catalytic performance of H_2SO_4-catalyzed C4 alkylation by formyl functional $[N_{1,1,1,1}][C_{10}SO_4]$ additive, AICHE J. 69 (2023) e18179.

TABLE 4.9 Calculated activation energies (E_a) and preexponential factors (k_0) for the used H_2SO_4 system.

		k_1	k_2	k_3	k_5	k_7
The pure used H_2SO_4 system	E_a (kJ mol^{-1})	−17.64	74.44	49.79	67.95	76.49
	k_0	6.05×10^{-3} (min^{-1})	6.93×10^{13} (min^{-1})	1.26×10^{12} (kg mol^{-1} min^{-1})	4.74×10^{16} (kg mol^{-1} min^{-1})	3.65×10^{3} (kg mol^{-2} min^{-2})

Adapted from reference Z. Ma, W. Zheng, W. Sun, L. Zhao, Enhanced catalytic performance of H_2SO_4-catalyzed C4 alkylation by formyl functional $[N_{1,1,1,1}][C_{10}SO_4]$ additive, AICHE J. 69 (2023) e18179.

4.2 Kinetics using ionic liquids as catalyst

Owing to the excellent advantages of negligible vapor pressure, strong chemical stability, adjustable Brønsted and Lewis acidity, easy separation from the products, and good capacity to dissolve a wide range of organic and inorganic compounds, ionic liquids (ILs) have attracted much attention for considerable applications in C4 alkylation [20,22–31]. The investigation of reaction kinetics is widely recognized for its fundamental significance, playing a crucial role not only in the design and optimization of existing chemical reactors but also in the development of innovative ones. However, to the best of our knowledge, there exists little literature focusing on the kinetics of ILs-catalyzed alkylation.

The kinetics of the IL-catalyzed C4 alkylation were investigated by Schilder et al., who assumed a pseudo-first-order reaction concerning the paraffin isobutane [32]. This assumption finds support in the studies conducted by Simpson et al. [33] and Langley and Pike [34], both of which demonstrated that C4 alkylation follows a first-order or pseudo-first-order kinetics. Fig. 4.10 presents the comprehensive concentration profile of 2-butene on both the organic and IL sides. The reduction in 2-butene concentration within the IL droplet is determined by solving the differential equation that describes simultaneous mass transport and reaction in a spherical catalyst (with radius r) for a first-order reaction [35].

After that, Zheng et al. investigated the kinetics of composite ionic liquid (CIL)-catalyzed alkylation [36]. Based on the carbonium ion mechanism, the kinetic model was further developed, in which three key groups were considered, i.e., TMPs, DMHs, and HEs. The C4 alkylation follows the carbonium ion mechanism, regardless of the catalyst used, such as concentrated H_2SO_4, HF, or ILs [20,37,38]. Based on the reaction mechanism and kinetics of C4 alkylation with H_2SO_4 as a catalyst [5], Zheng et al. formulated the reaction mechanism for composite ILs-catalyzed C4 alkylation.

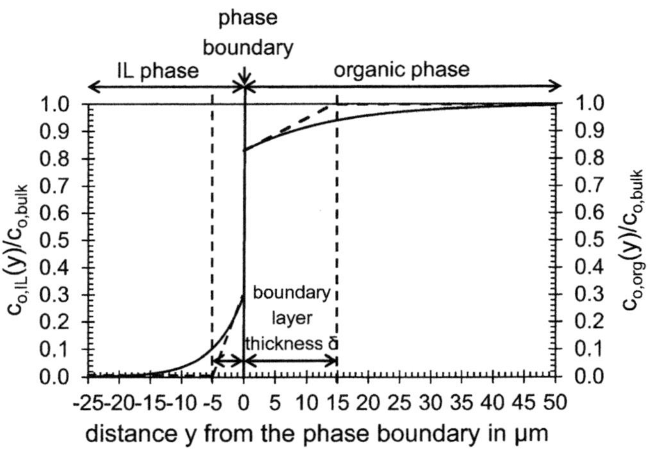

FIG. 4.10 Concentration profile of 2-butene in the ionic liquid phase and organic phase (schematically) for an Sh_{org} number of 25. *From reference L. Schilder, S. Maaß, A. Jess, Effective and intrinsic kinetics of liquid-phase isobutane/2-butene alkylation catalyzed by chloroaluminate ionic liquids, Ind. Eng. Chem. Res. 52 (2013) 1877–1885.*

In the initiation step, 2-butene undergoes rapid protonation, transforming into the *sec*-butyl carbonium ion (*sec*-$C_4H_9{}^+$),

$$2 - C_4H_8 + H^+ \rightarrow sec - C_4H_9{}^+ \tag{4.63}$$

Subsequently, a rapid hydride transfer takes place from isobutane to *sec*-$C_4H_9{}^+$, resulting in the formation of *tert*-butyl carbonium ion (*tert*-$C_4H_9{}^+$) and *n*-butane,

$$sec - C_4H_9{}^+ + i - C_4H_{10} \rightarrow tert - C_4H_9^+ + n - C_4H_{10} \tag{4.64}$$

The *tert*-$C_4H_9{}^+$ is inclined to undergo deprotonation to generate isobutene (*i*-C_4H_8) [9],

$$tert - C_4H_9{}^+ \rightarrow i - C_4H_8 + H^+ \tag{4.65}$$

Owing to a high energy barrier, the transformation of *tert*-$C_4H_9{}^+$ to 1-butene is difficult,

$$tert - C_4H_9{}^+ \rightarrow 1 - C_4H_8\ (\text{trace}) + H^+ \tag{4.66}$$

The carbonium ion-induced reactions mentioned above are intricate, and the corresponding intermediates are nearly impossible to detect. Thus, Zheng et al. assumed that *tert*-$C_4H_9{}^+$ is produced through the protonation of *i*-C_4H_8. The subsequent reaction pathways can be referenced from the previous kinetic model established by Sun et al. [5] for alkylation using H_2SO_4 as the catalyst. The schematic diagram of the alkylation reaction pathway network catalyzed by CIL is illustrated in Fig. 4.11.

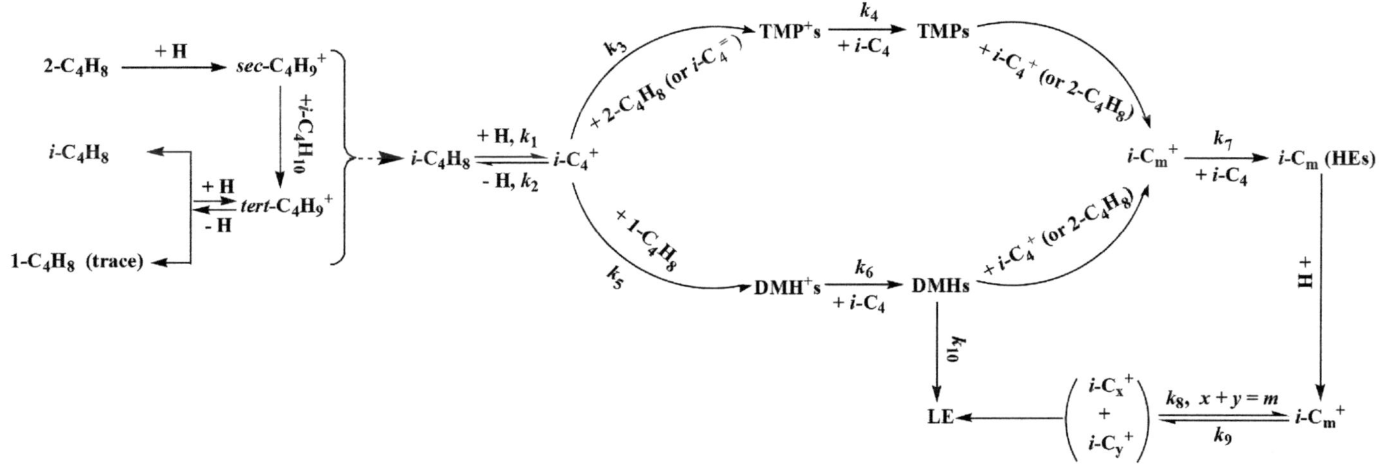

FIG. 4.11 Diagram of the CIL-catalyzed alkylation reaction pathway network. *From reference W. Zheng, D. Li, W. Sun, L. Zhao, Multi-scale modeling of isobutane alkylation with 2-butene using composite ionic liquids as catalyst, Chem. Eng. Sci. 186 (2018) 209–218.*

Due to the presence of more than 30 species in the isobutane alkylation system, including numerous isoparaffins with multiple isomers, along with the corresponding carbonium intermediates, the alkylation can be analyzed in only three or four groups. To prevent overfitting, TMPs, DMHs, HEs, and LEs, respectively, are treated as one single component, as a result of close molecular weight [5,13,21]. The rapid isomerization among the three butenes (1-butene, 2-butene, and isobutene) takes place under acidic conditions, and this can be computed using equilibrium Eq. (4.67). Consequently, the concentrations of the three butenes can be determined based on the overall concentration of butenes.

$$1-\mathrm{C_4H_8} \underset{}{\overset{K_1}{\rightleftarrows}} 2-\mathrm{C_4H_8} \underset{}{\overset{K_2}{\rightleftarrows}} i-\mathrm{C_4}^{=} \tag{4.67}$$

$$c_{1-\mathrm{C_4H_8}} = \alpha^* c_{\mathrm{C_4H_8}} \tag{4.68}$$

$$c_{i-\mathrm{C_4}^{=}} = \beta^* c_{\mathrm{C_4H_8}} \tag{4.69}$$

$$c_{2-\mathrm{C_4H_8}} = (1-\alpha-\beta)^* c_{\mathrm{C_4H_8}} \tag{4.70}$$

where α and β represent $1/(K_1+K_1*K_2+1)$ and $K_1/(K_1+K_1*K_2+1)$, respectively. According to the previous work [5], these values are 0.004718 and 0.1340, respectively, based on the thermodynamic equilibrium. Based on the aforementioned reaction steps and assumptions, Zheng et al. established the kinetic model of CIL-catalyzed C4 alkylation as follows:

$$\frac{dc_1}{dt} = -(1-\alpha-\beta)k_1c_1 - (1-\alpha)k_3c_1c_3 - \alpha k_5c_1c_3 - (1-\alpha)k_7c_1c_2c_4 + k_2c_3 \tag{4.71}$$

$$\frac{dc_2}{dt} = -k_4c_2c_4 - k_6c_2c_5 - (1-\alpha)k_7c_1c_2c_4 \tag{4.72}$$

$$\frac{dc_3}{dt} = -(1-\alpha-\beta)k_1c_1 + k_4c_2c_5 + k_6c_2c_5 - (1-\alpha)k_3c_1c_3 - \alpha k_5c_1c_3 - k_2c_3 \tag{4.73}$$

$$\frac{dc_4}{dt} = (1-\alpha)k_3c_1c_3 - k_4c_2c_5 - (1-\alpha)k_7c_1c_2c_4 \tag{4.74}$$

$$\frac{dc_5}{dt} = \alpha k_5c_{12}c_3 - k_6c_2c_5 \tag{4.75}$$

$$\frac{dc_6}{dt} = k_4c_2c_4 \tag{4.76}$$

$$\frac{dc_7}{dt} = k_6c_2c_5 - k_{10}c_7 \tag{4.77}$$

$$\frac{dc_8}{dt} = (1-\alpha)k_7c_1c_2c_4 + k_9c_9c_{10} - k_8c_8 \tag{4.78}$$

$$\frac{dc_9}{dt} = k_8c_8 - k_9c_9c_{10} \tag{4.79}$$

$$\frac{dc_{10}}{dt} = k_8c_8 - k_9c_9c_{10} \tag{4.80}$$

The initial conditions are $t=0$, $c_1=c_1^0$; $c_2=c_2^0$; $c_3=0$; $c_4=0$; $c_5=0$; $c_6=0$; $c_7=0$; $c_8=0$; $c_9=0$; $c_{10}=0$. In the above Eqs. (4.71)–(4.80), the corresponding species are 1, butene; 2, i-C_4; 3, i-C_4^+; 4, TMP^+; 5, DMH^+; 6, TMP; 7, DMH; 8, HE; 9, i-C_x^+; 10, i-$C_y^=$.

The concentration variations of three key groups in the CIL-catalyzed C4 alkylation at different temperatures are shown in Fig. 4.12. It can be seen that the obvious concentration changes in the three groups before

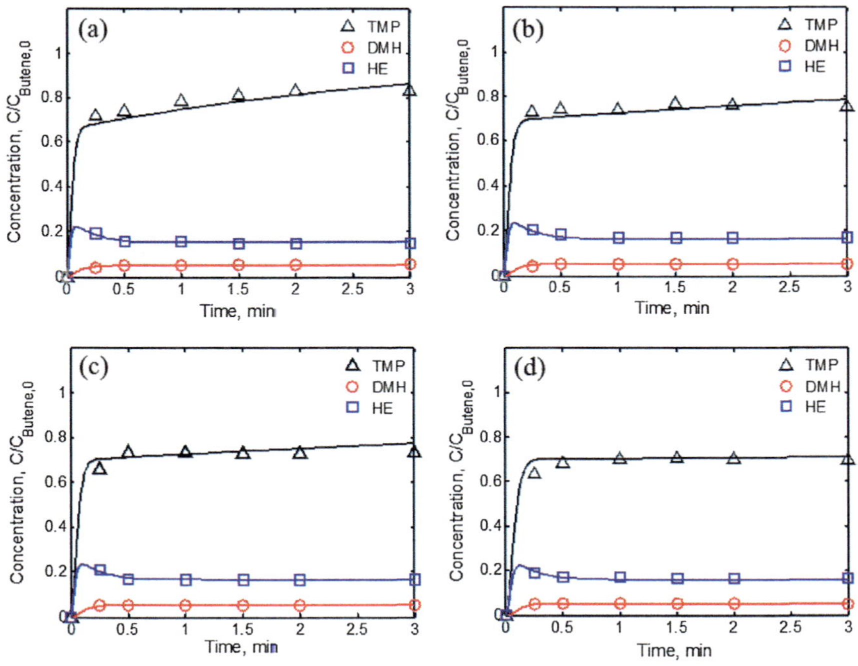

FIG. 4.12 Concentration profiles of key components in alkylate. Temperature: (A) 270.2K, (B) 279.2K, (C) 282.2K, (D) 288.2K. Symbols: experimental data; Line: calculated values by kinetic model. *From reference W. Zheng, D. Li, W. Sun, L. Zhao, Multi-scale modeling of isobutane alkylation with 2-butene using composite ionic liquids as catalyst, Chem. Eng. Sci. 186 (2018) 209–218.*

0.5 min and subsequently approaches equilibrium. In comparison to H_2SO_4, the shorter reaction time observed in the alkylation catalyzed by CIL can be attributed to the increased solubility of isobutane. The enhanced solubility of isobutane accelerates the hydride transfer from isobutane to TMP^+, a step considered to control the rate of C4 alkylation [28].

For details of the fitting and calculation process, please refer to the original text [36]. The estimated rate constants of k_1–k_{10} for CIL-catalyzed alkylation are shown in Table 4.10 with 95% confidence intervals.

Comparing the rate constants of CIL-catalyzed alkylation with those of H_2SO_4-catalyzed alkylation [5], it is evident that the rate constant k_4 for CIL alkylation is over two orders of magnitude greater than that of H_2SO_4-catalyzed alkylation. This difference can be attributed to the increased solubility and diffusion of isobutane in the CIL.

On the basis of the kinetic model, the model prediction of industrial CIL reactors was achieved [36]. The model prediction related to the conversion of olefin and yield of alkylate against mean residence time is presented in Fig. 4.13. This prediction result indicates that the mean residence time (τ) of the CIL-catalyzed alkylation satisfactorily ranges within 3 min [39], which is much lower than the residence time of H_2SO_4-catalyzed alkylation at the range of 20–30 min [12].

Liu et al. further established the kinetic model and predicted the concentration changes of four groups, such as the TMPs, DMHs, HEs, and LEs [40]. Different from the research of Zheng et al. [36], Liu et al. further investigated the reaction paths of HEs components and the self-alkylation of isobutane, and proposed the reaction route of CIL-catalyzed alkylation (see Fig. 4.14).

Based on the inclusion of the self-alkylation of isobutane and the oligomerization of 2-butene in the model, the formula is presented as follows:

$$nC_4^{=} \xrightarrow{k_1} iC_4^{=} \tag{4.81}$$

$$2iC_4^{0} + iC_4^{=} \xrightarrow{k_2} TMP + nC_4^{0} \tag{4.82}$$

$$iC_4^{=} + iC_4^{0} \xrightarrow{k_3} TMP \tag{4.83}$$

$$nC_4^{=} + iC_4^{0} \xrightarrow{k_4} DMH \tag{4.84}$$

$$3iC_4^{=} \xrightarrow{k_5} HE \tag{4.85}$$

$$HE \xrightarrow{k_6} 2LE \tag{4.86}$$

$$HE \xrightarrow{k_7} TMP + iC_4^{=} \tag{4.87}$$

TABLE 4.10 Estimated rate constants with 95% confidence intervals.

Temperature (K)	k_1 (min^{-1})	k_2 (min^{-1})	k_3,10^2 ($kg\ mol^{-1}\ min^{-1}$)	k_4, 10^2 ($kg\ mol^{-1}\ min^{-1}$)	k_5, 10^3 ($kg\ mol^{-1}\ min^{-1}$)	k_6, 10^3 ($kg\ mol^{-1}\ min^{-1}$)	k_7, 10^3 ($kg^2\ mol^{-2}\ min^{-1}$)	k_8 (min^{-1})	k_9, 10^2 ($kg\ mol^{-1}\ min^{-1}$)	k_{10}, 10^{-2} (min^{-1})
270.2	32.27 ± 2.16	0.36 ± 0.16	8.33	4.18 ± 0.47	3.84 ± 0.33	4.36	1.06 ± 0.08	0.63	1.08 ± 0.43	2.61
279.2	21.83 ± 0.37	0.16 ± 0.06								
282.2	19.65 ± 5.12	0.09 ± 0.06								
288.2	16.35 ± 1.13	0.07 ± 0.03								

Data from reference W. Zheng, D. Li, W. Sun, L. Zhao, Multi-scale modeling of isobutane alkylation with 2-butene using composite ionic liquids as catalyst, Chem. Eng. Sci. 186 (2018) 209–218.

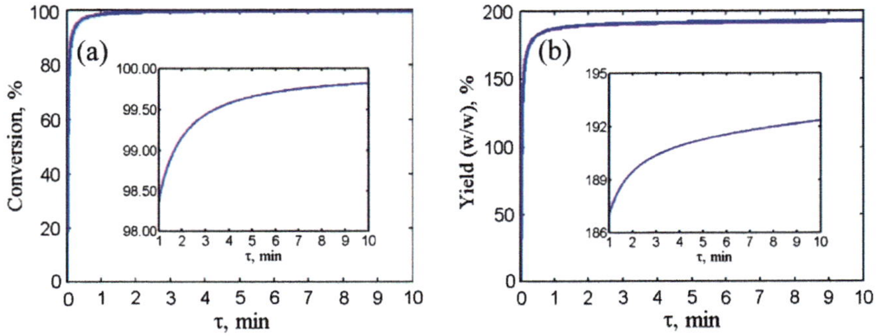

FIG. 4.13 Conversion of olefin (A) and alkylate yield (B) versus mean residence time. *From reference W. Zheng, D. Li, W. Sun, L. Zhao, Multi-scale modeling of isobutane alkylation with 2-butene using composite ionic liquids as catalyst, Chem. Eng. Sci. 186 (2018) 209–218.*

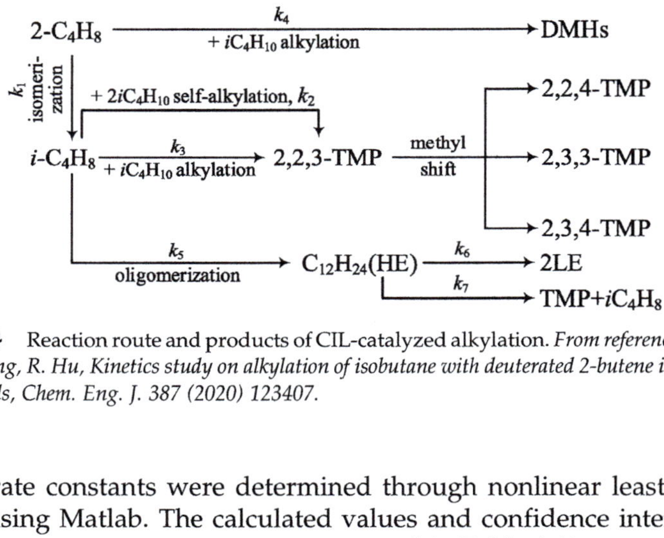

FIG. 4.14 Reaction route and products of CIL-catalyzed alkylation. *From reference Y. Liu, G. Wu, X. Pang, R. Hu, Kinetics study on alkylation of isobutane with deuterated 2-butene in composite ionic liquids, Chem. Eng. J. 387 (2020) 123407.*

The rate constants were determined through nonlinear least-squares fitting using Matlab. The calculated values and confidence intervals for k_1–k_7 at various temperatures are presented in Table 4.11.

At the same temperature, the value of k_5 is the largest, indicating that the oligomerization of 2-butene is much easier than the alkylation of isobutane and 2-butene. This result is consistent with previous findings [4,5,36]. The activation energy for DMH formation (60.4 kJ/mol) is higher than that for TMP formation (55.9 kJ/mol) (Table 4.12). Consequently, lower temperature is advisable for achieving a higher ratio of TMP/DMH. Moreover, the k_1 value for isomerization and the k_3 value for alkylation have similar orders of magnitude. This observation further underscores the intensity of isomerization from 2-butene to isobutylene in the CIL system.

TABLE 4.11 Kinetics parameters and confidence intervals at different temperatures.

Temperature (°C)	Values of k						
	k_1 (s^{-1})	k_2 ($mol^2\ mol^{-2}\ s^{-1}$)	k_3 ($mol\ mol^{-1}\ s^{-1}$)	k_4 ($mol\ mol^{-1}\ s^{-1}$)	k_5 ($mol^2\ mol^{-2}\ s^{-1}$)	k_6 (s^{-1})	k_7 (s^{-1})
5	0.13 ± 0.01	2.91×10^{-6}	0.12 ± 0.03	0.010 ± 0.002	0.83 ± 0.04	1.72×10^{-3}	9.74×10^{-4}
10	0.14 ± 0.03	4.75×10^{-6}	0.14 ± 0.01	0.012 ± 0.001	1.24 ± 0.02	1.48×10^{-3}	8.19×10^{-4}
15	0.19 ± 0.03	5.84×10^{-6}	0.17 ± 0.02	0.015 ± 0.002	1.48 ± 0.02	1.42×10^{-3}	7.74×10^{-4}
20	0.35 ± 0.05	6.55×10^{-6}	0.35 ± 0.04	0.024 ± 0.002	3.74 ± 0.03	1.18×10^{-3}	5.96×10^{-4}

Data from reference Y. Liu, G. Wu, X. Pang, R. Hu, Kinetics study on alkylation of isobutane with deuterated 2-butene in composite ionic liquids, Chem. Eng. J. 387 (2020) 123407.

TABLE 4.12 Activation energies and preexponential factors of the reactions.

Reactions	Activation energies (kJ/mol)	Pre-exponential factors
$nC_4^= \xrightarrow{k_1} iC_4^=$	65.2	4.99×10^5
$iC_4^= + iC_4^0 \xrightarrow{k_3} TMP$	55.9	8.80×10^3
$nC_4^= + iC_4^0 \xrightarrow{k_4} DMH$	60.4	5.16×10^3
$3iC_4^= \xrightarrow{k_5} HE$	53.9	3.01×10^4

Data from reference Y. Liu, G. Wu, X. Pang, R. Hu, Kinetics study on alkylation of isobutane with deuterated 2-butene in composite ionic liquids, Chem. Eng. J. 387 (2020) 123407.

4.3 Kinetics using sulfuric acid and additives as catalyst

To enhance the catalytic performance of H_2SO_4-catalyzed C4 alkylation on the industrial scale, introducing the additives is an effective approach. Surfactants, deep eutectic solvents (DESs), and ILs are among the additives that have been successfully developed to improve the catalytic efficiency of H_2SO_4-catalyzed C4 alkylation. The impact of ILs and DESs additives on H_2SO_4-catalyzed C4 alkylation has been extensively investigated by researchers such as Tang et al. [41], followed by studies conducted by Zhang et al. [42–44], Liu et al. [45], and Sun et al. [46–49]. However, there are few reliable parameters on the reaction kinetics of H_2SO_4-catalyzed C4 alkylation enhanced by additives.

Ma et al. estimated the rate constants of used H_2SO_4-catalyzed C4 alkylation enhanced by the formyl functional $[N_{1,1,1,1}][C_{10}SO_4]$ additive. The diagram of the whole reaction pathway network is shown in Fig. 4.15. The kinetic study of used H_2SO_4-catalyzed C4 alkylation has been introduced in Section 4.2. The concentration profiles of three key components in the catalytic system of the used H_2SO_4 with the formyl functional $[N_{1,1,1,1}][C_{10}SO_4]$ additive over time within the range of 277.15–289.15 K are plotted in Fig. 4.16. The fitting results about TMPs, DMHs, and HEs are correlated well with the experimental data.

The estimated parameters of k_1–k_{10} and k_{15} for C4 alkylation catalyzed by the used H_2SO_4 with the formyl functional $[N_{1,1,1,1}][C_{10}SO_4]$ additive are presented in Table 4.13 with 95% confidence intervals.

In the case of the $[N_{1,1,1,1}][C_{10}SO_4]$-assisting system with the formyl functional group, the apparent kinetic parameter k_4 for the rate-controlling step is higher than those of the used H_2SO_4 system and the fresh concentrated H_2SO_4 system. This is attributed to the exceptional capability of the

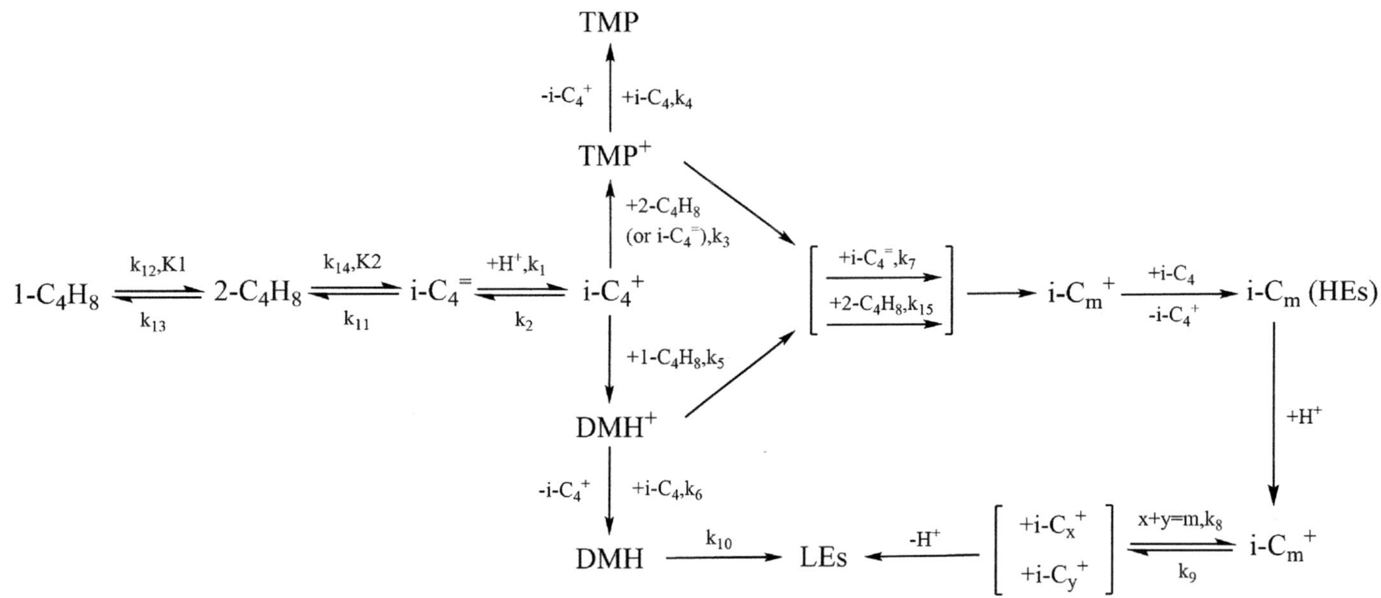

FIG. 4.15 Diagram of the alkylation reaction pathway network. *From reference Z. Ma, W. Zheng, W. Sun, L. Zhao, Enhanced catalytic performance of H_2SO_4-catalyzed C4 alkylation by formyl functional $[N_{1,1,1,1}][C_{10}SO_4]$ additive, AICHE J. 69 (2023) e18179.*

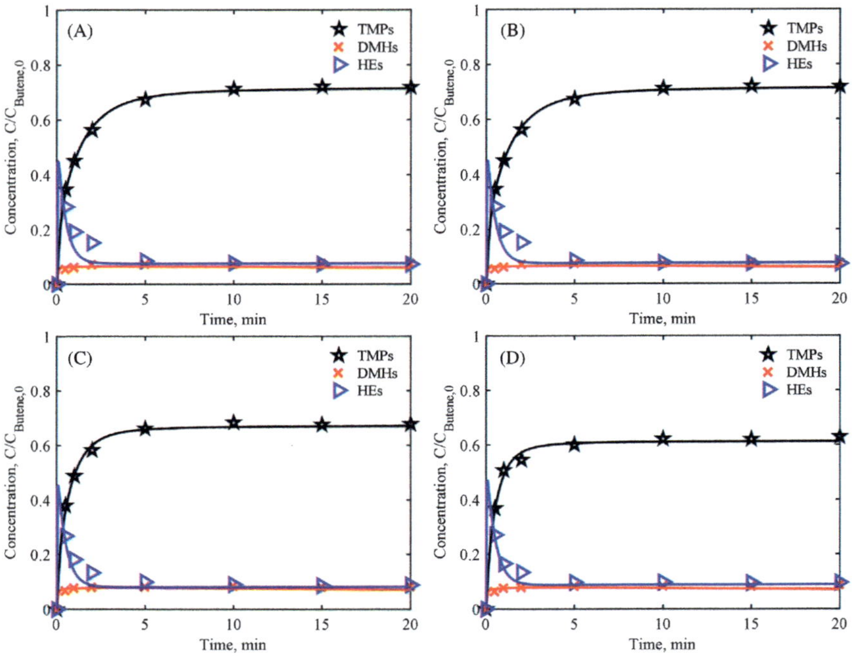

FIG. 4.16 Concentration profiles of key components in the used H_2SO_4-catalyzed alkylation with the formyl functional $[N_{1,1,1,1}][C_{10}SO_4]$ additive. Temperature: (A) 277.15K, (B) 281.15K, (C) 285.15K, (D) 289.15K. Symbols: experimental data. *Adapted from reference Z. Ma, W. Zheng, W. Sun, L. Zhao, Enhanced catalytic performance of H_2SO_4-catalyzed C4 alkylation by formyl functional $[N_{1,1,1,1}][C_{10}SO_4]$ additive, AICHE J. 69 (2023) e18179.*

TABLE 4.13 The estimated rate constants between the used H_2SO_4-catalyzed alkylation with formyl functional $[N_{1,1,1,1}][C_{10}SO_4]$ additive with 95% confidence intervals.

***T* (K)**	**k_1 (min^{-1})**	**k_2 (min^{-1})**	**k_3, 10^3 (kg mol^{-1} min^{-1})**	**k_4 (kg mol^{-1} min^{-1})**	**k_5, 10^4 (kg mol^{-1} min^{-1})**	**k_6 (kg mol^{-1} min^{-1})**
277.15	14.35 ± 0.19	0.89 ± 0.39	0.80 ± 0.04	1.69 ± 0.06	0.99 ± 0.03	4.98 ± 0.64
281.15	13.96 ± 0.33	1.04 ± 0.29	1.20 ± 0.59		1.21 ± 0.44	
285.15	12.27 ± 0.21	1.75 ± 0.44	1.42 ± 0.39		2.00 ± 0.68	
289.15	11.81 ± 0.47	3.26 ± 0.58	1.61 ± 0.20		2.20 ± 0.72	

***T* (K)**	**k_7 (10^2 kg mol^{-2} min^{-2})**	**k_8 (min^{-1})**	**k_9 (kg mol^{-1} min^{-1})**	**k_{10} (10^{-3} min^{-1})**	**k_{15} (kg mol^{-2} min^{-2})**
277.15	1.58 ± 0.38	1.72 ± 0.19	3.74 ± 0.60	6.33 ± 0.18	1.01 ± 0.13
281.15	1.63 ± 0.12				
285.15	1.68 ± 0.08				
289.15	1.76 ± 0.12				

Data from reference Z. Ma, W. Zheng, W. Sun, L. Zhao, Enhanced catalytic performance of H_2SO_4-catalyzed C4 alkylation by formyl functional $[N_{1,1,1,1}][C_{10}SO_4]$ additive, AICHE J. 69 (2023) e18179.

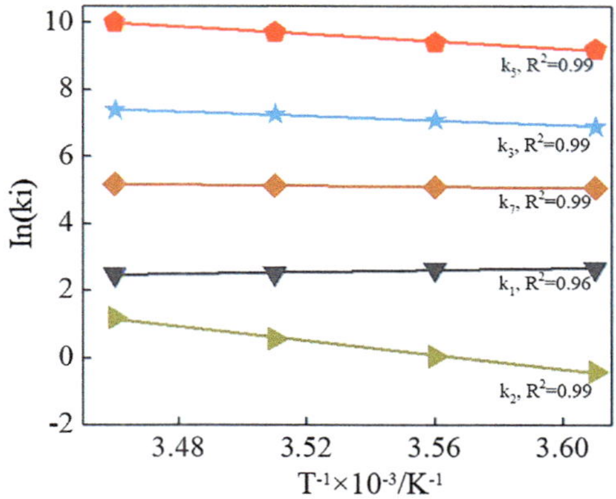

FIG. 4.17 Arrhenius relationship between $\ln(k_i)$ and $1/T$: the used H_2SO_4 system with formyl functional $[N_{1,1,1,1}][C_{10}SO_4]$. *From reference Z. Ma, W. Zheng, W. Sun, L. Zhao, Enhanced catalytic performance of H_2SO_4-catalyzed C4 alkylation by formyl functional $[N_{1,1,1,1}][C_{10}SO_4]$ additive, AICHE J. 69 (2023) e18179.*

TABLE 4.14 Calculated activation energies (E_a) and preexponential factors (k_0) for the used H_2SO_4 system with formyl functional $[N_{1,1,1,1}][C_{10}SO_4]$.

		k_1	k_2	k_3	k_5	k_7
$[N_{1,1,1,1}][C_{10}SO_4]$-assisting system	E_a (kJ mol^{-1})	−11.89	73.24	37.73	48.30	59.46
	k_0	8.33×10^{-2} (min^{-1})	5.05×10^{13} (min^{-1})	1.13×10^{10} (kg mol^{-1} min^{-1})	1.26×10^{13} (kg mol^{-1} min^{-1})	2.08×10^{3} (kg mol^{-2} min^{-2})

Data from reference Z. Ma, W. Zheng, W. Sun, L. Zhao, Enhanced catalytic performance of H_2SO_4-catalyzed C4 alkylation by formyl functional $[N_{1,1,1,1}][C_{10}SO_4]$ additive, AICHE J. 69 (2023) e18179.

formyl functional $[N_{1,1,1,1}][C_{10}SO_4]$ additive to enhance the solubility of isobutane in the acid phase. Additionally, there is a concurrent increase in the rate constants k_1–k_3 and k_5–k_7. Arrhenius relationships of k_1–k_3, k_5, and k_7 in the catalytic system of the $[N_{1,1,1,1}][C_{10}SO_4]$-assisting system are obtained by plotting the variation of $\ln(k_i)$ against $1/T$ in Fig. 4.17. The fitting results show good linear relationships ($R^2 > 0.95$). In the used H_2SO_4 system with the addition of the formyl functional $[N_{1,1,1,1}][C_{10}SO_4]$ additive, the activation energies for k_3, k_5, and k_7 are 37.73, 48.30, and 59.46 kJ/mol, respectively, as presented in Table 4.14. These values are lower than the corresponding activation energies observed in the pure used H_2SO_4 alkylation. This indicates that the incorporation of the formyl functional $[N_{1,1,1,1}][C_{10}SO_4]$ additive significantly decreases the activation energies for the crucial reaction pathways, resulting in an enhanced reaction rate and reduced reaction time.

References

[1] H. Bloch, H. Pines, L. Schmerling, The mechanism of paraffin isomerization, J. Am. Chem. Soc. 68 (1946) 153.

[2] G. Kramer, Hydride transfer reactions in concentrated sulfuric acid, J. Org. Chem. 30 (1965) 2671–2673.

[3] L.F. Albright, M.A. Spalding, J. Faunce, R.E. Eckert, Alkylation of isobutane with C4 olefins. 3. Two-step process using sulfuric acid as catalyst, Ind. Eng. Chem. Res. 27 (1988) 391–397.

[4] L.-m. Lee, P. Harriott, The kinetics of isobutane alkylation in sulfuric acid, Ind. Eng. Chem. Process. Des. Dev. 16 (1977) 282–287.

[5] W. Sun, Y. Shi, J. Chen, Z. Xi, L. Zhao, Alkylation kinetics of isobutane by C4 olefins using sulfuric acid as catalyst, Ind. Eng. Chem. Res. 52 (2013) 15262–15269.

[6] L.F. Albright, Mechanism for Alkylation of Isobutane with Light Olefins, ACS Publications, 1977.

[7] K. Kranz, Intro to Alkylation Chemistry Mechanisms, Operating Variables, and Olefin Interactions, DuPont™ STRATCO® Clean Fuel Technology, 2008.

[8] L.F. Albright, K. Li, Alkylation of isobutane with light olefins using sulfuric acid. Reaction mechanism and comparison with HF alkylation, Ind. Eng. Chem. Process. Des. Dev. 9 (1970) 447–454.

[9] A. Corma, A. Martinez, Chemistry, catalysts, and processes for isoparaffin–olefin alkylation: actual situation and future trends, Catal. Rev. 35 (1993) 483–570.

[10] I. Yucel, Atmospheric Model Applications, BoD—Books on Demand, 2012.

[11] N.V. Lebedeva, A. Nese, F.C. Sun, K. Matyjaszewski, S.S. Sheiko, Anti-Arrhenius cleavage of covalent bonds in bottlebrush macromolecules on substrate, Proc. Natl. Acad. Sci. 109 (2012) 9276–9280.

[12] L.F. Albright, Alkylation of isobutane with C3–C5 olefins to produce high-quality gasolines: physicochemical sequence of events, Ind. Eng. Chem. Res. 42 (2003) 4283–4289.

[13] P. Cao, L. Zheng, W. Sun, L. Zhao, Multiscale modeling of isobutane alkylation with mixed C4 olefins using sulfuric acid as catalyst, Ind. Eng. Chem. Res. 58 (2019) 6340–6349.

[14] E. Ivashkina, I. Dolganova, I. Dolganov, E. Ivanchina, A. Nurmakanova, A. Bekker, Modeling the H_2SO_4-catalyzed isobutane alkylation with alkenes considering the process unsteadiness, Catal. Today 329 (2019) 206–213.

[15] Y. Tian, Y. Wan, L. Zhang, G. Chu, A.C. Fisher, H. Zou, Optimized artificial neural network for evaluation: C4 alkylation process catalyzed by concentrated sulfuric acid, ACS Omega 7 (2021) 372–380.

[16] Y. Tian, Z. Li, S. Mei, M. Sheng, J.-f. Chen, G. Chu, L. Zhang, A.C. Fisher, H. Zou, Alkylation of isobutane and 2-butene by concentrated sulfuric acid in a rotating packed bed reactor, Ind. Eng. Chem. Res. 57 (2018) 13362–13371.

[17] Z. Xin, H. Jiang, Z. Zhang, Y. Chen, J. Wang, Kinetic model of olefins/isobutane alkylation using sulfuric acid as catalyst, ACS Omega 7 (2022) 9513–9526.

[18] B. Doshi, L.F. Albright, Degradation and isomerization reactions occurring during alkylation of isobutane with light olefins, Ind. Eng. Chem. Process. Des. Dev. 15 (1976) 53–60.

[19] L. Li, J. Zhang, C. Du, K. Wang, G. Luo, Kinetics study of sulfuric acid alkylation of isobutane and butene using a microstructured chemical system, Ind. Eng. Chem. Res. 58 (2018) 1150–1158.

[20] Z. Liu, X. Meng, R. Zhang, C. Xu, H. Dong, Y. Hu, Reaction performance of isobutane alkylation catalyzed by a composite ionic liquid at a short contact time, AICHE J. 60 (2014) 2244–2253.

[21] Z. Ma, W. Zheng, W. Sun, L. Zhao, Enhanced catalytic performance of H_2SO_4-catalyzed C4 alkylation by formyl functional $[N_{1,1,1,1}][C_{10}SO_4]$ additive, AICHE J. 69 (2023) e18179.

[22] S. Liu, C. Chen, F. Yu, L. Li, Z. Liu, S. Yu, C. Xie, F. Liu, Alkylation of isobutane/isobutene using Brønsted–Lewis acidic ionic liquids as catalysts, Fuel 159 (2015) 803–809.

[23] K. Yoo, V.V. Namboodiri, R.S. Varma, P.G. Smirniotis, Ionic liquid-catalyzed alkylation of isobutane with 2-butene, J. Catal. 222 (2004) 511–519.

[24] T.L.T. Bui, W. Korth, A. Jess, Influence of acidity of modified chloroaluminate based ionic liquid catalysts on alkylation of iso-butene with butene-2, Catal. Commun. 25 (2012) 118–124.

[25] T.L.T. Bui, W. Korth, S. Aschauer, A. Jess, Alkylation of isobutane with 2-butene using ionic liquids as catalyst, Green Chem. 11 (2009) 1961–1967.

[26] S. Aschauer, L. Schilder, W. Korth, S. Fritschi, A. Jess, Liquid-phase isobutane/butene-alkylation using promoted Lewis-acidic IL-catalysts, Catal. Lett. 141 (2011) 1405–1419.

[27] Y.N. Cong, Y. Liu, R.S. Hu, Determination of the presence of composite ions in catalysts and composite ionic liquids-catalyzed isobutane alkylation, Adv. Mater. Res. 287 (2011) 1666–1670.

[28] Y. Liu, L. Wang, R. Li, R. Hu, Reaction mechanism of ionic liquid catalyzed alkylation: alkylation of 2-butene with deuterated isobutene, J. Mol. Catal. A – Chem. 421 (2016) 29–36.

[29] H. Liu, S. Dai, D.-e. Jiang, Molecular dynamics simulation of anion effect on solubility, diffusivity, and permeability of carbon dioxide in ionic liquids, Ind. Eng. Chem. Res. 53 (2014) 10485–10490.

[30] W. Zheng, W. Xie, W. Sun, L. Zhao, Modeling of the interfacial behaviors for the isobutane alkylation with C4 olefin using ionic liquid as catalyst, Chem. Eng. Sci. 166 (2017) 42–52.

[31] T. Welton, Room-temperature ionic liquids. Solvents for synthesis and catalysis, Chem. Rev. 99 (1999) 2071–2084.

[32] L. Schilder, S. Maaß, A. Jess, Effective and intrinsic kinetics of liquid-phase isobutane/2-butene alkylation catalyzed by chloroaluminate ionic liquids, Ind. Eng. Chem. Res. 52 (2013) 1877–1885.

[33] M.F. Simpson, J. Wei, S. Sundaresan, Kinetic analysis of isobutane/butene alkylation over ultrastable HY zeolite, Ind. Eng. Chem. Res. 35 (1996) 3861–3873.

[34] J.R. Langley, R.W. Pike, The kinetics of alkylation of isobutane with propylene, AICHE J. 18 (1972) 698–705.

[35] E.W. Thiele, Relation between catalytic activity and size of particle, J. Ind. Eng. Chem. 31 (1939) 916–920.

[36] W. Zheng, D. Li, W. Sun, L. Zhao, Multi-scale modeling of isobutane alkylation with 2-butene using composite ionic liquids as catalyst, Chem. Eng. Sci. 186 (2018) 209–218.

[37] P. Wang, D. Wang, C. Xu, J. Gao, DFT calculations of the alkylation reaction mechanisms of isobutane and 2-butene catalyzed by Brönsted acids, Appl. Catal. A Gen. 332 (2007) 22–26.

[38] P. Wang, D. Wang, C. Xu, J. Liu, J. Gao, Ab initio calculations on the mechanism of isobutane and 2-butene alkylation reaction catalyzed by hydrofluoric acid, Catal. Today 125 (2007) 263–269.

[39] Z. Liu, C. Xu, C. Huang, Method for Manufacturing Alkylate Oil With Composite Ionic Liquid Used as Catalyst, US Patent 7,285,698, 2007.

[40] Y. Liu, G. Wu, X. Pang, R. Hu, Kinetics study on alkylation of isobutane with deuterated 2-butene in composite ionic liquids, Chem. Eng. J. 387 (2020) 123407.

[41] S. Tang, A.M. Scurto, B. Subramaniam, Improved 1-butene/isobutane alkylation with acidic ionic liquids and tunable acid/ionic liquid mixtures, J. Catal. 268 (2009) 243–250.

[42] Q. Huang, G. Zhao, S. Zhang, F. Yang, Improved catalytic lifetime of H_2SO_4 for isobutane alkylation with trace amount of ionic liquids buffer, Ind. Eng. Chem. Res. 54 (2015) 1464–1469.

[43] A. Wang, G. Zhao, F. Liu, L. Ullah, S. Zhang, A. Zheng, Anionic clusters enhanced catalytic performance of protic acid ionic liquids for isobutane alkylation, Ind. Eng. Chem. Res. 55 (2016) 8271–8280.

[44] H. Zhang, R. Liu, Z. Yang, F. Huo, R. Zhang, Z. Li, S. Zhang, Y. Wang, Alkylation of isobutane/butene promoted by fluoride-containing ionic liquids, Fuel 211 (2018) 233–240.

[45] Y. Liu, R. Li, H. Sun, R. Hu, Effects of catalyst composition on the ionic liquid catalyzed isobutane/2-butene alkylation, J. Mol. Catal. A – Chem. 398 (2015) 133–139.

[46] W. Zheng, P. Cao, Y. Yuan, C. Huang, Z. Wang, W. Sun, L. Zhao, Experimental and modeling study of isobutane alkylation with C4 olefin catalyzed by Brønsted acidic ionic liquid/sulfuric acid, Chem. Eng. J. 377 (2019) 119578.

[47] P. Cao, Y. Yuan, C. Huang, W. Sun, L. Zhao, Promoting the sulfuric acid catalyzed isobutane alkylation by quaternary ammonium ionic liquids, AICHE J. 66 (2020) e16979.

[48] W. Zheng, Z. Ma, W. Sun, L. Zhao, Target high-efficiency ionic liquids to promote H_2SO_4-catalyzed C4 alkylation by machine learning, AICHE J. 68 (2022) e17698.

[49] Z. Ma, J. Sha, W. Zheng, W. Sun, L. Zhao, Effects of deep eutectic solvents on H_2SO_4-catalyzed alkylation: combining experiment and molecular dynamics simulation, AICHE J. 68 (2022) e17556.

CHAPTER

5

Reactors, catalyst deactivation, and regeneration

In recent years, the global production of alkylate has predominantly relied on traditional processes involving H_2SO_4 and HF [1]. H_2SO_4 processes mainly include the DuPont Stratco process and the Lummus CDAlky process. Meanwhile, the HF process primarily employs the ConocoPhillips ReVAP process. Both H_2SO_4 and HF processes are technologically mature. However, the H_2SO_4 process is characterized by strong corrosiveness, high acid consumption, regeneration challenges, and significant environmental pollution. On the other hand, HF is highly volatile and extremely toxic, posing severe hazards in case of leaks. Due to safety and environmental considerations, newly constructed alkylation units in recent years have predominantly favored the H_2SO_4 process [2].

Solid acid alkylation processes are represented by various technologies developed by different companies, including the AlkyClean process by Lummus, the KBR process, the Alkylene process by UOP, the FBA process by Topsoe, and the Eurofuel technology by Lurgi [3]. Each of these processes has unique advantages, but as of now, only the AlkyClean process has achieved industrial-scale implementation.

5.1 Industrial reactors in the H_2SO_4 process

The advancements in H_2SO_4-catalyzed C4 alkylation technology have primarily concentrated on improving the mass transfer within reactors [4–7]. Simultaneously, researchers have been dedicated to enhancing and optimizing processes related to product separation and the transfer of reaction heat. Depending on the mixing methods employed, reactors for the H_2SO_4 process can be divided into various types, such as stirred mixing, static mixing, injection mixing, and shear mixing, among others. These developments are geared toward enhancing the efficiency of the

Aliphatic Alkylation
https://doi.org/10.1016/B978-0-443-21984-9.00006-4

alkylation process. These developments aim to enhance the efficiency and effectiveness of the alkylation process.

5.1.1 Stirred mixing reactor

The Stratco Contactor reactor is the most mature stirred mixing reactor in terms of technology, as shown in Fig. 5.1. The Stratco reactor is a horizontal reactor equipped with an inner circulation tube, a heat exchanger tube bundle to remove the heat of reaction and stirring impellers. The hydrocarbon feed and recycled acid are introduced at the suction side of the impeller within the circulation tube, ensuring a finely dispersed acid-continuous emulsion. This minimizes temperature variations and suppresses side reactions.

The reactor utilizes a product coolant system, where the effluent from the reaction is depressurized and cooled after passing through an acid settler, which generates low-temperature and low-pressure conditions. This cooled effluent serves as a coolant, circulating through the reactor tube bundle to remove the heat generated during the alkylation reaction. Compared to closed-loop coolant circulation or self-cooling processes, the cooling effect of the hydrocarbon effluent can maintain a higher concentration of isobutane inside the reactor, while the lowest recycled isobutane quantity comes from the deisobutanizer tower.

Cai et al. conducted the computational fluid dynamics (CFD) simulations analysis on the fluid flow inside the Stratco reactor, employing the multiple reference frames (MRF) method combined with the standard k-ε turbulence model [8]. Building upon the examination of the basic flow

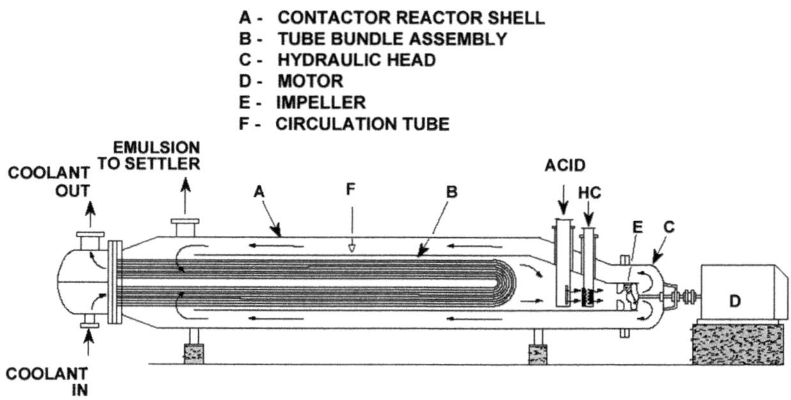

FIG. 5.1 Stratco Contactor reactor used in H_2SO_4-catalyzed alkylation. *From A. Feller, Reaction Mechanism and Deactivation Pathways in Zeolite Catalyzed Isobutane/2-Butene Alkylation, Technische Universität München, 2003.*

patterns in the entire field, the velocity distribution and turbulence parameters in the impeller mixing and tube heat exchanger areas were analyzed. Structural optimizations were performed on the reactor. According to the simulation results, the main issues identified in the Stratco reactor include: (1) potential sulfuric acid leakage and corrosion at the agitator shaft; (2) high viscosity of sulfuric acid at low temperatures, impeding mass transfer and leading to a sharp increase in the power required for agitation; (3) difficulty in separating the acid-hydrocarbon emulsion, requiring extended settling time and large settling vessels for efficient mass transfer; (4) high acid consumption, resulting in high acid regeneration costs.

5.1.2 Static mixing reactor

Static mixing reactors mainly come in two forms: one achieves static mixing through internal components, and the other accomplishes static mixing through the use of packing materials.

The H_2SO_4 alkylation reactor designed by Cunningham et al. with a static mixer is a vertical cylindrical structure with a conical bottom [9]. Inside the reactor, there are perforated baffles and multiple emulsion injection pipelines, as shown in Fig. 5.2. The emulsion inlet pipelines

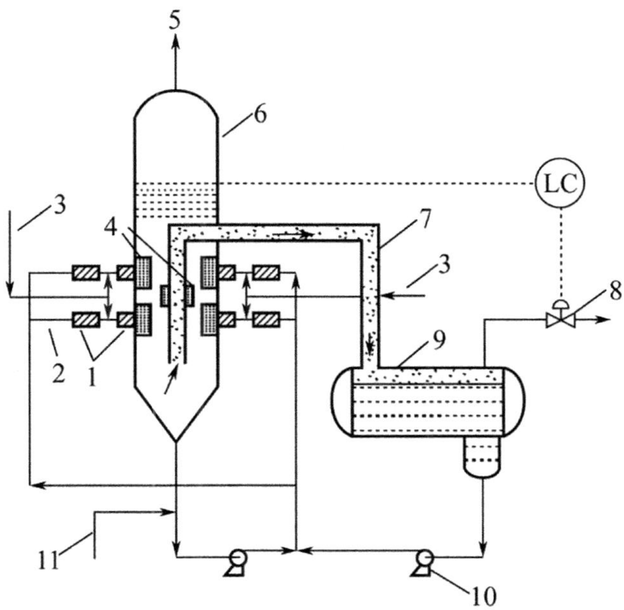

FIG. 5.2 H_2SO_4 alkylation reactor with static mixer. *From B.A. Cunningham, Sulfuric Acid Alkylation Reactor System With Static Mixers, US5785933A.*

are equipped with static mixers, and the liquid enters the reactor tangentially, making a circular motion, which facilitates the separation of acid-hydrocarbon components in the reaction product. The reactor and settling tank are connected by a transfer pipe with a vertical portion suspended within the central part of the reactor vessel and a sealable horizontal portion extending through the cylindrical wall of the reactor vessel. The transfer pipe has one end near the conical bottom portion of the reactor vessel and another end at the top portion of the settling vessel. Multiple emulsion injection inlet pipes are tangentially positioned and communicate with the interior of the reactor vessel for injecting emulsified reactants into the reactor vessel. This type of reactor is not prone to leakage, minimizing the environmental impact of reactants. The reaction temperature in the reactor ranges from 2°C to 10°C, and the reaction heat is carried away by the evaporation of alkanes from the top. The evaporated isobutane is compressed, cooled, and then recirculated back to the reactor. Although this reactor replaces traditional agitators, whether its mixing efficiency is superior to agitation mixing requires further investigation. Additionally, the external multistage static mixers and internal perforated baffles not only complicate the structure but also significantly increase energy consumption, which is not conducive to industrial applications.

CDTech's CDAlky process, designed by Smith et al., centers around a large vertical downflow reactor. As depicted in Fig. 5.3, the reactor is equipped with patented packing, and a distributor is located at the top. Sulfuric acid and raw materials enter the reactor from the top, undergo thorough mixing and react in the packing section, eliminating the need for conventional agitation equipment. Butane undergoes vaporization at the bottom of the reactor, carrying away the heat generated during the alkylation and thereby achieving direct cooling of the reactor. After compression and separation, a portion of these vaporized gases is recirculated to the top of the reactor to control the reactor temperature, while another portion can be further separated to obtain normal and isomeric paraffins.

This process is not constrained by the viscosity of the acid, and the reaction temperature can be maintained at −3°C, improving the quality of the alkylate. Additionally, the two-stage aggregation separator employed in this process not only exhibits high separation efficiency but also yields dry reaction products. This eliminates acid washing, alkali washing, and water washing, reducing acid consumption by 50% and alleviating downstream equipment corrosion. The current issue is the development of the packing material. Some of the facilities where the CDAlky process has been implemented include Shandong Shencheng Chemical Group Co., Ltd. (200,000 tons/year), Ningbo Haiyue New Materials Co., Ltd. (600,000 tons/year), Guangxi Qinzhou Tianheng Petrochemical Co., Ltd. (200,000 tons/year), and Yuntianhua Group Co., Ltd. (240,000 tons/year).

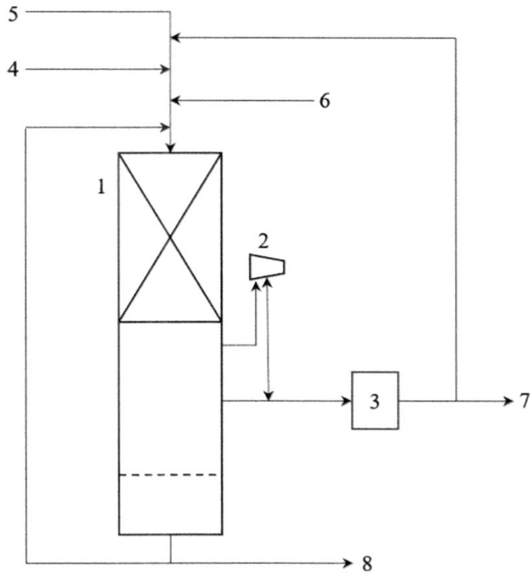

FIG. 5.3 Simplified process flow diagram of the CDAlky process. (1) Reactor; (2) compressor; (3) separator; (4) olefins; (5) raw materials; (6) sulfuric acid; (7) hydrocarbon components; (8) waste acid. *Modified from J. Lawrence, A. Smith, A.P. Gelbein, J. William M. Cross, H_2SO_4 Alkylation by Conversion of Olefin Feed to Oligomers and Sulfate Esters, US7977525.*

Liquid-phase C4 alkanes and concentrated H_2SO_4 are thoroughly emulsified outside the device using a stirring device. They enter the device through the liquid-phase feed pipe from the top of the tower via a raw material pump, with a feed temperature of 2–10°C. Gaseous C4 olefins enter the device from the top of the tower and flow downward together with the liquid phase after mixing. As they pass through the packing section, liquid-phase acid/hydrocarbons react simultaneously with gaseous alkene hydrocarbons, and acid/hydrocarbons are separated. Using a nozzle-type feed pipe, fresh liquid C4 olefins are introduced between the packing sections. After thorough mixing through the self-mixing device, the gas-liquid phase enters the next level of packing for further reaction. The temperature inside the tower is maintained at 2–18°C during the reaction. The raw material gas entering the top of the tower and the evaporated gaseous alkene hydrocarbons from the heated reaction process flow out from the bottom of the tower. The alkylate and H_2SO_4 after the reaction also flow out from the bottom of the tower. The device utilizes the self-evaporation of liquid-phase C4 alkanes during the reaction to remove the reaction heat, maintaining the reaction at an appropriate temperature. However, the phase separation of acid and hydrocarbons in the packing materials affects the emulsification of acid hydrocarbons, which limits its industrial applications.

The N-type static mixing reactor employed in SINOALKY alkylation technology [10–13] is a typical three-stage reactor, with each reactor comprised of the combination of three static mixers. The I and II stage mixers are designed for the initial rapid reaction of the alkylation process, providing high mixing intensity to rapidly disperse the olefins entering the reactor, thereby reducing the occurrence of side reactions. The III stage mixer, with lower mixing intensity compared to the I and II stages, matches the slow hydride transfer reaction of the alkylation. It plays a role in maintaining the droplet size of the acid-hydrocarbon mixture and prolonging the reaction residence time. From the perspective of phase separation, the horizontal flow direction favors the separation of the acid and hydrocarbon phases, while a vertical arrangement hinders the separation of the acid and hydrocarbon phases. Therefore, a vertical arrangement is more conducive to preventing the acid and hydrocarbon phases from layering after the dispersion of acid-hydrocarbon droplets through the premixer. Moreover, the N-type configuration efficiently decreases the vertical reactor height, leading to a reduced head for the acid-hydrocarbon pump in circulation. This results in decreased energy consumption and a smaller footprint compared to a horizontal setup, simplifying on-site pipeline layout optimization.

5.1.3 Injection mixing reactor

The injection mixing reactor includes water mist nozzles, a distributor (a bubbling device for the sprayer), Venturi tube mixers, etc., which replace mechanical stirring to achieve the rapid mixing of acid and hydrocarbon.

The injection reactor designed by Bakshi from RHT company is a vertical vessel, as shown in Fig. 5.4. This reactor primarily replaces mechanical mixing devices with injection devices and utilizes coalescers for gas-liquid separation and acid-hydrocarbon separation. Sulfuric acid enters the nozzle axially, while the mixture of olefins, recycled isobutane, and refrigerant (isobutane) enters the nozzle axially in the vertical direction. Subsequently, the raw materials and acid are introduced into the reactor, initiating the preliminary separation of acid hydrocarbons. The sulfuric acid from the bottom persists in entering the nozzle for further reaction facilitated by the acid circulation pump, thereby achieving a continuous acid circulation. The temperature inside the reactor is controlled at −3°C, and the top C4 gas undergoes the absorption/desorption process for cold regeneration, reducing the operating costs compared to compression refrigeration. The reactor is equipped with a defoamer, and the reaction products enter the coalescers through the defoamer, achieving the separation of the acid phase and hydrocarbon phase.

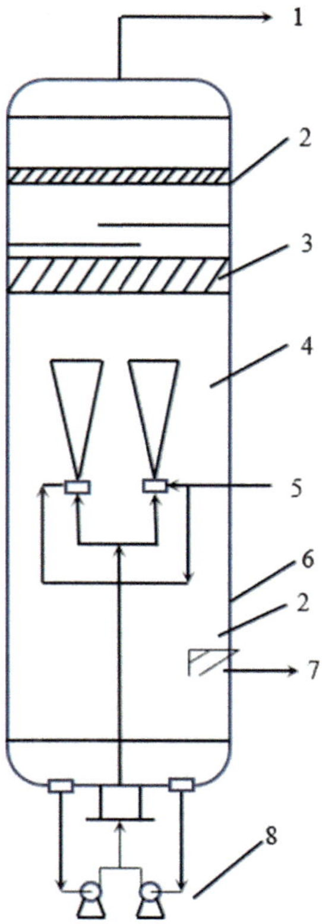

FIG. 5.4 Novel mixing device reactor of alkylation. (1) Removal of C4 adsorption cycle; (2) foam breaker; (3) filler; (4) injector; (5) hydrocarbon feed; (6) reactor shell; (7) decoalescence separation of reaction effluent; (8) acid cycle. *Modified from A.S. Bakshi, Sulfuric Acid Alkylation Process, US7652187.*

The main issues with this reactor are: the complex structure of the nozzle, making it challenging to achieve uniform injection mixing of acid and hydrocarbon; at low temperatures, the high viscosity of H_2SO_4 is not conducive to injection; compared to the Stratco and CDAlky processes, the nozzle design in this process is difficult, and the short reaction contact time makes it challenging to control the reaction.

Alexanyan et al. designed an injection-type circulating alkylation reactor, as illustrated in Fig. 5.5 [14]. The reactor is divided into a premixing zone and reaction zone. Isobutane and acid are fed in a volumetric ratio of 1:1. In the premixing zone, isobutane enters the emulsification area through a nozzle, rapidly mixing with acid in the annular gap to form an emulsion. The emulsification zone and the reaction zone are connected by a throat, with a diameter ratio of 1:2. On the one hand, the emulsion enters the reaction zone through the throat at a higher velocity; on the

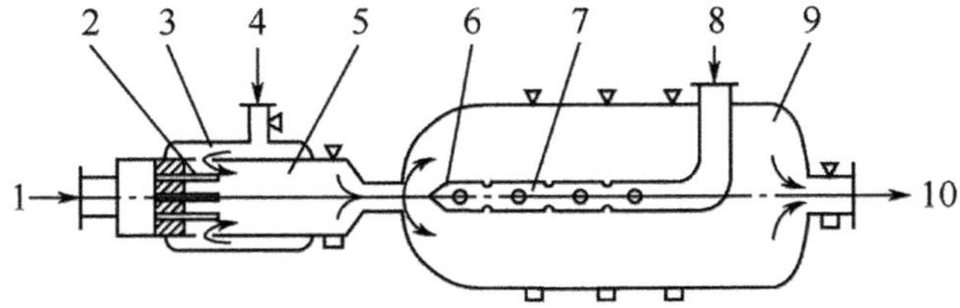

FIG. 5.5 Injection-type circulating alkylation reactor. (1) Isobutane feed; (2) nozzle; (3) ring gap; (4) sulfuric acid feed; (5) premixing zone; (6) conical closed end; (7) nozzle; (8) olefin feed; (9) reaction zone; (10) de-settling device. *From G.G. Alexanyan, N.B. Librovitch, Y.A. Prochukhane, Process for the Alkylation of Olefins, US5777189.*

other hand, olefins are injected into the emulsion in the opposite direction through nozzles in the reaction zone, initiating contact reactions. The volume of the reaction zone is significantly larger than that of the premixing zone. After the premixture of isobutane and acid enters the reaction zone, the volume abruptly increases, and adiabatic expansion absorbs the reaction heat. The reaction does not require stirring or mixing devices, reducing maintenance and operating costs. However, the reactor has not been widely adopted in industry, and reasons include the high viscosity of sulfuric acid, resulting in poor injection mixing effects during the scale-up process. Additionally, the significant difference in density between sulfuric acid and isobutane causes the emulsion formed in the emulsification zone to quickly stratify in the reaction zone due to insufficient turbulence, leading to olefin self-polymerization and affecting the quality of alkylate.

Gershuni proposed a parallel-nozzle vertical reactor, as shown in Fig. 5.6 [15]. The reactor consists of five parts: mixing zone, emulsification zone, diffusion zone, prereaction zone, and reaction zone, arranged from bottom to top. Isoparaffins are rapidly sprayed through nozzles, and sulfuric acid in the emulsification zone thoroughly mixes to form an emulsion. The mixture then enters the prereaction zone to react with olefins. The prereaction zone and the reaction zone are separated by baffles, and nozzles are placed on the baffles. These nozzles make an angle of 0–30 degrees with the reactor axis. The fluid exiting the prereaction zone continues to be sprayed upward into the reaction zone, where it thoroughly reacts with olefins from multiple parallel nozzles. The products and unreacted materials flow into the gas-liquid separator at the bottom of the reaction zone through multiple outlet pipes. The gas is directed to subsequent compression sections, while the liquid undergoes static separation in the separator. The lower liquid mainly contains partial sulfuric acid and unreacted alkanes, which are pumped back to the bottom mixing zone for further participation in the reaction. The upper liquid primarily consists of the main reaction product, which is sent to subsequent sections for purification and refinement. This reactor has nozzles in all three parts, eliminating the need for a stirring device and effectively promoting the

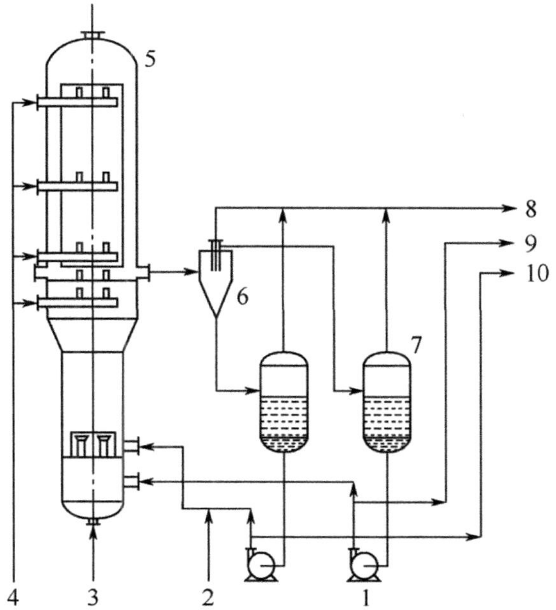

FIG. 5.6 The schematic representation parallel-nozzle vertical reactor. (1) Pump; (2) sulfuric acid; (3) isobaric hydrocarbons; (4) olefins; (5) reactor; (6) gas-liquid separator; (7) static separator; (8) gas decompression; (9) product deseparation and purification; (10) used acid. *From S. Gershuni, Method and Device for the Production of Alkylates, US7435864.*

mixing of acid and hydrocarbon, making it advantageous for the alkylation. However, compared to the Stratco reactor, this reactor has a complex structure, increasing energy consumption and making equipment maintenance and cleaning challenges.

5.1.4 Shear mixing reactor

The alkylation reaction process proposed by Hassan et al. is mainly composed of three stages: mixing zone, reaction zone, and separation zone, as illustrated in Fig. 5.7 [16]. In the mixing zone, a unique shear mixing device is employed, as illustrated in Fig. 5.8, which has a metal shell to control internal temperature and pressure. By high-speed rotation of three-stage gears, the fluid is sheared to form an emulsion with a diameter smaller than 5 μm, achieving the mixing of reactants. The mixed emulsion then enters a stirred-tank reactor for alkylation under suitable conditions. After the reaction, the effluent undergoes separation, with the bottom acid being recycled and waste acid periodically discharged. Unreacted isobutane from the top flows back to the shear device for recycling. In the shear mixing device, the acid and hydrocarbon phases form an emulsion through shear, not only increasing the reaction area and yield of alkylate but also

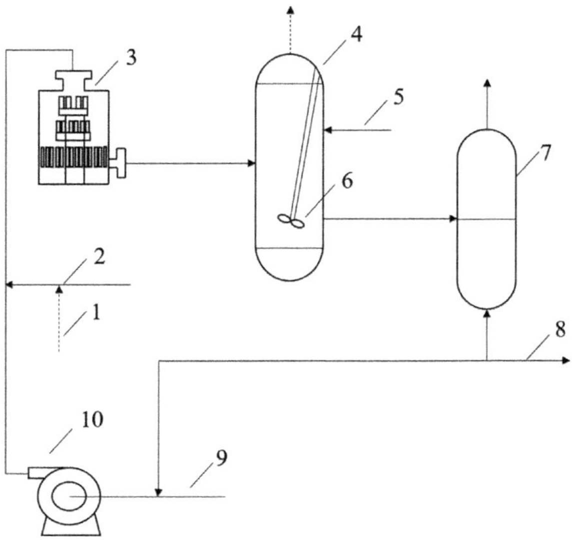

FIG. 5.7 Process flow diagram of alkylation process proposed by Hassan et al. [16]. (1) Cyclic isobutane; (2) hydrocarbon feed; (3) shear mixer; (4) stirred-tank reactor; (5) acid or water; (6) agitator paddle; (7) separator; (8) waste acid outlet; (9) new acid inlet; (10) pump. *Modified from A. Hassan, E. Bagherzadeh, R.G. Anthony, et al., System and Process for Alkylation, US8269057.*

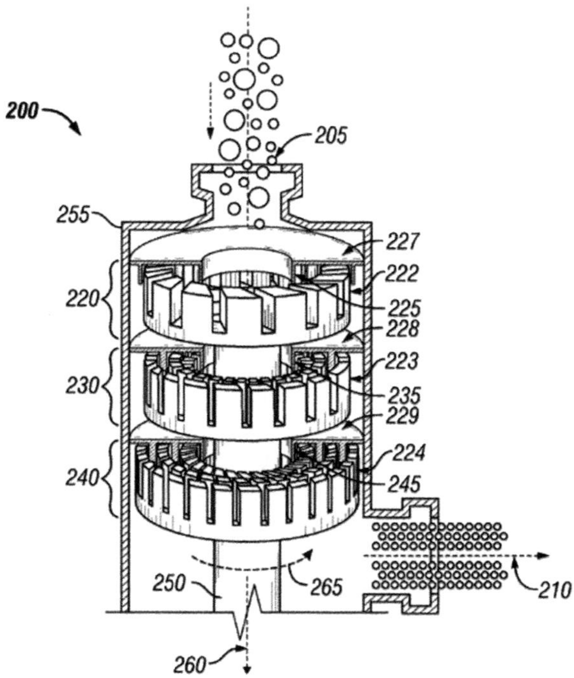

FIG. 5.8 Longitudinal cross-sectional view of a multistage high shear device. *From A. Hassan, E. Bagherzadeh, R.G. Anthony, et al., System and Process for Alkylation, US8269057.*

reducing the usage of catalyst and the regeneration load of downstream catalysts. Furthermore, it can maintain a uniform temperature distribution in the reactor, inhibiting the side reactions. Additionally, the high-speed shear device can surpass the limits of mass transfer, allowing the reaction to approach reaction control. However, this shear mixing reactor faces challenges and has not been industrially applied. The complex three-stage shear device not only raises equipment costs but also requires higher energy consumption. Subsequent stirring devices are still necessary.

5.1.5 Microsystem reactor

Luo et al. designed a microsystem reactor, as shown in Fig. 5.9 [17]. In the catalytic system of the microsystem reactor, H_2SO_4 and hydrocarbon feed (previously stored in a piston storage tank) are transported into a microreactor via two feeding pipes, utilizing two metering pumps. The microreactor serves as a membrane dispersion reactor with an average pore diameter of 5 μm [18,19]. The micro-structured system significantly enhances the quality of alkylate by utilizing a reactor of smaller droplet

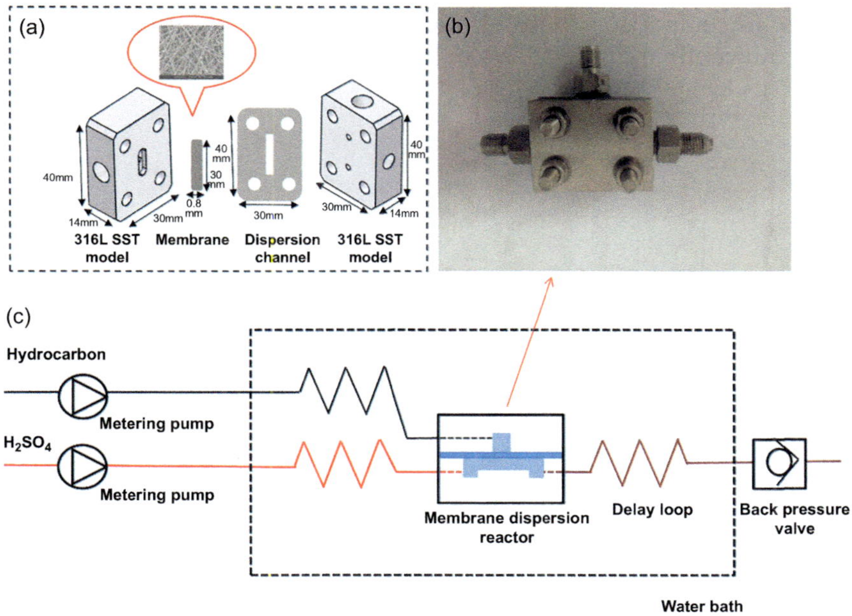

FIG. 5.9 Schematic overview of the experimental setup. (A) Detailed structures of the membrane dispersion reactor. (B) Appearance structures of the membrane dispersion reactor. (C) The schematic overview of the micro-structured chemical system. *From L. Li, J. Zhang, C. Du, et al., Process intensification of sulfuric acid alkylation using a microstructured chemical system, Ind. Eng. Chem. Res. 57 (2018) 3523–3529.*

size and notably shorter reaction time, thereby improving transport performances. Olefin conversion can reach up to 100% within 30s. The selectivity of the C_8 product reaches 71% with the RON of 94.4 with the acid concentration of 94wt% at 4°C using the I/O ratio of 8:1. Within this reaction system, the acid concentration is reduced to only 0.1% per cycle, suggesting effective reduction in acid consumption in this micro-reaction process. Moreover, Luo et al. investigated the alkylation kinetics employing the micro-structured system [19]. In this system, the kinetic parameter for HEs formation is the largest, attributed to easy polymerization of butene. The activation energy for TMP formation is the lowest, whereas that for HE is the highest, suggesting that the reaction is preferable at a lower reaction temperature. A rate constant in the micro-structured system is 100 times higher than that obtained in batch reaction systems.

Recently, a microreactor configured with staggered herringbone grooves was adopted by Wang et al. for intensifying the mass and heat transfer of H_2SO_4-catalyzed C4 alkylation. The diagram of a staggered herringbone groove microreactor is presented in Figs. 5.10 and 5.11. The hydrodynamic and mixing behaviors of the reactant and catalyst phases in the microreactor with different channel sizes were simulated using CFDs. In comparison to the stirred batch reactor, the reaction time can be significantly reduced to 29s in the 9.3mL designed microreactor with

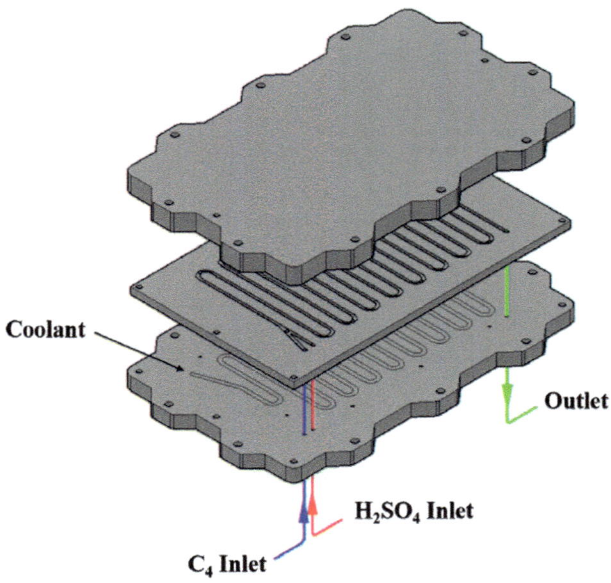

FIG. 5.10 The staggered herringbone groove microreactor. *From D. Wang, T. Zhang, Y. Yang, et al., Simulation and design microreactor configured with micromixers to intensify the isobutane/1-butene alkylation process, J. Taiwan Inst. Chem. Eng. 98 (2019) 53–62.*

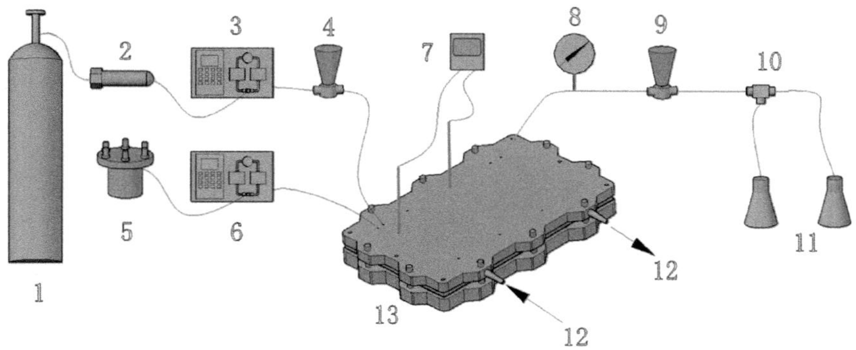

FIG. 5.11 Experimental setup of the isobutane alkylation microreactor system. (1) Feed storage tank; (2) dryer; (3) double piston pump; (4) back pressure regulator; (5) H_2SO_4 tank; (6) double piston pump; (7) temperature indicator; (8) pressure gauge; (9) back pressure regulator; (10) three-way valve; (11) sample collection; (12) cold medium inlet and outlet; (13) staggered herringbone groove microreactor. *From D. Wang, T. Zhang, Y. Yang, et al., Simulation and design microreactor configured with micromixers to intensify the isobutane/1-butene alkylation process, J. Taiwan Inst. Chem. Eng. 98 (2019) 53–62.*

a 1-butene conversion of 91.4% at constant reaction temperature. Under the optimal reaction conditions, the 1-butene conversion can reach 94.9%, and the selectivity of C_8 components is 91.08% with RON of 99.7.

5.1.6 Rotating packed bed reactor

Rotating packed bed (RPB) acts as a process intensification technique in the industry, which stands out as an outstanding heterogeneous reactor that notably improves the mixing of liquid-liquid two-phase systems [20–25]. The investigation of a novel RPB reactor was firstly developed by Tian et al. to enhance the alkylation catalyzed by H_2SO_4, as shown in Fig. 5.12) [12]. Under the optimal conditions, the RON reaches 98.85 with yields of 90.65% for C_8 and 85.47% for TMPs components. Importantly, the RPB achieves high RON at an I/O ratio of 30:1 with low internal I/O, which indicates that the high mass transfer and micro-mixing efficiencies of the RPB reactor enable the production of high-quality alkylate with a low inner I/O, reducing the massive energy required for compressors and pumps to cycle unreacted isobutane and mix with fresh feed to increase the inner I/O in the reactor. Additionally, an empirical correlation model is proposed for predicting multiproduct yields and RON, with a deviation within $\pm 10\%$.

To further strengthen the catalytic performance of the RPB reactor, three homologs of sulfates and two hydrophilic surfactants were employed as additives to modify the properties of the acid solution in Tian's work [26] (Fig. 5.13).

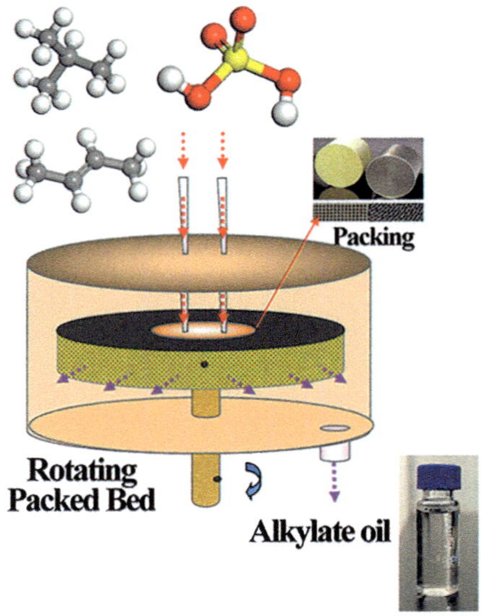

FIG. 5.12 Diagram of a novel RPB reactor. *From Y. Liu, G. Liu, G. Wu, et al., Alkylation of isobutane and 2-butene in rotating packed-bed reactors: using ionic liquid and solid acid as catalysts, Ind. Eng. Chem. Res. 59 (2020) 14767–14775.*

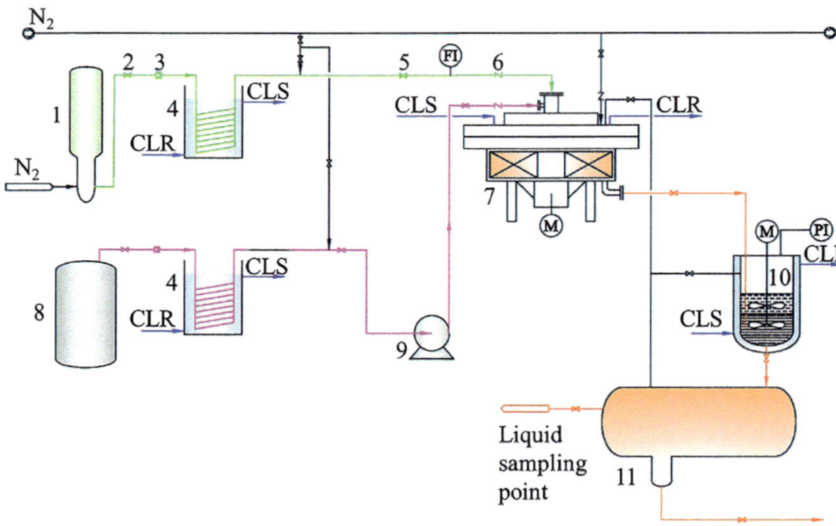

FIG. 5.13 Schematic diagram of the rotating packed bed (RPB) alkylation process. (1) Mixed hydrocarbon tank; (2) cutoff valve; (3) filter; (4) thermostatic tank; (5) plug valve; (6) one-way valve; (7) rotating packed bed; (8) storage tank of acid; (9) pump; (10) stirred tank; (11) product tank; *CLR*, cooling liquid return; *CLS*, cooling liquid supply. *From Y. Tian, S. Mei, L.L. Zhang, et al., Improved H_2SO_4-catalyzed alkylation reaction in a rotating packed bed reactor by adding additives, Can. J. Chem. Eng. 100 (2022) 3395–3407.*

The RPB reactor can be employed to enhance heterogeneous mixing between the acid and hydrocarbon phases. On the other hand, by utilizing additives to regulate the solubility of isobutane in an acid solution, the content of isobutane in the reaction zone can be adjusted. Through this method, an outstanding alkylate with RON of approximately 99 was achieved under relatively mild conditions (the reaction at 8°C and an isobutane/butene ratio of 30:1). Table 5.1 lists the alkylation performances catalyzed by different acid solutions with various additives for comparison.

5.2 Industrial reactor in the HF process [33]

The design of the HF alkylation reactor has remained stagnant with no significant changes since the early 1970s. One prevalent reactor type utilizes a combination of a shell and tube heat exchanger with a sparger system. In this particular design, the HF alkylation acid undergoes continuous circulation through the exchanger-reactor, passing baffles designed to facilitate a homogeneous flow. Simultaneously, the hydrocarbon reactants are introduced into the liquid in the acid flow direction through spray nozzles positioned along several spargers. These spargers protrude into the reactor at intervals along its length from an external supply manifold, as illustrated in Fig. 5.14.

The effectiveness of the process relies on the formation of small droplets of hydrocarbon reactants within the flowing acid stream. Less than optimal liquid dispersion and the presence of large droplets can lead to secondary reactions and a decline in the quality of alkylate.

5.3 Industrial reactor in the ILs process

In recent decades, ionic liquids (ILs) have been extensively investigated as alternative catalysts for alkylation production [13,34]. ILs are ionic, salt-like materials that are liquid at less than 100°C. ILs have found applications as solvents and homogeneous catalysts, due to their negligible vapor pressure, excellent solubility to a wide range of organic and inorganic compounds, and the capacity for recycling and reuse [27,28,35,36]. The reactors of the ILs alkylation process can be classified into two types: cyclone reactor and RPB reactor.

As early as 2006, a new composite ionic liquid (CIL) alkylation process has been proven in a pilot plant and retrofitted into an existing 65,000-tons/year H_2SO_4 alkylation unit in China. However, the Stratco reactor used for the IL alkylation process faces challenges, including low selectivity of TMPs and difficulties in removing the catalysts once the reaction is completed. On the other hand, alkylation catalyzed by the liquid acid catalyst is a heterogeneous system, which takes place near or on the contact

TABLE 5.1 The alkylation performance catalyzed by different catalysts in the stirred tank reactor (STR).

Catalyst	Content	N (r/min)	T (°C)	t (min)	I/O	A/HC	y_{cs} (%)	RON
$[N_{12,1,1,1}]Cl/H_2SO_4$ [1]	40 mmol/L	1000	10	8	15:1	0.6	87.5	97.8
$[OPSIm][HSO_4]/H_2SO_4$ [27]	200 mmol/L	3000	5	15	6.75	1	/	98.0
Caprolactam/H_2SO_4 [28]	1 wt%	1000	12	10.5	100:1	1.5	88.1	/
Graphene oxide/H_2SO_4 [29]	0.1 wt%	1500	10	10	13:1	0.5	88.7	96.7
Trifluoroacetic acid/H_2SO_4 [19]	20 wt%	1000	12	10.5	8:1	1.5	95.7	/
Trifluoroacetic acid/H_2SO_4 [19]	1 wt%	1400	5	10	9:1	1	92.26	98.55
$[Bmim][SbF_6]/H_2SO_4$ [30]	0.5 wt%	1000	/	10	8.5:1	0.8	90.07	98.00
ADM-C_{12}-SbF_6/H_2SO_4 [31]	0.5 wt%	1000	1.5	10	10:1	0.8	/	98.6
$[N_{222}H][CF_3SO_3(CF_3SO_3H)_x]$ (x=0, 1, 2) [32]	/	/	5	10	8.5:1	0.6	86.23	97.3

Data from Y. Tian, S. Mei, L.L. Zhang, et al., Improved H_2SO_4-catalyzed alkylation reaction in a rotating packed bed reactor by adding additives, Can. J. Chem. Eng. 100 (2022) 3395–3407.

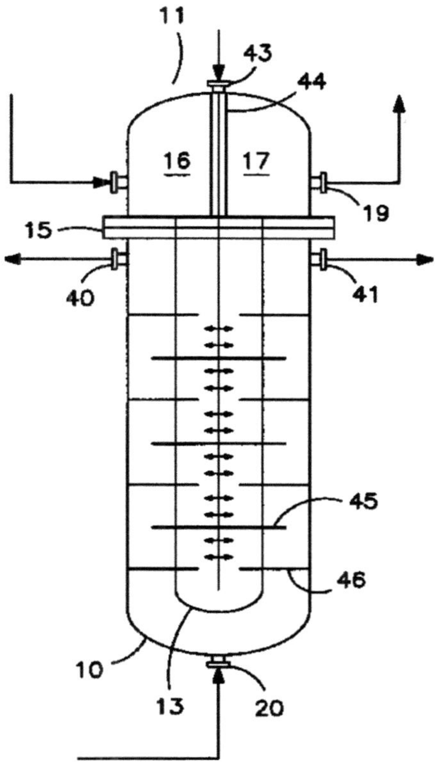

FIG. 5.14 The reactor of HF alkylation. *From R.A. Strauss, R.R. Hemrajani, HF Alkylation Reactor, US7947232B2, 2011.*

interface between the catalyst and the hydrocarbon phases [29,37]. If the reaction time is too long, it results in more side reactions in the reactor and the separator. To solve these problems, a novel liquid-liquid cyclone reactor (LLCR) is designed to improve the mass transfer of the reactants and accelerate separation of the acid-hydrocarbon emulsion. Zhang et al. conducted in-depth studies on the LLCR reactor, including flow field simulations and design of light phase inlet structures [10,30–32]. Fig. 5.15 shows a schematic diagram of the proposed novel LLCR. The innovative LLCR features separate inlets for the light phase and heavy phase, where the light phase is introduced through two symmetric tangential slots in the inlet, while the heavy phase enters axially with the assistance of a guided vane. The operational principle of the LLCR involves the heavy phase gradually undergoing high-speed rotation along the guided vane upon entry. Consequently, a layer of the heavy phase forms near the wall surface. As the light phase enters the LLCR through the tangential slots, it moves toward the axis due to centrifugal force [38], facilitating the mixing

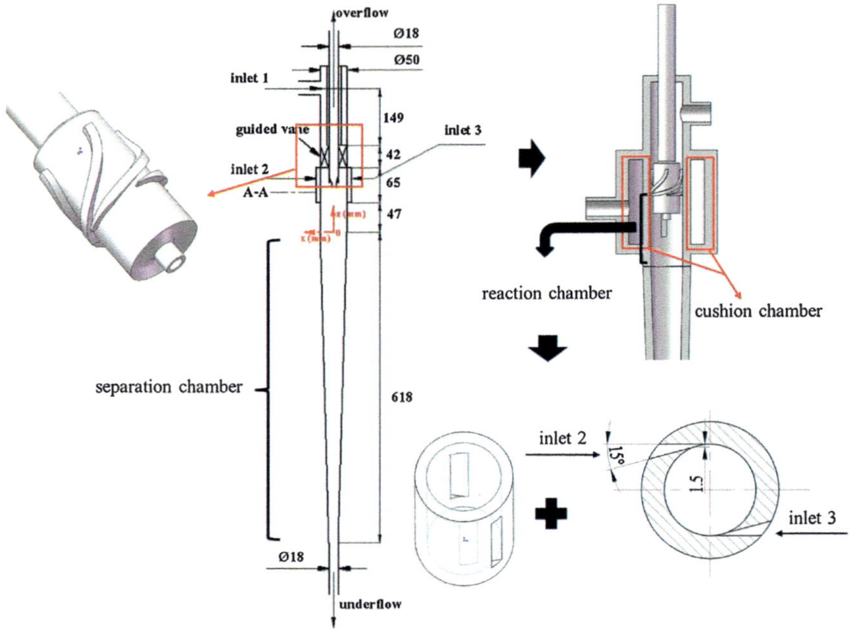

FIG. 5.15 Schematic diagram of the novel LLCR structure. *From M. Zhang, L. Wang, L. Zhu, et al., Phase holdup distribution and dispersion performance in a novel liquid–liquid cyclone reactor of isobutane alkylation catalyzed by ionic liquid, Chem. Eng. Res. Des. 125 (2017) 257–264.*

process between the two phases. Ultimately, the product exits through the overflow outlet, while the ionic liquid is discharged through the underflow outlet. The typical size of cyclone reactor is listed in Table 5.2.

TABLE 5.2 Size of the cyclone reactor.

Name	Number
Column section diameter (mm)	50
Overflow port diameter (mm)	12
Underflow port diameter (mm)	18
Tangential inlet diameter (mm)	4
Column section height (mm)	165
Cone section height (mm)	620

Data from J. Bi, P. Yang, J. Duan, et al., Numerical investigation of secondary flow in a cyclone reactor for alkylation catalyzed by ionic liquid, Ind. Eng. Chem. Res. 61 (2022) 16461–16469.

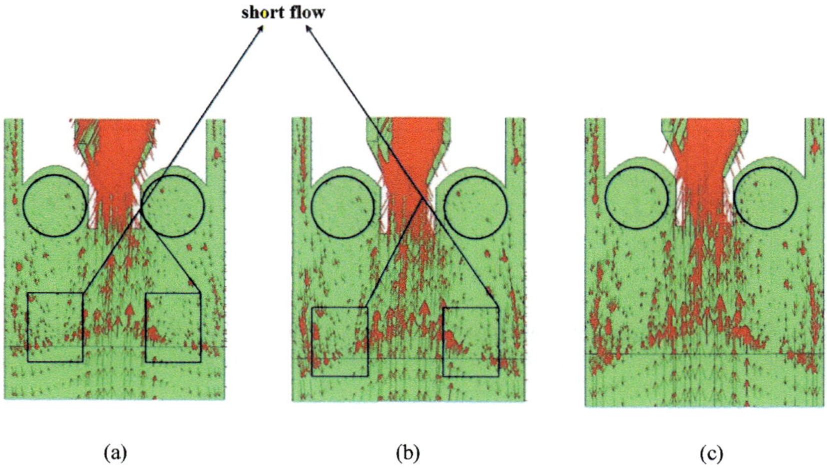

FIG. 5.16 Axial velocity vector of kerosene in the reaction chamber under different total inlet flows: (A) $2\,m^3/h$, (B) $3\,m^3/h$, (C) $4\,m^3/h$. *From M. Zhang, L. Zhu, Z. Wang, et al., Flow field in a liquid–liquid cyclone reactor for isobutane alkylation catalyzed by ionic liquid, Chem. Eng. Res. Des. 125 (2017) 282–290.*

Through the Eulerian-Eulerian (CFD) multiphase flow model and Reynolds Stress Model (RSM), the dispersion performance in the reaction chamber and flow behavior in the separation chamber were simulated (Fig. 5.16) [10].

The effects of the number and width of light phase inlet structures in LLCR on mixing and separation performance were conducted using the Euler-Euler model and Reynolds Stress Model [31]. Three designs of light phase inlet, namely Type O, Type A, and Type B, are shown in Fig. 5.17. The simulation results indicated improvements in both mixing and separation performance with an increasing number of light phase inlets. Reducing the width of the light phase inlet is found to contribute to a decrease in droplet breakup, an increase in mass transfer time, and separation efficiency.

The axial velocity and streamline distribution in the cyclone reactor for ionic liquid alkylation are modeled using the Reynolds stress and Eulerian multiphase flow models [32]. Notably, clear secondary flows are identified in the zero axial velocity wave zone (ZAVWZ). Optimal back-mixing is achieved in the inlet mixing zone with a high secondary flow rate. Meanwhile, primary phase separation occurs in the cone zone with limited secondary flow. Fig. 5.18 shows the changes in the secondary flow of the light phase at different overflow ratios in the cyclone reactor. Increasing the total flow rate enhances the secondary flow of two phases. An appropriate feed ratio facilitates secondary flow formation. The secondary flow of the heavy phase intensifies and concentrates in the column

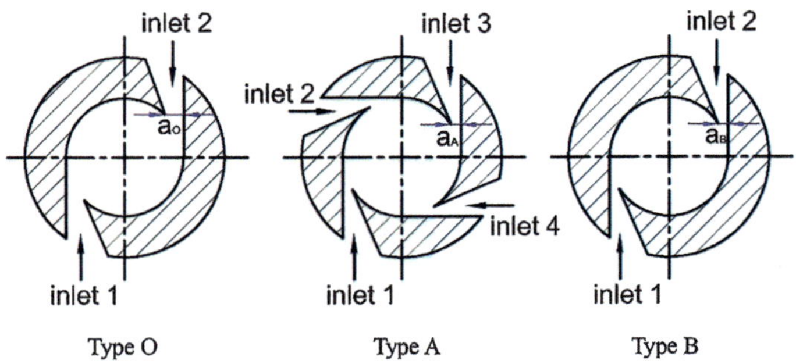

FIG. 5.17 Three types of light phase inlet. *From S. Duan, M. Chen, R. Zhang, et al., Effect of light phase inlet structure on liquid-liquid cyclone reactor for isobutane alkylation catalyzed by ionic liquid, Chem. Eng. Process. 140 (2019) 22–28.*

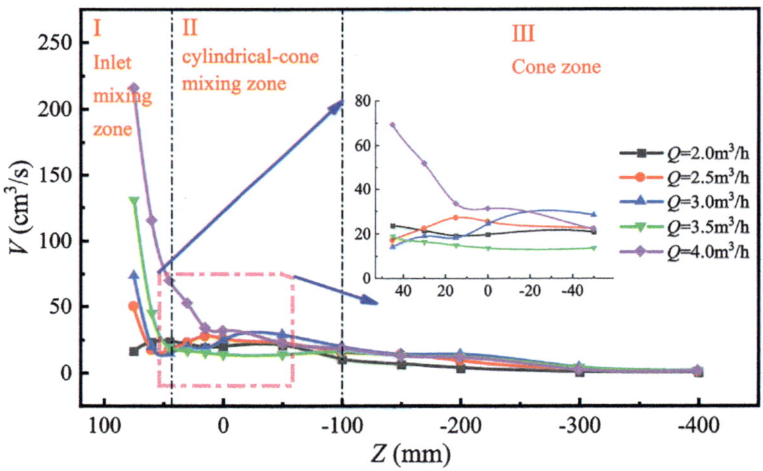

FIG. 5.18 Distribution of the secondary flow rate in the cyclone reactor with different total inlet flows. *From J. Bi, P. Yang, J. Duan, et al., Numerical investigation of secondary flow in a cyclone reactor for alkylation catalyzed by ionic liquid, Ind. Eng. Chem. Res. 61 (2022) 16461–16469.*

with elevated operation parameters (overflow ratio, F, feed ratio, E, and total inlet flow, Q). Additionally, the secondary flow of the light phase increases in the inlet mixing area, promoting effective mixing between the two phases. Conversely, an increase in the overflow ratio diminishes the secondary flow of the light phase in the inlet mixing area, impeding back-mixing. Moreover, an excessively high inlet flow disrupts flow field stability, causing an undesirable rise in the secondary flow rate in the cone area, detrimental to separation. A high feed ratio leads to a rapid concentration of the light phase through the overflow outlet, limiting the overall phase contact. The optimal mixing of two phases in the reactor is achieved at $F=0.40$, $E=1.00$, and $Q=4.0\,m^3/h$.

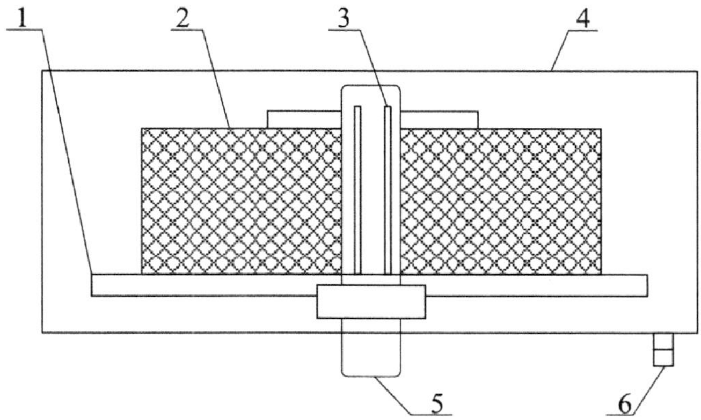

FIG. 5.19 Schematic of RPB: (1) Rotor; (2) packing; (3) liquid inlet; (4) glass window; (5) shaft; and (6) liquid outlet. *From Y. Liu, G. Liu, G. Wu, et al., Alkylation of isobutane and 2-butene in rotating packed-bed reactors: using ionic liquid and solid acid as catalysts, Ind. Eng. Chem. Res. 59 (2020) 14767–14775.*

The RPB proves to be an effective multiphase reactor. Within the RPB, the presence of a centrifugal force enhances mass transfer [39]. While RPB finds numerous industrial applications, the designs for these applications are currently tailored on a case-by-case basis. The alkylation of isobutane with 2-butene catalyzed by ILs/solid acid (SA) was studied based on the RPB reactors by Tian et al. [40], as shown in Fig. 5.19. Residence time, reaction temperature, the I/O ratio, and the acid-to-hydrocarbon ratio exert a significant influence on alkylate quality. Under optimal conditions, the RON can achieve 99.8, with TMPs components constituting 87.1 wt%. The alkylation data are collected in Table 5.3.

The droplets in the RPB allow the exploration of the correlation between the average droplet diameter and interfacial area. The droplet diameter and interfacial area proved to be highly dependent on physical parameters such as acid-to-hydrocarbon ratio, rotational speed, and interface tension of acid-hydrocarbon. For liquid flow patterns, the liquid from the inner packing layer ultimately transforms into droplets in the cavity area. Given the brief duration of the liquid passage through the packing layer, the liquid shape is assumed to be a spherical droplet during the rotation process. Following the suggestions of Albright et al. [41], the apparent volume ratio of acid-to-hydrocarbon (ϕ) is employed to signify the IL to hydrocarbon phase ratio in the reaction zone. Clearly, the average droplet diameter significantly impacts the interfacial area of hydrocarbons. A representative image of droplets in RPB is depicted in Fig. 5.20. The interfacial area of acid-hydrocarbon phases can be increased with a high acid-to-hydrocarbon ratio and small droplet diameter. The large interfacial area is beneficial to the increase of the yield and quality of alkylate.

TABLE 5.3 Product distribution using different packing layers.[a]

Packing layers Product distribution (wt%)	IL@SA	SA	MCM-41	Pure SA[b]	Stirred reactor[c]
2-Methylbutane	4.9	5.5	4.1	6.5	5.9
2,3-Dimethylbutene	2.9	3.3	5.1	6.7	6.3
C_6–C_7 (others)	3.0	4.1	3.9	7.2	9.1
Total light ends	10.8	12.9	13.1	20.4	21.3
2,2,4-Trimethylpentane	39.2	36.4	32.4	20.1	22.8
2,2,3-Trimethylpentane	2.3	2.5	2.4	0.4	1.2
2,3,4-Trimethylpentane	16.4	17.4	18.4	6.5	10.7
2,3,3-Trimethylpentane	13.0	12.3	12.1	3.4	11.2
Total TMP	70.9	68.6	65.3	30.4	45.9
2,5-Dimethylhexane	4.5	5.4	2.2	2.7	3.0
2,4-Dimethylhexane	3.0	3.2	1.9	2.3	2.5
2,3-Dimethylhexane	1.2	1.6	0.5	0.5	2.3
2,2-Dimethylhexane	0.1	0.2	0.1	0.1	0.2
Total C_8	79.7	79	70	48.5[d]	53.9
2,2,5-Trimethylhexane	6.5	4.5	3.8	5.8	8.3
C_{9+} (Others)	1.9	2.4	11.4	15.1	10.1
C_{12+}	1.1	1.2	1.7	10.2	6.4
Total heavy ends	9.5	8.1	16.9	31.1	24.8
Calculated RON	96.0	95.8	95.1	85.2	92.2
Calculated MON	93.4	93.0	92.7	83.7	90.5
Yield on 2-butene (wt%)	144.8	160.1	158.2	78.6	194.9

[a] *Single alkylation procedure, rotational speed = 800 rpm, I/O = 7, reaction temperature = 25°C.*
[b] *Using SA as the catalyst without the presence of IL.*
[c] *Data of a stirred reactor: pure [BMIm]–$AlCl_3$ was used, I/O = 7, agitation = 800 rpm, residence time = 15 min, and reaction temperature = 25°C.*
[d] *Containing 12.5 wt% isooctene.*
Data from Y. Liu, G. Liu, G. Wu, et al., Alkylation of isobutane and 2-butene in rotating packed-bed reactors: using ionic liquid and solid acid as catalysts, Ind. Eng. Chem. Res. 59 (2020) 14767–14775.

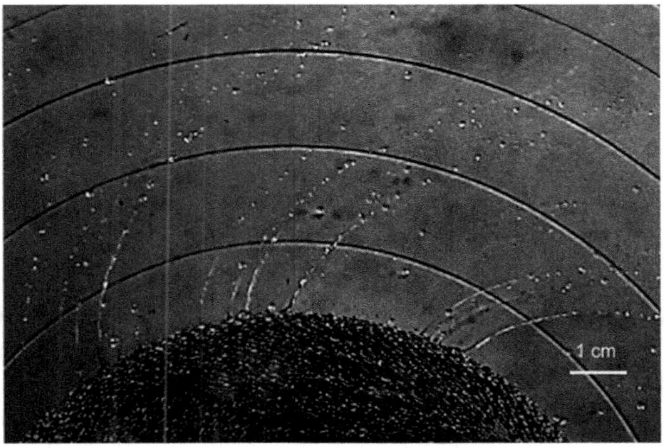

FIG. 5.20 Image of droplets in the cavity zone (Isobutane, rotational speed = 800 rpm). *From Y. Liu, G. Liu, G. Wu, et al., Alkylation of isobutane and 2-butene in rotating packed-bed reactors: using ionic liquid and solid acid as catalysts, Ind. Eng. Chem. Res. 59 (2020) 14767–14775.*

5.4 Industrial reactors in the solid acid process

Li et al. simulated the isobutane/2-butene alkylation process to optimize the fixed-solid-acid-bed reactor using a β-zeolite catalyst [42]. They employed the Lagrange method to neutralize the numerical problems that originate from nonlinear reactions and Eulerian grids. The results show that the paraffin/olefin and alkylate reflux ratios are two critical parameters in designing SCA processes. For batch runs without alkylate reflux, the paraffin/olefin ratio should be larger than 400 to achieve selectivity above 80% (2-butene to C_8). Alkylate reflux is necessary for continuous operations to reduce the burden in distillation, and the reflux ratio should be determined simultaneously with the input paraffin/olefin ratio. The temperature in fixed-bed reactors can be neglected as long as the mixer temperature is set to be steady and the reaction is terminated before the catalyst is completely spent. Multiple inlet points can intensify the process by reducing the number of reactors, and the radical bed can handle a large liquid flux. Finally, according to the techno-economic analysis, in the optimal design, material consumption and equipment purchase are two main components in the total cost, whereas the utility cost in distillation represents only a minor portion of the total cost.

In 2015, a milestone was reached with the successful commissioning of the world's first industrial-scale solid acid alkylation unit with a capacity of 100,000 tons/year [43]. This unit was established at Shandong Huifeng Petrochemical Company and employed the Lummus process package along with catalysts from Albemarle company [32]. However, it is worth noting that solid acid alkylation technologies currently face significant challenges, particularly in terms of catalyst deactivation and the complexities associated with catalyst regeneration.

5.5 Deactivation mechanism

5.5.1 Deactivation process of H_2SO_4

In C4 alkylation, when the concentration of H_2SO_4 falls below 85 wt%, the catalytic efficiency decreases significantly. The typical composition of used H_2SO_4 from alkylation is listed in Table 5.4 [44,45].

The abovementioned water-insoluble materials primarily refer to hydrocarbon polymers (C_{12}-C_{20}) formed during the alkylation. These polymers are generally referred to as acid-soluble oil or "red oil," which contains a significant amount of conjunct polymers. Conjunct polymers are byproducts formed during C4 alkylation that dilute the acid catalyst so that it eventually loses effectiveness [46]. They contain C_5 and maybe C_6 ring groups with molecular weights mainly in the range 300–500 and are highly unsaturated with both conjugated and nonconjugated double bonds [47,48]. Sulfonic acid are complex substances similar to the aforementioned water-insoluble material but tend to be slightly more reactive. Organic sulfates mainly include dibutyl sulfate and acidic monobasic sulfate. Moreover, sulfurous acid may be formed by the reaction of sulfur dioxide with water [45].

The concentration of concentrated sulfuric acid, as a catalyst in the alkylation reaction, has a significant impact on the reaction. The appropriate acid concentration is essential to obtain high-quality alkylate. When the concentration of H_2SO_4 exceeds 99%, it can directly react with isobutane, leading to the increased acid consumption. Conversely, when the concentration of H_2SO_4 falls below 85%, the activity of H_2SO_4 as a catalyst decreases significantly, and it also tends to cause severe corrosion to the

TABLE 5.4 The compositions of used H_2SO_4.

Component	Content (wt%)
Titratable acidity	90.0
H_2SO_4	88.5
Water-insoluble materials	3.0–4.0
H_2O	2.0–3.0
Sulfonic acid	1.0–3.0
Organic sulfates	1.0–3.0
SO_2	0.1–0.3

equipment. Therefore, the suitable sulfuric acid concentration for alkylation is typically in the range of 95%–96%.

In industrial alkylation, fresh H_2SO_4 with a concentration of 98–99 wt% is commonly added to the reactor. However, due to the presence of water in the raw materials, dilution by water generated as a byproduct, and the formation of sulfate ester and acid-soluble polymers during the reaction, the concentration of H_2SO_4 gradually decreases during the operation. To ensure the quality of alkylate and prevent corrosion of the equipment, it is necessary to discharge it as used H_2SO_4 when the acid concentration drops to 88–90 wt%.

The reasons for the decrease in acidity of concentrated H_2SO_4 primarily come from two aspects: the influence of impurities in alkylation raw materials and the formation of byproducts during the alkylation reaction. This can be analyzed in more detail from several perspectives.

5.5.1.1 *Impact of raw material*

The C4 feedstock for alkylation is obtained through catalytic cracking separation. The complexity of raw material sources results in relatively complex impurities in the raw materials. Water can dilute H_2SO_4 and, in subsequent alkylation, serve as a reactant to generate other byproducts [49]. Butadiene present in the raw materials is a major impurity contributing to acid consumption. Butadiene has a higher solubility in H_2SO_4 compared to isobutane, making it dissolve more readily in acid, thereby forming alkyl sulfates and reducing the concentration of H_2SO_4. Additionally, alkyl sulfates can also affect the yield and quality of the alkylate. Typically, the primary method for removing butadiene involves selective catalytic hydrogenation of the feed gas to convert diene hydrocarbons into monoene hydrocarbons, thus avoiding the generation of byproducts and reducing acid consumption [50]. In the C4 fraction, there may also be a small amount of ethylene, which can react with H_2SO_4 to produce isohydrogen sulfate, as shown in the following reaction equation:

$$CH_2 = CH_2 + H_2SO_4 \rightarrow CH_3{-}CH_2OSO_2 \quad (5.1)$$

Furthermore, ethylene as an impurity leads to an accumulation effect in the reaction phase. Even with a small amount of ethylene present in the feedstock, it can result in the daily production of several hundred kilograms of ethylene in the acid phase, leading to a significant generation of used H_2SO_4. Currently, industrial units include a pretreatment system to address ethylene impurities. This system employs a dehydrogenation tower to remove olefins such as ethylene and propylene. Sulfides have a substantial diluting effect on H_2SO_4, with each ton of sulfides causing the generation of several tens of tons of used H_2SO_4. Therefore, sulfides are typically removed to reduce their content to below 10 μg/g [51] before entering the reactor. In addition, sulfides can promote side reactions such as polymerization, further accelerating the depletion of H_2SO_4. Dimethyl ether and methanol are also significant acid-consuming impurities that can reduce the yield and RON of alkylate. These impurities should likewise be removed before the alkylation reaction [51].

5.5.1.2 Impact of side reactions

Due to the occurrence of numerous side reactions, alongside the production of alkylate as the primary reaction product, a portion of hydrocarbons soluble in acid, known as ASO, is also generated. As sulfuric acid is continuously recycled, ASO accumulates and participates in subsequent alkylation as reactants. With a deepening understanding of the alkylation process, it has been found that ASO primarily forms through the following four pathways:

(1) Polymerization of olefins

In the early stage of alkylation, butenes have a relatively high solubility in H_2SO_4, and the concentration of olefins is excessive compared to alkanes at the acid-hydrocarbon reaction interface. Therefore, olefins are prone to polymerization reactions, leading to the formation of high molecular weight polymers. Carbonium ions generated in this reaction continue to react, forming alkanes. Albright proposed that this reaction is one of the primary pathways for ASO formation [47]. The reaction equation is represented as follows in Eq. (5.2):

$$4C_4H_8 \rightarrow 16(-C_{1.0}H_{1.75}-) + [H] \tag{5.2}$$

(2) Reversible sulfates

In alkylation, a small amount of olefins reacts with H_2SO_4 to form sulfates. This reaction is mostly reversible, and most sulfates are unstable under acidic conditions, releasing olefins, which then undergo alkylation with isobutane. However, in certain regions of the alkylation reactor and acid settler, where the concentration of isobutane is lower at the acid-hydrocarbon interface, the olefins released from sulfates can further react, leading to the formation of ASO and alkylate with lower RON. The specific reaction equations are as follows:

$$C_3H_6 + H_2SO_4 \rightleftharpoons (C_3H_7)HSO_4 \tag{5.3}$$

$$(C_3H_7)HSO_4 + C_3H_6 \rightleftharpoons (C_3H_7)_2SO_4 \tag{5.4}$$

$$C_4H_8 + H_2SO_4 \rightleftharpoons (C_4H_9)HSO_4 \tag{5.5}$$

$$C_4H_8 + (C_4H_9)HSO_4 \rightleftharpoons (C_4H_9)_2SO_4 \tag{5.6}$$

(3) Oxidative polymerization

Due to the strong oxidative and hygroscopic properties of concentrated H_2SO_4, a small fraction of olefins can undergo oxidative polymerization reactions. Apart from forming polyolefins, this process also generates SO_2 and water.

$$4C_4H_8 + 2H_2SO_4 \rightleftharpoons 16(-C_{1.0}H_{1.75}-) + 4H_2O + 2SO_2 \qquad (5.7)$$

(4) Degradation reactions

TMPs and other isomeric alkanes can undergo degradation reactions upon contact with H_2SO_4, leading to the formation of C_4-C_{16} isomeric alkanes and ASO. Higher acidity of H_2SO_4 leads to the generation of a higher ASO content. Olefins react with protonated CF_3SO_3H to form tert-butyl carbonium ions, and tert-butyl carbonium ions react with butenes to generate C_8^+. The C_8^+ then loses an H^+ to form the corresponding C_8 component, as shown in Eq. (5.8):

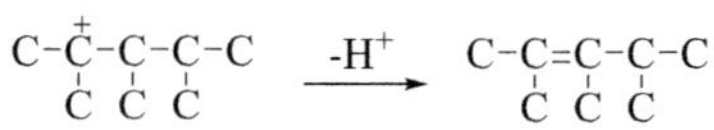
(5.8)

In the trifluoromethanesulfonic acid-catalyzed alkylation reaction, the generated olefins, as shown in Eq. (5.9), lose an H^- to form the corresponding carbonium ions. The carbonium ions undergo addition and hydride transfer reactions, leading to the formation of complex olefin molecules. Finally, the products undergo protonation and cyclization reactions, resulting in the formation of ASO.

(5.9)

Ipatieff and Grosse were the pioneering researchers who initially observed that ASO results from polymerization reactions [52]. Subsequently, the composition of ASO was characterized. Bloch employed UV and FT-IR absorption spectroscopy to analyze ASO [53]. It is revealed that ASO primarily consists of hydrocarbon compounds characterized by rings containing five carbon atoms along with branched structures. On average, each ASO molecule contains 2–4 double bonds, with two of these bonds typically being conjugated. These conjugated double bonds are likely distributed across both the ring hydrocarbons and their branches. However, it is worth noting that the actual complexity of ASO molecules in terms of compound size and quantity is likely far greater than the individual average cyclic compounds studied by Bloch. This complexity arises due to the broad boiling point range and intricate composition of ASO.

As early as 1963, Miron and Lee concluded the properties of the conjunct polymers and hydrogenated polymers [54], as listed in Tables 5.5 and 5.6. They investigated seven distinct ASO samples extracted from different acids: HF, H_2SO_4, and $AlCl_3$. Conjunct polymers are incidental byproducts of acid-catalyzed hydrocarbon reactions, whose structures were determined by physical property correlations, ultraviolet, infrared, and mass spectroscopy, along with various chemical methods, including hydrogenation. The conjunct polymers are identified as cyclic polyolefinic hydrocarbons featuring a substantial proportion of conjugated double bonds. Although five-membered ring systems predominate, the presence of larger, and possibly smaller, rings are speculated. Both condensed and uncondensed ring systems are considered. On average, the ASO molecule can be regarded as dicyclic, possessing one pair of conjugated double bonds and one-half to one and one-half isolated double bonds. Approximately 60% of the carbon atoms are found in side chains. Polymer distillate fractions with as many as five rings and four double bonds per mole have been identified.

Figs. 5.21–5.25 depict instances of molecular types that might exist in conjunct polymer I in Miron and Lee's work [54], categorized based on their molecular size. The tricyclic structures make up 60%–70% of polymer I, comprising approximately 14–15 ring carbon atoms and 4–5 alkyl carbon atoms. The properties of compositions in different ASO are listed in Table 5.7.

Albright used the same method of ASO extraction to study H_2SO_4-ASO samples with different storage time [60,61]. It is found that there is a direct correlation between the extended storage time of used H_2SO_4 and the increase in both the molecular weight and the degree of unsaturation within ASO molecules. In-depth analyses utilizing IR and ^{H}NMR spectroscopy reveal striking similarities in the functional groups present across the three different ASO types.

In an acidic environment, ASO molecules undergo esterification and sulfonation reactions. These chemical transformations result in the creation of molecules bearing sulfone and sulfonic acid groups, alongside other pertinent functional groups. Albright postulated that hydrocarbons soluble in acid undergo ionization processes, generating carbonium ions. These carbonium ions play a vital role in alkylation, manifesting in three primary aspects [60–62]:

1. Acidity decline with ASO accumulation: As ASO accumulates within H_2SO_4, the acidity of the H_2SO_4 gradually diminishes. Consequently, this decline in acidity leads to a reduction in catalytic activity. It becomes imperative to replace the H_2SO_4 once its acidity falls below a certain critical threshold.
2. Altering physical properties: The presence of ASO in the reaction system exerts a discernible influence on its physical properties.

TABLE 5.5 Properties of conjunct polymers.

	Conjunct polymer oil						
	HF-catalyzed reactions				H_2SO_4-catalyzed reactions		$AlCl_3$-catalyzed reaction
Property	**Polymer I**	**Polymer II**	**Polymer III**	**Polymer IV**	**Polymer V**	**Polymer VI**	**Polymer VII**
°API (60°F)	21.2	35.1	28.1	27.5	22.1	…	…
Density, 25°C, computed	0.9195	0.8413	0.8798	0.8831	0.9064	0.9057[a]	0.8787[a]
Refractive index, 25°C	1.5179	1.4800	1.4984	1.4978	1.5010	1.4967	1.4903
Specific dispersion, 25°C	130.2	134.0	134.6	132.4	106.4	[b]	…
Bromine no.[c]	188	162	131	157	179	194	187
Iodine no.[d]	437	379	402	419	314	381	393
Maleic anhydride value[e]	354	351	362	299	253	298	322
Molecular weight[f]	265	302	295	302	362	265	285
% Carbon, found	88.12	87.00	87.75	87.60	81.58	82.12	86.84
% Hydrogen, found	11.29	12.74	12.22	12.30	11.45	11.33	12.38
Total, % C+% H, found	99.41	99.74	99.97	99.90	93.03[g]	93.45[h]	99.22
% C, corrected[i]	88.64	87.23	87.78	87.69	87.69	87.88	87.52
% H, corrected[j]	11.36	12.77	12.22	12.31	12.31	12.12	12.48
Molecular formula	$C_{19.6}H_{29.9}$	$C_{21.9}H_{38.3}$	$C_{21.6}H_{35.8}$	$C_{22.0}H_{36.9}$	$C_{26.4}H_{44.2}$	$C_{19.4}H_{31.9}$	$C_{20.8}H_{35.3}$
Type formula	$C_nH_{2n-9.3}$	$C_nH_{2n-5.5}$	$C_nH_{2n-7.4}$	$C_nH_{2n-7.1}$	$C_nH_{2n-8.8}$	$C_nH_{2n-6.9}$	$C_nH_{2n-6.3}$

[a] *By pycnometer.*
[b] *Oil too dark for accurate determination.*
[c] *By Francis method [55].*
[d] *By mercuric acetate-catalyzed Wijs method [56].*
[e] *Expressed as mg. maleic anhydride reacting with 1 g material under reflux in toluene solution, determined by back-titration with standard alkali.*
[f] *Ebullioscopic, in benzene [57–59].*
[g] *Contained 1.00% sulfur and 2.35% ash.*
[h] *Contained 2.90 % sulfur.*
[i] *Corrected %C = 100(%C)/%C + %H.*
[j] *Corrected %H = 100(%H)/%C + %H.*
Data from S. Miron, R.J. Lee, Molecular structure of conjunct polymers, J. Chem. Eng. Data 8 (1963) 150–160.

TABLE 5.6 Properties and structural composition of hydrogenated polymers.

Property	Polymer I	Polymer II	Polymer III	Polymer IV	Polymer V
Density, 20°C[a]	0.9029	0.8323	0.8554	0.8660	0.8673
Refractive index, 20°C	1.4872	1.4580	1.4670	1.4727	1.4734
Specific dispersion (F-C), 20°C	99.7	98.3	97.5	101.9[b]	96.0
Bromine no.[c]	3.2	0.2	0.7	2.2	0.4
Molecular weight[d]	273	310	302	303	323
Corrected % C[e]	86.67	85.84	86.28	86.43	86.17
Corrected % H[f]	13.33	14.16	13.72	13.57	13.83
Molecular formula	$C_{19.70}H_{36.10}$	$C_{22.16}H_{43.66}$	$C_{21.70}H_{41.11}$	$C_{21.81}H_{40.79}$	$C_{23.17}H_{44.32}$
Type formula	$C_nH_{2n-3.30}$	$C_nH_{2n-0.77}$	$C_nH_{2n-2.29}$	$C_nH_{2n-2.83}$	$C_nH_{2n-2.02}$
Average no. of rings per mole	2.65	1.38	2.14	2.41	2.01
Average ring size, C atoms	5.8	4.4	4.6	4.7	5.7
Average no. of C atoms per ring[g]	5.0	4.1	4.0	4.1	5.0
Average no. of C atoms per mole	19.7	22.2	21.7	21.8	23.2
Average no. of ring C atoms per mole	13.3	5.7	8.6	9.9	10.1
Average no. of alkyl C atoms per mole	6.4	16.5	13.1	11.9	13.1
% Naphthenic C atoms	67.5	25.7	39.6	45.4	43.6
% Alkyl C atoms	32.5	74.3	60.4	54.6	56.4

[a] *By pycnometer.*
[b] *This value of specific dispersion indicates that polymer IV may have been not fully hydrogenated. However, difference does not appear great enough to affect conclusions.*
[c] *Francis method [55].*
[d] *Ebullioscopic, in benzene.*
[e] *Corrected %C = 100(%C)/%C + %H.*
[f] *Corrected %H = 100(%H)/%C + %H.*
[g] *Assuming 50% condensation.*
Data from S. Miron, R.J. Lee, Molecular structure of conjunct polymers, J. Chem. Eng. Data 8 (1963) 150–160.

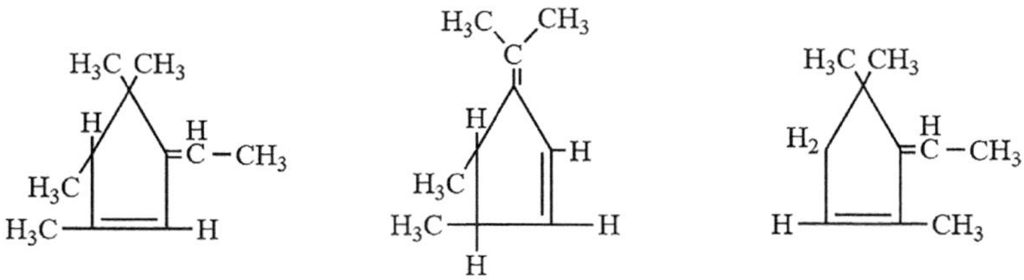

FIG. 5.21 Probable structures of typical monocyclic molecules in polymer I. *From S. Miron, R.J. Lee, Molecular structure of conjunct polymers, J. Chem. Eng. Data 8 (1963) 150–160.*

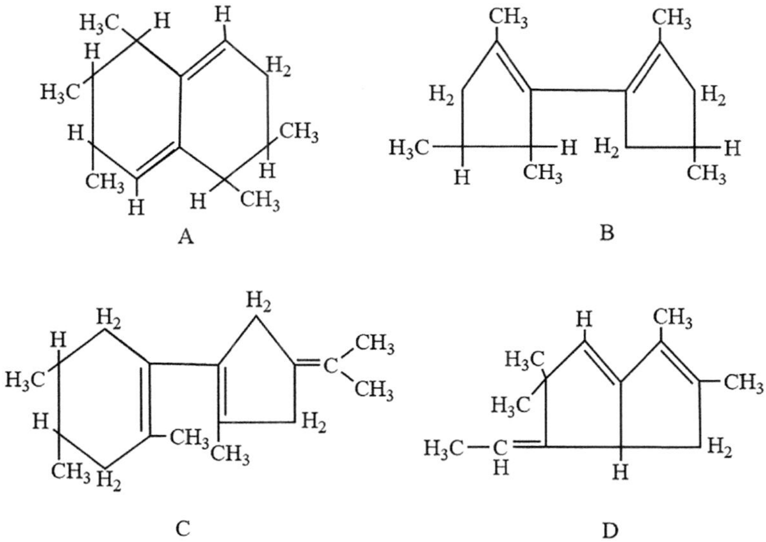

FIG. 5.22 Probable structures of typical dicyclic molecules in polymer I. *From S. Miron, R.J. Lee, Molecular structure of conjunct polymers, J. Chem. Eng. Data, 8 (1963) 150–160.*

This alteration can impact the phase behavior and fluid dynamics of the system, further complicating the interplay of chemical reactions.

3. Chemical effects on alkylation: ASO present within H_2SO_4 has a significant effect on the alkylation reaction system. These chemical effects can modify reaction pathways, intermediate species, and overall reaction kinetics, thus necessitating a comprehensive understanding of their role in the alkylation process.

Xin et al. verified the influence of ASO on the reaction [63]. First, on the basis of the UV absorption spectra, the structures of typical ASO molecules were examined, which consist of five-membered or six-membered ring hydrocarbons with two conjugated double bonds. These molecules can possess complex structures involving simultaneous features like conjugated dienes, homoannular dienes, exocyclic double bonds, and various substituents such as alkyl, acyl, and ring residues [53,54,64,65] (Fig. 5.26).

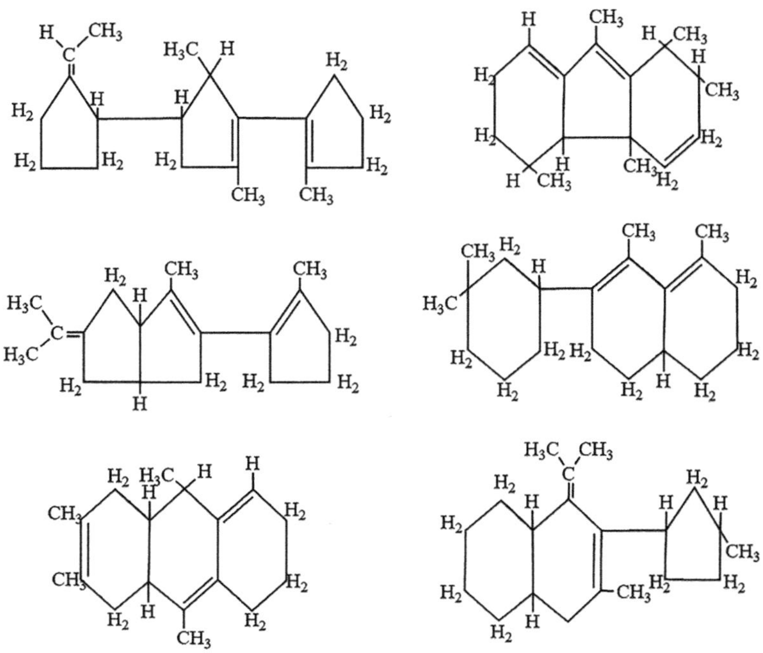

FIG. 5.23 Probable structures of typical tricyclic molecules in polymer I. *From S. Miron, R.J. Lee, Molecular structure of conjunct polymers, J. Chem. Eng. Data 8 (1963) 150–160.*

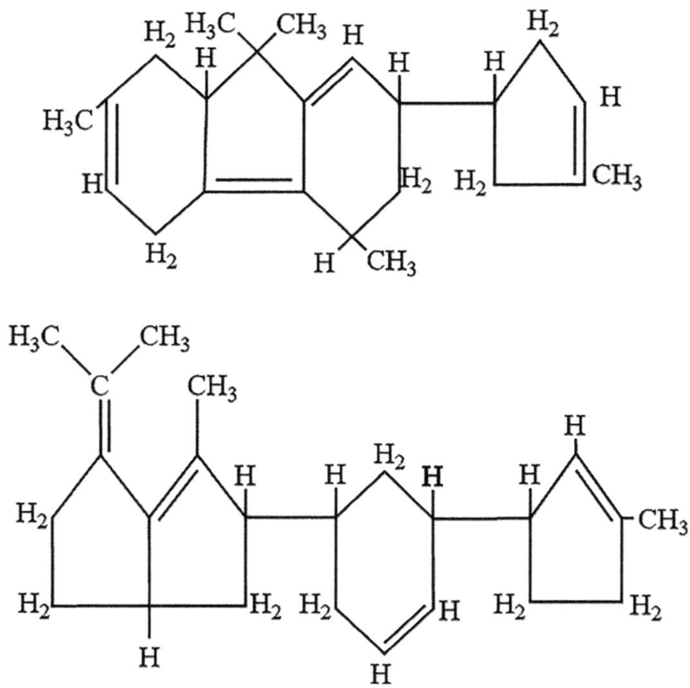

FIG. 5.24 Probable structures of typical tetracyclic molecules in polymer I. *From S. Miron, R.J. Lee, Molecular structure of conjunct polymers, J. Chem. Eng. Data 8 (1963) 150–160.*

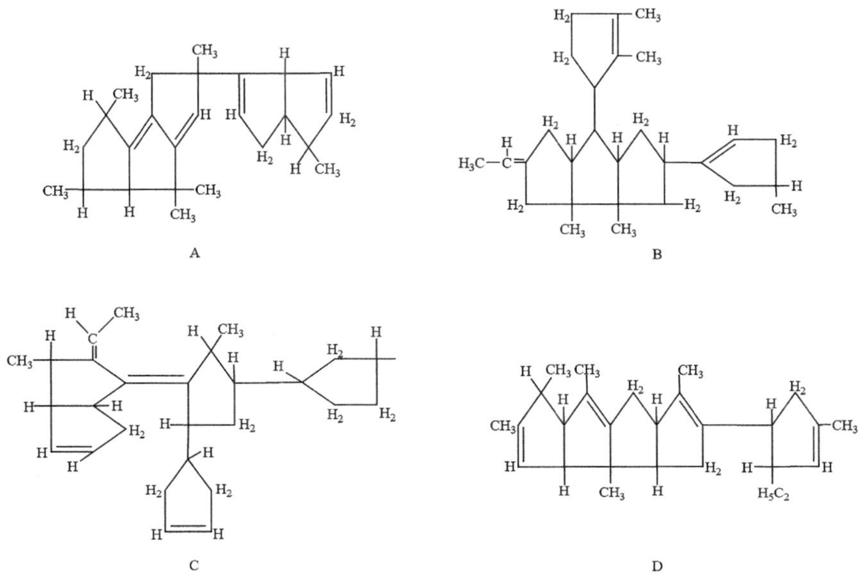

FIG. 5.25 Probable structures of typical pentacyclic molecules in polymer I. *From S. Miron, R.J. Lee, Molecular structure of conjunct polymers, J. Chem. Eng. Data 8 (1963) 150–160.*

TABLE 5.7 The properties of compositions in different ASO.

Acid-soluble hydrocarbon	**HF(IV)**	**H_2SO_4(VI)**	**$AlCl_3$(VII)**
Average molecular weight	302	265	285
Molecular formula	$C_{22.0}H_{36}$	$C_{19.4}H_{31.9}$	$C_{20.8}H_{35.3}$
General molecular formula	C_nH_{2n-7}	$C_nH_{2n-6.9}$	$C_nH_{2n-6.3}$

Data from S. Miron, R.J. Lee, Molecular structure of conjunct polymers, J. Chem. Eng. Data 8 (1963) 150–160.

From GC-MS, Fig. 5.27 illustrates an increase in the carbon number of molecules from less than 8 to more than 20 with increasing retention time, along with a transition in ring types from five-membered to six-membered rings. Additionally, the number of rings increases from one to two or more. Furthermore, heteroatoms, especially oxygen, and long-chain branched substituents become more prevalent with longer retention times. In addition, a majority of the molecules are oxygenated, possibly containing hydroxyl, ether, aldehyde, or ketone groups. Table 5.8 categorizes the compounds with higher concentrations, providing valuable

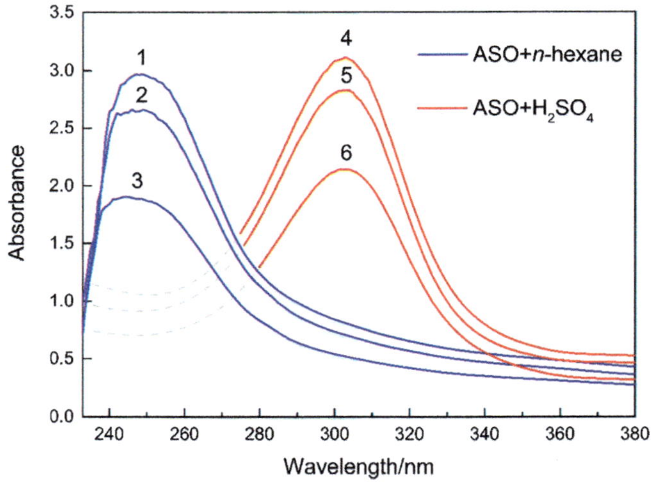

FIG. 5.26 UV absorption spectra of ASO sample ($C_{ASO+n\text{-hexane}}:C_{ASO+H_2SO4}=5:1$). *From Y.Y. Xin, Y.F. Hu, L. Mengchen, et al., Isobutane alkylation catalyzed by H_2SO_4: Effect of H_2SO_4 acid impurities on alkylate distribution, Energ. Fuel 35 (2021) 1664–1676.*

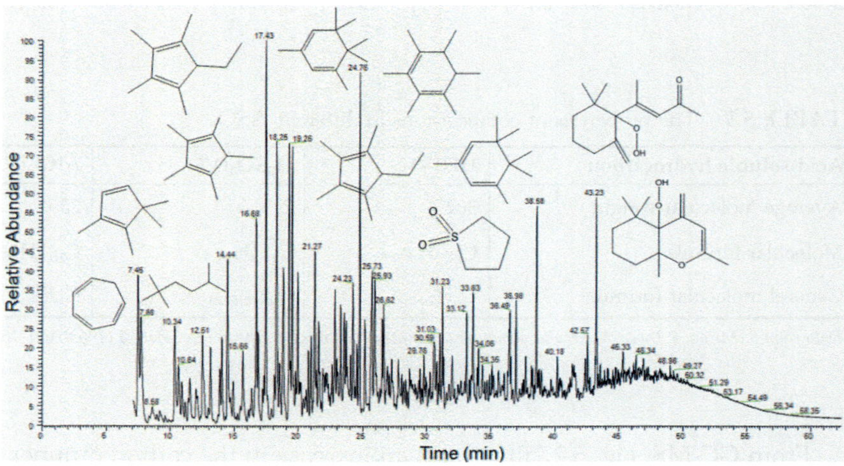

FIG. 5.27 GC-MS spectra of ASO. *From Y.Y. Xin, Y.F. Hu, L. Mengchen, et al., Isobutane alkylation catalyzed by H_2SO_4: Effect of H_2SO_4 acid impurities on alkylate distribution, Energ. Fuel 35 (2021) 1664–1676.*

insights into the composition of ASO. Therefore, the molecular complexity of ASO compounds increases as retention time increases.

Table 5.8 illustrates that a significant portion of ASO molecules are cyclic conjugated olefins, encompassing monocyclic, bicyclic, tricyclic olefins, and more. These molecules exist as uncopolymerized or copolymerized five-membered or six-membered cyclic compounds, characterized by

TABLE 5.8 Main compounds in ASO.

Category	Structure's formula	Compound's name	Content (%)	
1	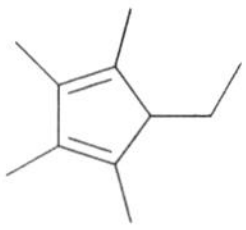	ASO-1: 5-ethyl-1,2.3,4-tetramethyl-1,3-cyclopentadiene	7.80	15.67
	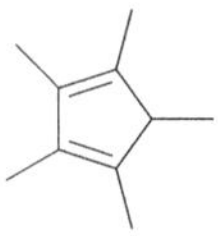	ASO-2: 1,2,3,4,5-pentamethyl-cyclopentadiene	4.37	
	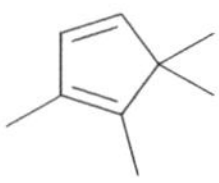	ASO-3: 1,2,5,5-tetramethyl-1,3-cyclopentadiene	2.13	
	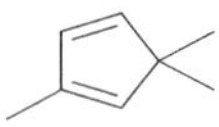	ASO-4: 2,5,5-trimethyl-1,3-cyclopentadiene	1.37	
2	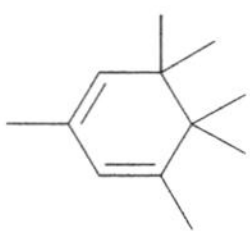	ASO-5: 1,3,5,5,6,6-hexamethyl-1,3-cyclohexadiene	7.58	14.12
	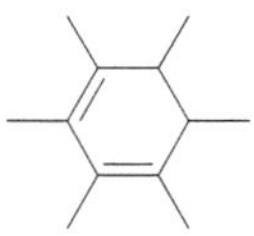	ASO-6: 1,2,3,4,5,6-hexamethyl-1,3-cyclohexadiene	6.54	
3		ASO-7: cycloheptatriene	2.02	2.02
4	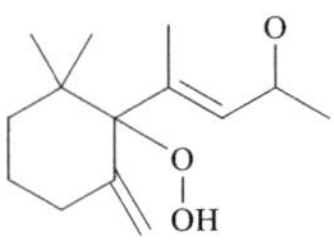	ASO-8: 4-(1-hydroperoxy-2,2-dimethyl-6-methylene-cyclohexyl)-pent-3-en-2-one	1.77	4.08
	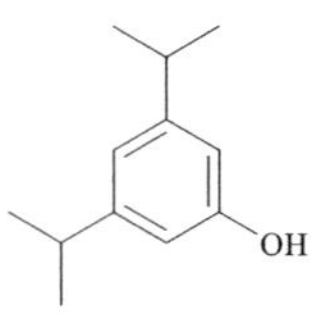	ASO-9: phenol,3,5-bis(1-methylethyl)-	1.41	
	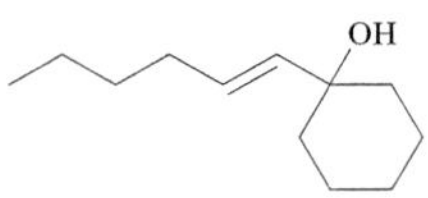	ASO-10: cyclohexanol,1-(1-hexenyl)-, (E)-	0.9	

Continued

TABLE 5.8 Main compounds in ASO—cont'd

Category	Structure's formula	Compound's name	Content (%)	
5	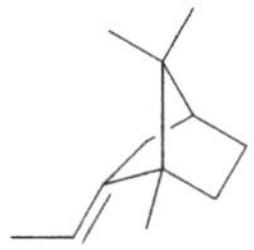	ASO-11: (E)-2-ethyl-1,7,7-trimethyl-dicyclic [2.2.1] heptane	3.02	5.26
	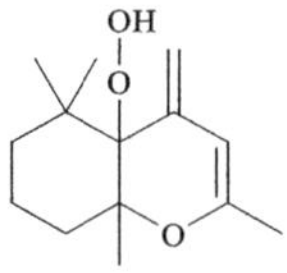	ASO-12: 2,2,5,8a-tetramethyl-4-methylene-6,7,8,8a-tetrahydro-4H,5H-chromen-4a-yl hydro peroxide	1.37	
	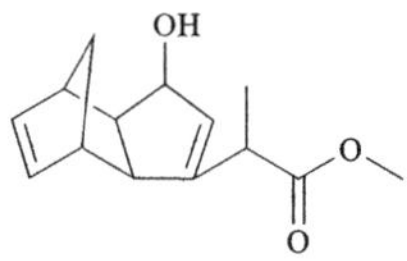	ASO-13: 2-(5-hydroxytricyclo [5.2.1.0 (2,6)]deca-3,8-dien-3-yl) propionic acid, methyl ester	0.87	
6	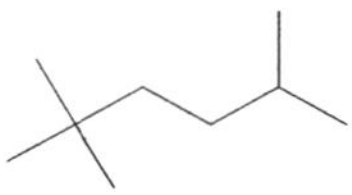	ASO-14: 2,2,5-trimethylhexane	1.13	2.18
	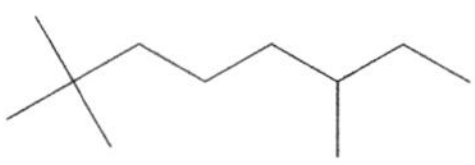	ASO-15: 2,2,6-trimethyloctane	1.05	
7	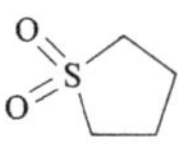	ASO-16: sulfolane	2.02	2.02
8	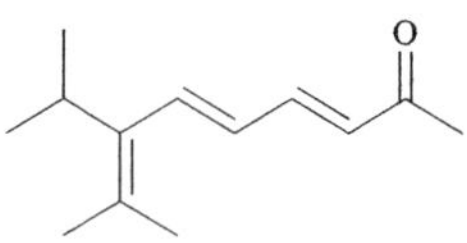	ASO-17: 3,5,7-nonatrien-2-one,8-methyl-7-(1-methylethyl)-, (E,E)-	1.46	3.13
	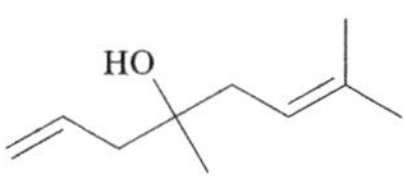	ASO-18: 4,7-dimethyl-1,6-octadien-4-ol	1.19	
	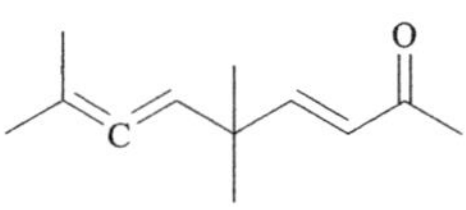	ASO-19: 5,5,8-trimethyl-nona-3,6,7-trien-2-one	1.08	

Categories: 1. cyclopentadiene; 2. cyclohexadiene; 3. cycloheptatriene; 4. monocyclic olefins; 5. bicyclic olefins; 6. alkanes with branched chains; 7. sulfolane; 8. olefins with branched chains.
Data from Y.Y. Xin, Y.F. Hu, L. Mengchen, et al., Isobutane alkylation catalyzed by H_2SO_4: Effect of H_2SO_4 acid impurities on alkylate distribution, Energ. Fuel 35 (2021) 1664–1676.

numerous conjugated double bonds and branched alkane structures distributed along the side chains of the ring core. Additionally, some ASO molecules consist of long-chain alkanes, long-chain olefins, naphthenes with olefin substituents, and heteroatomic compounds containing elements like sulfur and oxygen.

The content of the compounds detailed in Table 5.8 amounts to 49.08%. Among these, the most abundant compounds are cyclopentadiene (15.67%) and cyclohexadiene (14.12%). Other prominent species include olefins (34.83%), alkanes (2.18%), and heteroatomic compounds containing oxygen elements (12.07%). GC-MS analysis has also unveiled specific fine structures. Notably, sulfones exemplified by sulfolane are present in noteworthy quantities within ASO. These compounds likely originate from the reactions between cis-configurational dienes and acid molecules, potentially forming hydroxyl groups through ester hydrolysis, while molecules containing ether, aldehyde, or ketone groups may result from the interactions of olefin polymers [47].

The carbon number distribution of these compounds is depicted in Fig. 5.28. The retention times for compounds within specific carbon number ranges are approximately as follows: C_8–C_9, 7.15–12.51 min; C_{10}–C_{11}, 12.62–19.25 min; C_{12}, 19.38–29.38 min; C_{13}, 29.45–30.09 min; C_{14}–C_{15}, 30.17–39.78 min; and C_{16}–C_{20}, 39.83–61.72 min. Comparing the chromatographic peaks, it is evident that carbon numbers primarily fall within the range of C_{10}–C_{15}. Notably, C_{12} exhibits the highest content at 34.14%. The contents of C_{10}–C_{11} and C_{14}–C_{15} are 24.39% and 19.08%, respectively. Compounds with the lowest (C_8 and C_9) and highest (C_{16} and C_{20}) molecular weights have relatively lower contents. C_{13} is the least abundant at 1.39%. These findings suggest that the production sequence favors the formation of C_8, C_{12}, C_{16}, and C_{20} [66,67].

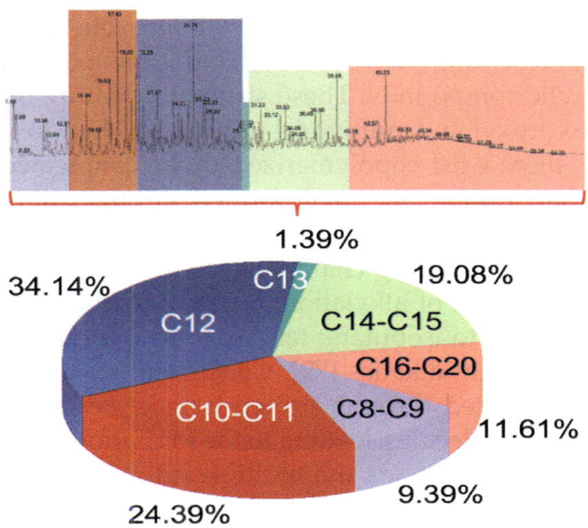

FIG. 5.28 Carbon number distribution of ASO compounds. *From Y.Y. Xin, Y.F. Hu, L. Mengchen, et al., Isobutane alkylation catalyzed by H_2SO_4: Effect of H_2SO_4 acid impurities on alkylate distribution, Energ. Fuel 35 (2021) 1664–1676.*

$$C_4^+ \xrightarrow{+C_4^=} C_8^+ \xrightarrow{+C_4^=} C_{12}^+ \xrightarrow{+C_4^=} C_{16}^+ \xrightarrow{+C_4^=} C_{20}^+ \tag{5.10}$$

(5.11)

The primary generation of C_{12} fractions stems from the polymerization reactions involving olefin and C_8. Initially, C_8 serves as the main product, with a significant presence of C_8^+ observed during the alkylation processes. The polymerization process (refer to reaction 5.10) is notably more efficient and rapid compared to the hydride transfer between isobutane and the C_8 carbonium ion [68].

Moreover, the hydrocarbon cracking of larger molecules like C_{16} and C_{20} is relatively simpler and quicker than that of smaller molecules such as C_8 and C_{12}, leading to a relatively higher production of C_{12}. Other components are likely formed through various reactions including polymerization, cracking, isomerization, hydride transfer, disproportionation, self-alkylation, among others. An instance of disproportionation capable of generating C_5–C_{11} is one example of such reactions [69,70].

$$C_8 + C_8 \rightarrow C_7 + C_9/C_6 + C_{10}/C_5 + C_{11} \tag{5.12}$$

Additionally, the ASO was extracted by *n*-hexane from used acid to analyze the specific structures of the complex compounds in ASO with the quantitative analysis [71]. The results are similar to those obtained by Xin et al. [63]. Table 5.9 lists these compounds with a higher content in ASO, totaling 42.83%. Among these, monocyclic conjugated olefins dominate, representing uncopolymerized five-membered and six-membered cyclic compounds. These structures feature branched alkane chains with hydroxyl and sulfhydryl groups attached to the side chains. Additionally, there exist copolymerized cyclic conjugated olefins, comprising less than 1.0% of the total, formed through the polymerization of monocyclic olefins within a highly acidic environment. The remaining components encompass long-chain alkanes, long-chain olefins, and sulfate esters. At the start of alkylation, olefins exhibiting higher solubility in H_2SO_4 compared to isobutane tend toward polymerization, yielding high molecular weight alkanes [72]. Concurrently, carbocations such as C_4^+ and C_8^+, generated through proton transfer reactions, engage with one or more olefin molecules, leading to the formation of cyclic monomers with conjugated double bonds and multiple side chains via deprotonation and cyclization processes. Moreover, a portion of the side chains undergoes oxidation by H_2SO_4, resulting in the formation of carbonyl and sulfonyl groups. This phenomenon elucidates the predominance of monocyclic conjugated olefins within ASO, characterized by diverse functional groups along the side chains (Fig. 5.29).

TABLE 5.9 Main compounds in ASO.

No.	Compounds	Structure	Content (wt%)
1	*p*-Isopropylbenzyl alcohol	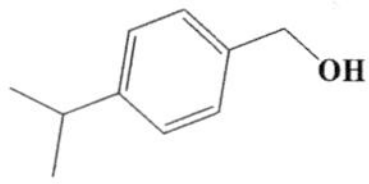	9.72
2	4-Methoxy-benzenethiol	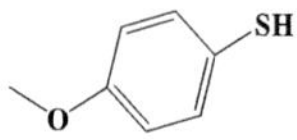	8.92
3	4-Tert-butylbenzyl alcohol	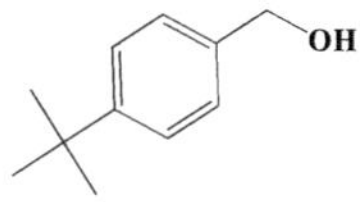	6.17
4	5-Ethyl-1,2,3,4-tetramethyl-1,3-cyclopentadiene	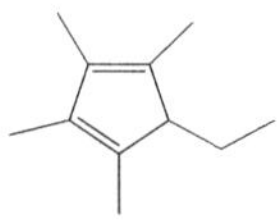	2.55
5	1,2,3,4,5,6-Hexamethyl-1,3-cyclohexadiene	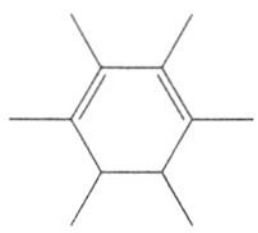	2.04
6	1-(1-Cyclohexen-1-yl)-ethanone	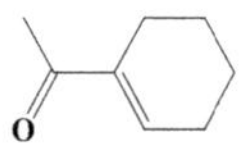	1.77
7	1,2,3,4,5-Pentamethyl-1,3-cyclopentadiene	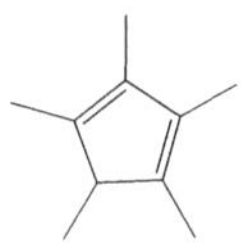	1.58
8	2,5-Dimethylphenyl acetate	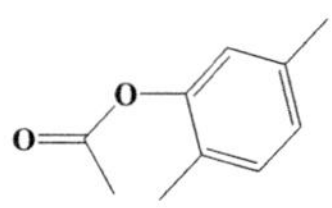	1.42
9	2,2,4-Trimethyl-hexane	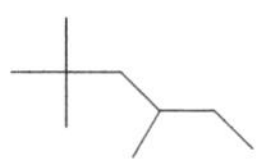	1.26
10	(4E,6Z)-2,6-Dimethyl-2,4,6-octatriene	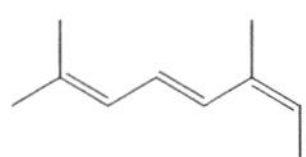	1.16
11	1-Butyl-4-methoxy-benzene	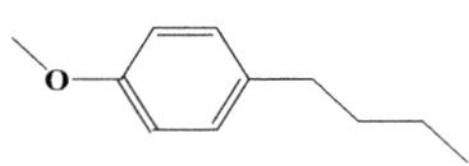	1.15
12	Methyl-cyclopentane	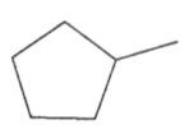	0.98
13	1,2,3-Trimethyl-cyclopentene	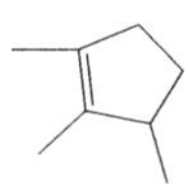	0.73

Continued

TABLE 5.9 Main compounds in ASO—cont'd

No.	Compounds	Structure	Content (wt%)
14	2-Ethyl-4,5-dimethyl-phenol	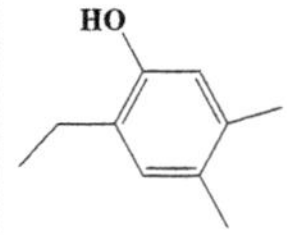	0.70
15	2-(4-Methoxyphenyl)-2-methyl-1,3-dioxolane	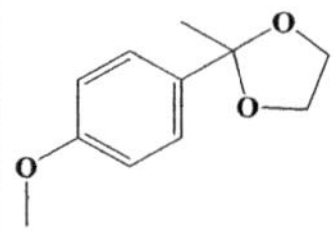	0.67
16	1,3-Dimethyl-4,5,6,7,8,8a-hexahydro-2(1*H*)-azulenone	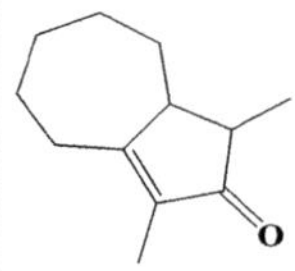	0.61
17	2,4-Dimethyl-3-hexene(c,t)	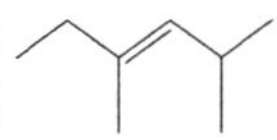	0.61
18	1,3-Cyclohexadiene, 1,5,5,6-tetramethyl-	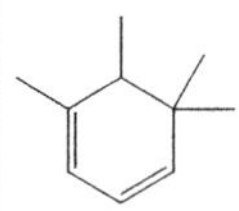	0.59
19	Sulfurous acid, decyl pentyl ester	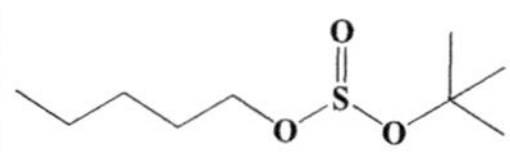	0.17
20	2H-Inden-2-one, 1,4,5,6,7,7a-hexahydro-7a-methyl-, (S)-	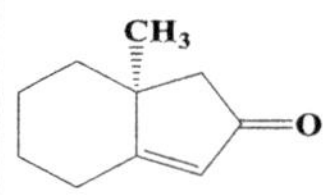	0.03
Total content			42.83

Data from Ref. Z.H. Ma, W.Z. Zheng, W.Z. Sun, et al., Enhanced catalytic performance of H_2SO_4-catalyzed C4 alkylation by formyl functional $[N_{1,1,1,1}][C_{10}SO_4]$ additive, AIChE J., 69 (2023) e18179.

In Xin's research, the recyclability of the catalyst was assessed by examining the distribution, selectivity, RON, and TMP/DMH ratio of the alkylates [63]. The RON value and TMP/DMH ratio serve as crucial metrics for evaluating the efficacy of alkylation catalysts. A higher RON value or greater TMP/DMH ratio is preferred. It can be inferred that the impact of ASO on the alkylation process primarily hinges on its components with relatively elevated concentrations. The effects of ASO on the alkylate distribution and the catalytic performances are shown in Fig. 5.30.

In Fig. 5.30, during the high catalytic activity stage, there is a notable increase in the proportions of C_8 and TMP, while the fraction of C_9^+

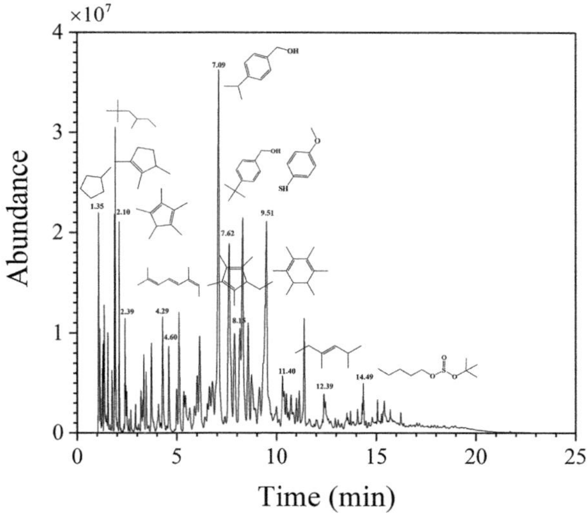

FIG. 5.29 GC-MS spectra of ASO. *From Z.H. Ma, W.Z. Zheng, W.Z. Sun, et al., Enhanced catalytic performance of H_2SO_4-catalyzed C4 alkylation by formyl functional [N1,1,1,1][$C_{10}SO_4$] additive, AIChE J. 69 (2023) e18179.*

decreases as the quantity of ASO rises. The presence of ASO exhibits positive effects on the alkylation process when its concentration in the acid is relatively low. Acting as a surfactant and a reservoir of hydrogen ions, ASO facilitates the acceleration of alkylation reactions, leading to enhanced selectivity of TMP components [66]. Additionally, ASO has the potential to improve the solubility of isobutane in H_2SO_4. The involvement of ASO in various reaction sequences can be summarized briefly. Initially, upon entering the acidic phase, isobutane could interact with ASO ions to generate tert-butyl ions through Eq. (5.13).

ASO^+ +  $\rightleftharpoons$ ASO + $C4^+$ (5.13)

The generated tert-butyl ions then promptly react with isobutylene, yielding $TMPs^+$, which may undergo isomerization. Ultimately, TMPs are formed through the transfer of hydride ion from appropriate ASO groups to $TMPs^+$ as depicted in Eq. (5.14):

$$ASO + C_8{}^+ \rightleftharpoons ASO^+ + C_8 \tag{5.14}$$

Secondly, both the acid and ASO ions participate in the dehydrogenation of a portion of isobutane, yielding isobutylene at the same time. The generated isobutylene then reacts with tert-butyl ions to form TMPs. Thirdly, the tert-butyl cation and C4 olefin can undergo a reaction, resulting in the cation with the C_8 side group attached to ASO [73]. It is worth

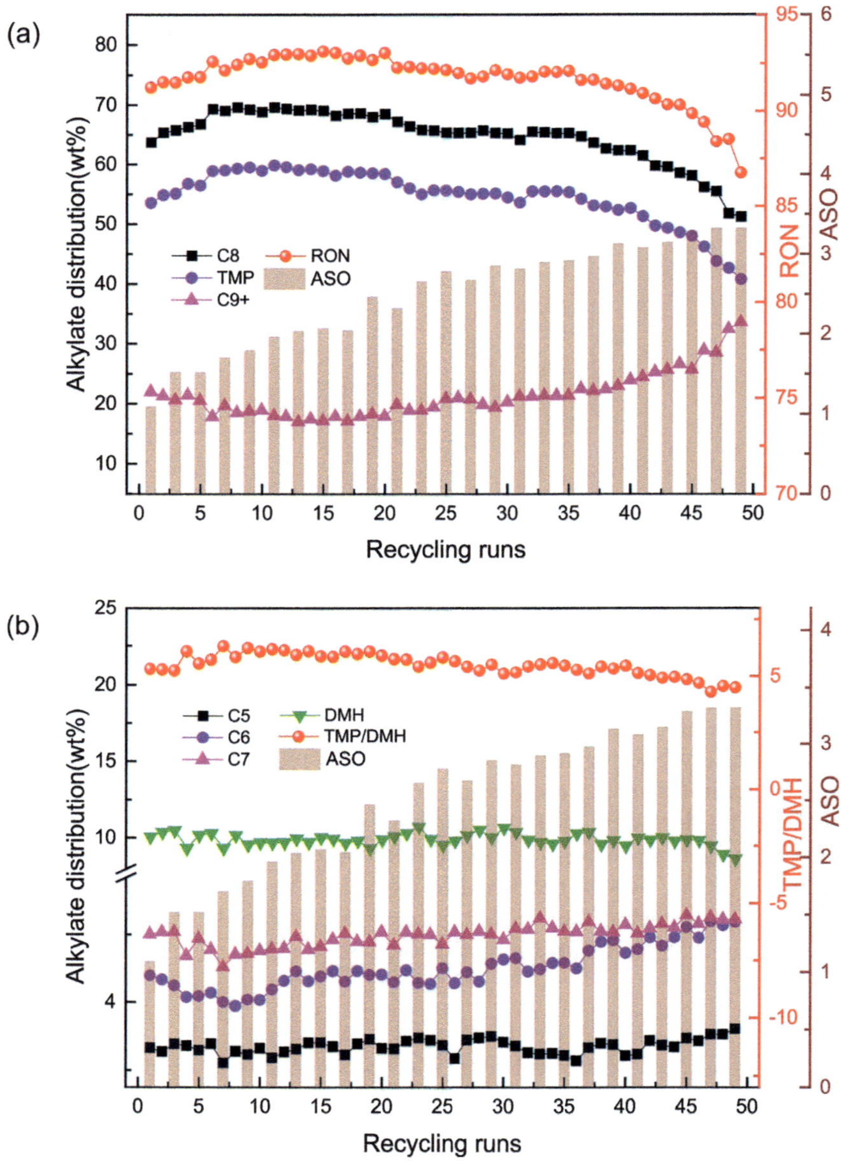

FIG. 5.30 Effects of ASO content on alkylate quality. *From Y.Y. Xin, Y.F. Hu, L. Mengchen, et al., Isobutane alkylation catalyzed by H_2SO_4: Effect of H_2SO_4 acid impurities on alkylate distribution, Energ. Fuel 35 (2021) 1664–1676.*

noting that detachment of this side group can yield the TMP molecule. Due to their higher solubility in the acid compared to isobutane, olefins should dissolve more rapidly in the acid from their mixtures with isobutane. Thus, at the reaction interface, protons transfer from the acid to various unsaturated groups, including olefin groups and ASO hydrocarbons, producing different types of hydrocarbon cations such as ASO^+ [73]. When this transfer occurs within the acid phase, the rate of hydride transfer may be accelerated. As isobutane enters the acid phase later in the reaction, the predominant flow is from isobutane to ASO through Eq. (5.13). Additionally, some ASO molecules contain tertiary C–H bonds, including ASO 1, 2, 6, 9, 13–15, 17, etc., as listed in Table 5.8. It is believed that TMP formation occurs via the transfer of hydride ions from appropriate tertiary C–H bonds through Eq. (5.15). Under these conditions, the net flow of hydride ions is presumed to be from ASO to alkyl cations.

$$\geq C{-}H + C_8H_{17}^{+} \rightleftharpoons \geq C^{+} + C_8H_{18} \tag{5.15}$$

ASO can function as a reservoir for hydride ions, with initial outward and subsequent inward flows observed during the alkylation process. Hydride ion movement between different portions of ASO is also noted as the alkylation progressed.

A direct correlation exists between alkylate quality and isobutane solubility in the presence of ASO, with even small traces of ASO compounds sufficient to eliminate the induction effect in the alkylation process [62,74]. ASO is believed to possess surface-active properties in the two-phase system, enhancing the solubility of isobutane in the acid phase [62,74–76]. ASO molecules containing hydroxyl groups, such as ASO 8, 9, 10, 12, 13, and 18 in Table 5.8, exhibit both hydrophilic and lipophilic characteristics, potentially positively impacting isobutane solubility in the acid phase.

In the gradual deactivation stage, the concentrations of C_5, C_6, C_7, and C_9^+ increase as ASO continues to form in the acid phase. The rise in light ends (C_5, C_6, and C_7) and heavy ends (C_9^+) is notably influenced by ASO growth, with heavy ends showing a significant increase alongside ASO accumulation. Oligomerization reactions proceed at a rate 4 times faster than hydrogen transfer, suggesting that heavy ends are predominantly formed through polymerization reactions involving ASO olefins [68,77].

During the final deactivation stage, a sharp decline in selectivity of high-octane TMPs components is observed after the 45th recycle run. This decline impacts the RON value and TMP/DMH ratio while HEs (C_9^+) and LEs (C_5–C_7) increase noticeably. ASO accumulation primarily causes catalyst deactivation by diluting the acid and reducing its strength, thereby accelerating the formation of C_9^+ and ASO. These changes ultimately compromise the alkylate quality and shorten the lifetime of the catalyst.

In the optimal concentration range (approximately 1%–2%), ASO can enhance the generation of tert-butyl ions and quicken the hydride ion transfer rate. Moreover, olefins may interact with ASO to generate tert-butyl cations instead of reacting with other olefins, thereby reducing the formation of undesired HEs during the initial stage of the alkylation process [73]. Additionally, ASO can impact the viscosity and surface tension of H_2SO_4, facilitating the solubility of isobutane within the catalytic system. Nevertheless, as depicted in Fig. 5.30, surpassing an ASO concentration of 2.5% notably affects the quality of the alkylate. Therefore, it is advisable to confine the appropriate ASO quantity within the range of 1.5%–2.5% for optimal performance.

Albright et al. recognized and concluded the impact of water on alkylation [78]. Acid containing 0–5 wt% water and 0–15 wt% ASO were used as the catalyst. Through the continuous process, the optimal water content is determined to be 0.5%–1.0% with high-quality alkylate at acid strengths as low as 85 wt%. The presence of water in the acid phase of alkylation affects both the viscosity and solubility of isobutane. The solubility of isobutane decreases slightly as the water content raises.

The presence of water in the acid phase of alkylation affects both the viscosity and solubility of isobutane. It is crucial to maintain a specific water content within the acid, as deviations from this narrow range can lead to undesirable outcomes. Insufficient water can hinder the ionization reaction, resulting in a lower yield of alkylate. Conversely, excessive water promotes undesired reactions, leading to increased levels of LEs, DMHs, and HEs, which compete with the formation of the desired components of TMPs. Xin et al. made a comprehensive explanation about the role of water in alkylation [63]. The effects of H_2O on the alkylate distribution and catalytic performances are shown in Fig. 5.31.

Owing to the strong ability of water absorption, the existence of water mainly decreases the acid strength of H_2SO_4. However, on the other side, the presence of H_2O plays a positive role in alkylation during the catalytic activity strengthening stage with the water content ranging between 2.0 and 4.0 wt%. The minor amount of H_2O is beneficial for the ionization of H_2SO_4 to H^+, which is a critical step for enhancing the acid strength of H_2SO_4 [78]. On the other side, the presence of a small quantity of H_2O accelerates the formation of carbocation and proton transfer between isobutane and the carbocation [79]. Meanwhile, the presence of H_2O contributes to a reduction in the viscosity of H_2SO_4, thereby facilitating mass transfer within the system [79]. In addition, it is noteworthy that the H_2O content surpasses that of ASO and esters, particularly in the initial stage.

With the alkylation proceeds, water gradually accumulates in the acid and accelerates the decrease of the acid strength, in which the water content is about 4.5–5.4 wt%. The further decrease of acid strength significantly affects the normal main reactions, leading to an increase in the byproduct content. Therefore, a small amount of H_2O is advantageous in the initial stages of the alkylation reaction process; numerous adverse effects become more pronounced as the H_2O content exceeds a certain

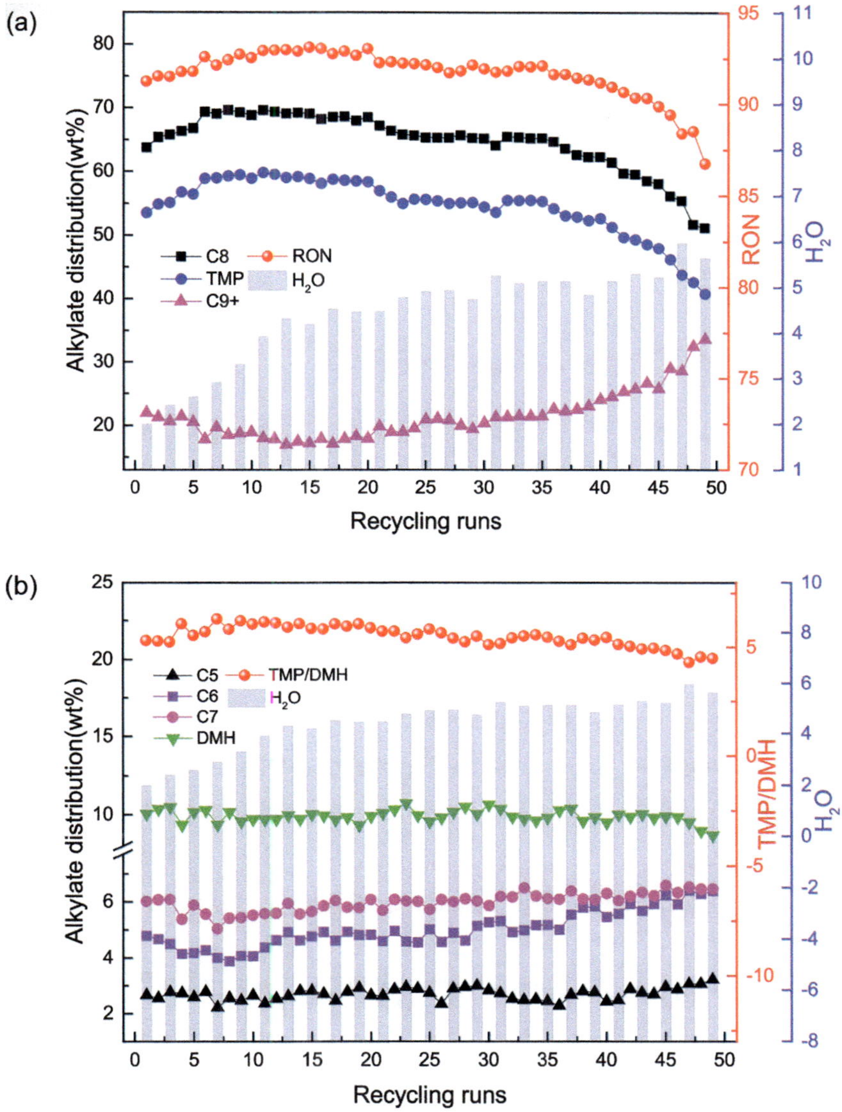

FIG. 5.31 Effects of H_2O content on alkylate quality. *From Y.Y. Xin, Y.F. Hu, L. Mengchen, et al., Isobutane alkylation catalyzed by H_2SO_4: Effect of H_2SO_4 acid impurities on alkylate distribution, Energ. Fuel 35 (2021) 1664–1676.*

threshold. The present study indicates that the optimal amount of H_2O ranges from 2 to 4 wt%.

The impurities existing H_2SO_4 mainly include ASO, H_2O, and esters. The effects of esters on the alkylate distribution and catalytic performances are shown in Fig. 5.32. According to the analysis of composition

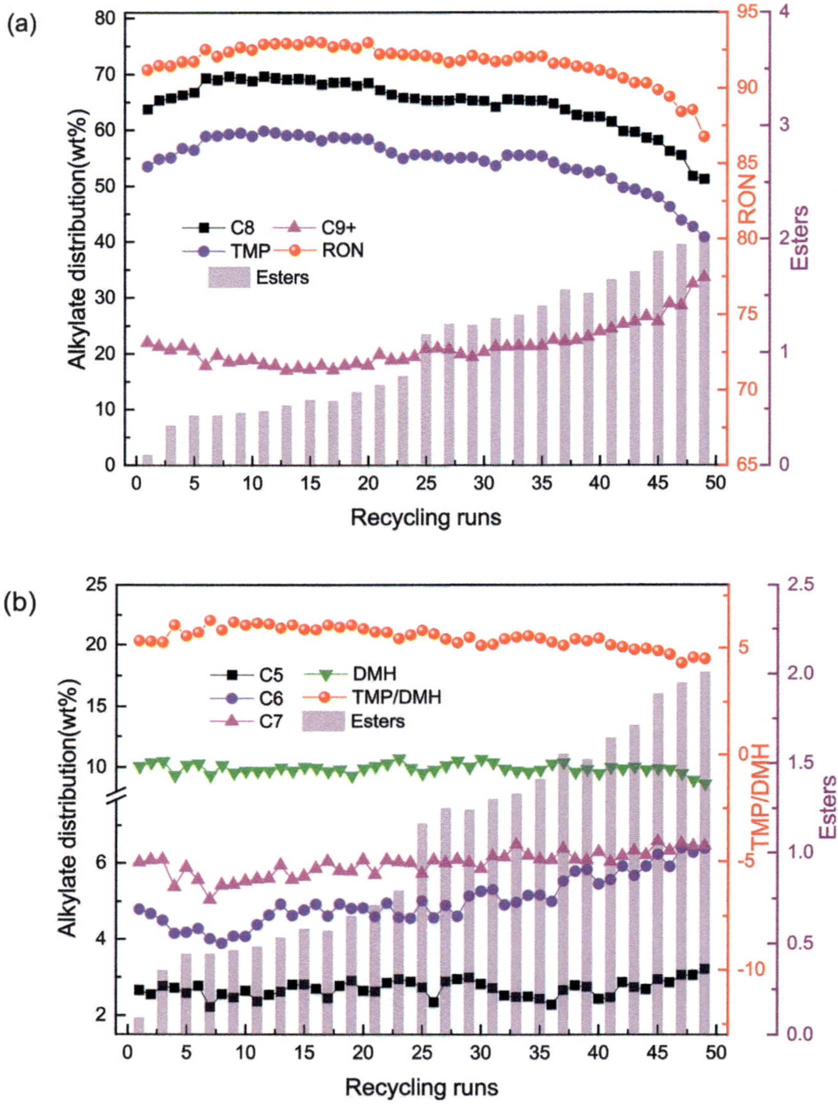

FIG. 5.32 Effects of ester content on alkylate quality. *From Y.Y. Xin, Y.F. Hu, L. Mengchen, et al., Isobutane alkylation catalyzed by* H_2SO_4*: Effect of* H_2SO_4 *acid impurities on alkylate distribution, Energ. Fuel 35 (2021) 1664–1676.*

of used acid, the content of esters in acid is smaller than the contents of H_2O and ASO in the initial stage. Besides, compared to H_2O and ASO, the accumulation rate of the esters content is slower.

In the initial phase, olefin exhibits a significantly higher dissolution rate compared to alkane. Consequently, an olefin-rich environment develops above the surface of the acid phase. Esters are generated in this setting, likely through the initial protonation of olefin. This protonation leads to the formation of a secondary carbonium ion, which rapidly combines with the acid, resulting in the production of esters [60,72,80–83]. These esters are soluble in both alkanes and acid, which can act as anionic surfactants with both hydrophilic and lipophilic properties. This feature facilitates the solubility of reactants, representing the primary advantageous effect of esters in the alkylation process. Second, the production of ester intermediates is a part of the initial reactions. The higher yields of esters play a crucial role in preventing oligomerization reactions, consequently minimizing the production of byproducts such as conjunct polymers to a negligible extent. The notable presence of ester intermediates, attributed to the absence of free acid, significantly influences the selectivity toward the desired alkylate for gasoline.

In the gradual deactivation stage, the alkylate quality is decreased with the increase in the content of esters. As depicted in Fig. 5.32, the ester content notably increases compared to other stages. This can be attributed to the gradual decrease in the acid strength, which enhances ester stability and favors their accumulation due to a shifted equilibrium in the reversible reaction with acid [44].

The rapid ester formation, surpassing carbocation isomerization rates, leads carbocations to react with esters, forming heavier carbocations, contributing to the growth of HEs. These HEs enter the alkylate phase or the acid phase as ASO. Furthermore, the presence of esters not only deactivates the acid catalyst but also hinders acid regeneration [81]. Studies indicate that higher ASO levels, along with poor-quality gasoline, SO_2, and H_2O, are produced when acid-containing esters are stored at room temperature [47,84]. Ester decomposition rates increase with temperature, potentially forming hydroxyl groups during hydrolysis, promoting the solubility of ASO and undesired products in the acid phase.

To summarize, in the initial stages, a small quantity of esters (approximately 0.4–0.7 wt%) positively influences the alkylation process, surpassing the impact of H_2O and ASO. However, in later stages, when ester levels surpass 1.5 wt%, their presence significantly raises catalyst consumption and complicates the regeneration of acid.

5.5.2 Deactivation process of Ionic liquids

As a novel catalyst for C4 alkylation, chloroaluminate ionic liquids (ILs) have been identified as a potential replacement for conventional liquid acid catalysts [85]. Chloroaluminate ILs exhibit high solubility for transition-metal salts, thereby offering the potential for catalyzing numerous organic reactions within these ILs [28]. Researchers at China University of Petroleum have achieved success in creating a composite ionic liquid (CIL) using triethylamine hydrochloride, anhydrous aluminum chloride, and cuprous chloride. This CIL demonstrates excellent catalytic efficacy in the C4 alkylation process [86,87]. Similar to the alkylation catalyzed by H_2SO_4, Ma et al. found the formation of solid and ASO during CIL alkylation, which can change the structure of CIL and affect the catalytic performance of CIL [88].

5.5.2.1 Deactivation of Brønsted acid in CIL

Indeed, as early as 1945, Schmering et al. noted that chlorohydrocarbons act as intermediates in the formation of carbonium ions during alkylation processes. Hence, the production of chlorohydrocarbons serves as intermediates in Eq. (5.16). HCl plays a crucial role as a proton donor in this context. The chlorobutanes derived from the reaction between HCl and butene are pivotal intermediates for the formation of carbonium ions. Consequently, the loss of these intermediates from the reaction system, due to their departure with the alkylate, leads to a reduction in Brønsted acidity. Consequently, this reduction contributes to the deactivation of CIL since these intermediates cannot be transformed into carbonium ions without the assistance of Lewis acidic $Al_2Cl_7^-$.

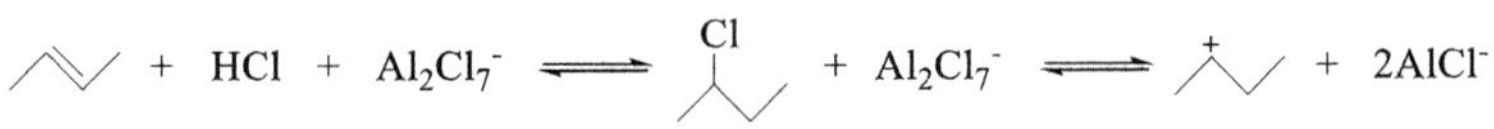

(5.16)

During several separation stages, Ma et al. separated utilized CIL into liquid and solid phases [88]. Various analytical techniques including nuclear magnetic resonance, X-ray diffraction, X-ray photoelectron spectroscopy, and elemental analyses were employed to characterize these fractions. Furthermore, the concentration of ASO in the solid phase was examined. It is revealed that the solid phase primarily comprises CuCl with approximately 1 wt% of ASO. The reduction in CIL acidity and product selectivity is attributed to the loss of aluminum chloride and the separation of CuCl in solid form from CIL.

Zhang et al. investigated the deactivation mechanism and the activity recovery approach of CIL [89]. It is pointed out that the deactivation of CIL occurred due to the loss of Brønsted acidity and/or Lewis acidity. The decrease in Brønsted acidity is attributed to the loss of trace amounts of chlorohydrocarbons formed during the outflowing alkylate. On the other hand, the deactivation of Lewis acidity mainly stems from the complexation of the formed ASO with the anions of CIL (Fig. 5.33).

As shown in Table 5.10, alkylate contains certain levels of chlorides. Treatment with 0.1 M NaOH solution removed 4%–7% of the total chloride, indicating minimal inorganic chloride presence. Thus, organic chlorides constitute the primary component. Table 5.11 displays chloride levels in alkylate boiling fractions postcaustic washing, with approximately 91% of total organic chlorides concentrated in the narrow fraction

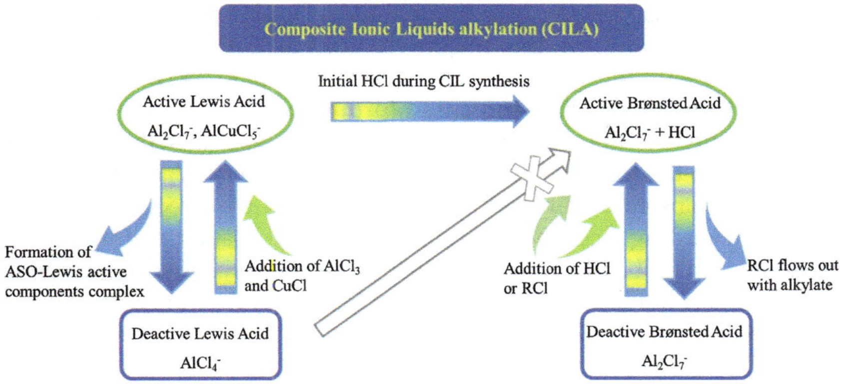

FIG. 5.33 Deactivation mechanism and activity-recovery approach of CIL. *From X. Zhang, R. Zhang, X. Meng, et al., Deactivation mechanism and activity-recovery approach of composite ionic liquids for isobutane alkylation, Appl. Catal. A-Gen. 557 (2018) 64–71.*

TABLE 5.10 Content of the chlorides in alkylate.[a]

Chloride content	A	B	C	D	E
Before caustic wash (mg/L)	326	274	236	250	240
After caustic wash (mg/L) (organic chlorides)	309	262	220	240	223
Removal percentage (%) (inorganic chlorides)	5.3	4.6	6.8	4.3	7.1

[a] *20 mL alkylate washed with 25 mL 0.1 mol/L NaOH for 0.5 h at 25°C; samples from alkylation reaction with different amounts of C_4 intake.*
Data from X. Zhang, R. Zhang, X. Meng, et al., Deactivation mechanism and activity-recovery approach of composite ionic liquids for isobutane alkylation, Appl. Catal. A-Gen. 557 (2018) 64–71.

TABLE 5.11 Chloride content of alkylate boiling fractions.[a]

Boiling range (°C)	Fraction percentage (wt%)	Chloride content (mg/L)	Chloride percentage (wt%)
<58	5.8	3500	91.2
58–90	5.2	49	1.1
90–95	16.5	4	0.3
95–109	65.0	10	2.9
>109	7.5	129	4.3

[a] *The chloride content of alkylate after washing with caustic is 223 mg/L (sample E from Table 5.10).*
Data from X. Zhang, R. Zhang, X. Meng, et al., Deactivation mechanism and activity-recovery approach of composite ionic liquids for isobutane alkylation, Appl. Catal. A-Gen. 557 (2018) 64–71.

below 58°C, suggesting predominantly volatile chlorohydrocarbons. GC analysis identifies the major chlorohydrocarbons in alkylate as 2-chloro-2-methylpropane (15 wt%), 2-chlorobutane (42 wt%), 1-chloro-2-methylpropane (9 wt%), and chloroform (18 wt%).

Consequently, the deactivation of Brønsted acid in CIL is primarily attributed to the loss of chlorohydrocarbons as the alkylate exits the system. The Brønsted acidity of CIL can be restored by directly introducing HCl or by adding a compound that produces HCl, such as minute quantities of methanol and water, or by incorporating chlorohydrocarbons like *t*-butyl chloride [90,91] or the chlorohydrocarbon-rich (e.g., low) boiling fraction of alkylate [92].

5.5.2.2 Deactivation of Lewis acid in CIL

The active Lewis acid components in CIL, such as $Al_2Cl_7^-$ and $AlCuCl_5^-$, undergo reactions with water and methanol in the feed, leading to the formation of solids and HCl [93]. Therefore, it is anticipated that drying and deoxygenating the feed can prevent solid formation. However, when the feed is dried and deoxygenated but includes Brønsted acid additives like HCl and *t*-butyl chloride, abundant solids are produced. This observation suggests that solid formation is not primarily due to impurities in the C4 feed but is associated with the deactivation or alteration of the Lewis acid in the CIL.

In Fig. 5.34, it is evident that C_8 selectivity and alkylate RON experience a significant decline when butene conversion is below 100%. At the same

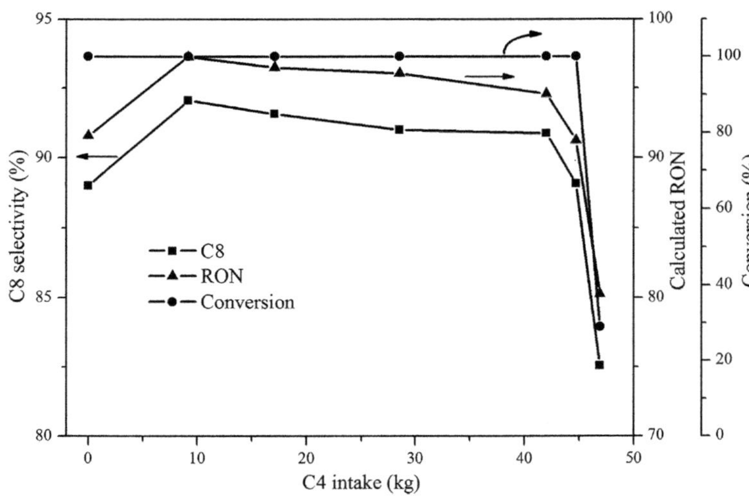

FIG. 5.34 C_8 selectivity, alkylate RON and butene conversion as function of C4 intake. (CIL intake 200 g, I/O ratio 20:1, C4 flow rate 750 mL/h, stirring speed 1200 rpm, 47.7 mg HCl/kg C4 feed). *From X. Zhang, R. Zhang, X. Meng, et al., Deactivation mechanism and activity-recovery approach of composite ionic liquids for isobutane alkylation, Appl. Catal. A-Gen. 557 (2018) 64–71.*

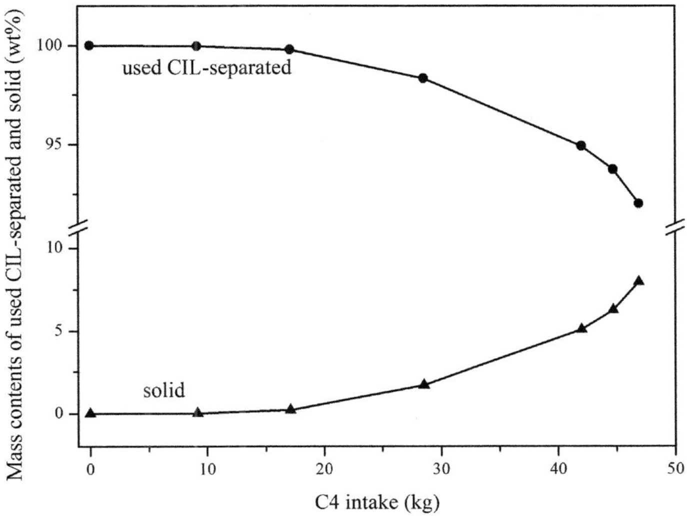

FIG. 5.35 Amount of the used CIL-separated and solid as function of C4 intake. (CIL intake 200 g, I/O ratio 20:1, C4 flow rate 750 mL/h, stirring speed 1200 rpm, 47.7 mg HCl/kg C4 feed). *From X. Zhang, R. Zhang, X. Meng, et al., Deactivation mechanism and activity-recovery approach of composite ionic liquids for isobutane alkylation, Appl. Catal. A-Gen. 557 (2018) 64–71.*

time, CIL is consumed, and solid accumulates with an increasing C4 intake (Fig. 5.35). The quantity of ASO rises sharply as CIL approaches deactivation (Fig. 5.36). Studies indicate that the main cause of alkylation catalyst deactivation is the formation of ASO. ASO is predominantly found in the used CIL-separated phase, with minimal ASO present in the solid. However, the formation of solid increases noticeably as the ASO content in the used CIL-separated phase rises [88,94,95].

Thus, as the C4 intake increases, there is a noticeable rise in the chloride content in the alkylate, solid content in the reaction system, and ASO content in the used CIL-separated phase. The presence of ASO can deactivate CIL through electron donor/acceptor interactions, forming complexes with $Al_2Cl_7^-$. This interaction leads to the formation of ASO-$Al_2Cl_7^-$ complexes, which hinders the conversion of chlorohydrocarbons with $Al_2Cl_7^-$ into carbonium ions, ultimately leading to the deactivation of CIL.

5.5.3 Deactivation process of solid acid

The preceding text has already introduced the entire process of alkylation. The solid acid-catalyzed C4 alkylation reaction mainly includes elementary reactions such as protonation [96,97], addition reaction [98],

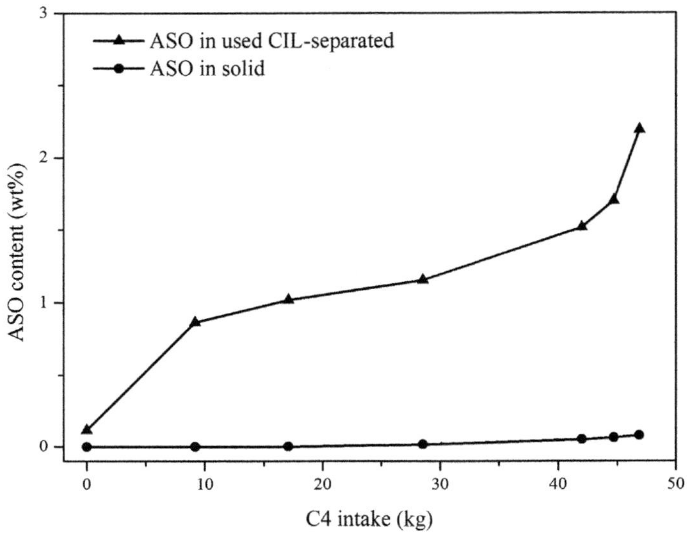

FIG. 5.36 ASO content as function of C4 intake. (CIL intake 200 g, I/O ratio 20:1, C4 flow rate 750 mL/h, stirring speed 1200 rpm, 47.7 mg HCl/kg C4 feed). *From X. Zhang, R. Zhang, X. Meng, et al., Deactivation mechanism and activity-recovery approach of composite ionic liquids for isobutane alkylation, Appl. Catal. A-Gen. 557 (2018) 64–71.*

intermolecular hydrogen ion transfer reaction, intramolecular hydrogen transfer reaction, intramolecular methyl transfer reactions, β-scission reaction [99,100], and deprotonation reaction. When using zeolites as catalysts, the process is as follows [101–103]: Protonation of olefins takes place via a carbocation transition state. The positively charged central carbon atom of the carbocation interacts with the negatively charged active oxygen atom in the zeolite framework, resulting in the formation of an alkoxyl group. Because the central carbon atom of the alkoxyl group still carries a positive charge, it can undergo subsequent reactions with electron-donating groups. This involves hydride transfer reactions with isobutane or addition reactions with olefins. Specifically, the detailed process of hydride transfer between the tert-butyl alkoxyl group (TBA) and isobutane is as follows: free isobutane molecules first enter the zeolite channels to form adsorbed isobutane molecules. After overcoming a certain energy barrier, TBA forms the transition state of tert-butyl carbocation. Subsequently, this transition state of tert-butyl carbocation reacts with isobutane to form an intermediate product sharing hydrogen atoms undergoing rotation. The entire process of hydride transfer is symmetrically completed. The specific process of the addition reaction between TBA and isobutene is as follows: free isobutene molecules first enter the zeolite channels to form adsorbed isobutene molecules. TBA overcomes a certain energy barrier, forming the transition state of tert-butyl carbocation. The transition state of tert-butyl carbocation reacts with isobutene, leading to the formation of C8 carbocation, which rotates at the active site, adsorbing onto the active oxygen atom with a new central C atom, ultimately generating the C8 alkoxyl group, thereby completing the addition reaction.

Subsequently, the C8 alkoxyl group undergoes skeletal isomerization and hydrogen ion transfer with isobutane, resulting in the formation of the alkylate product, trimethylpentanes (TMPs). Moreover, the C8 alkoxyl group can further participate in addition reactions with olefins, leading to the formation of higher molecular weight alkoxyl groups. These alkoxyl groups may undergo deprotonation reactions, resulting in the formation of high-molecular-weight olefins that adsorb onto active sites, effectively covering these sites. In contrast, they can undergo hydride transfer reactions with isobutane, producing large-molecule alkanes that cannot diffuse out of the catalyst pores, thereby obstructing the catalyst pores. Carbonaceous precursors cannot rapidly desorb from the catalyst surface, resulting in their coverage on active sites or the blockage of pores and a gradual decline and eventual deactivation of catalyst activity.

5.5.3.1 Characterization of deactivated catalysts

The main reason for the rapid deactivation of solid-acid catalysts of C4 alkylation is the competing reaction of conjugate oligomerization of olefins during their interaction with strong acids. Deactivation can be slowed down by choosing appropriate reaction conditions and reactor design, but eventually all catalysts have to be regenerated. To characterize the deposits and the deactivated catalysts, a wide variety of methods have been employed, ranging from ^{13}C CP MAS NMR, ^{1}H NMR to mass spectrometry, IR-spectroscopy, Solid-state UV/Vis spectroscopy, GC-MS, MALDI-TOF MS, Confocal fluorescence microscopy, In situ UV/Vis microspectroscopy, and Space- and Time-resolved in situ spectroscopy. MALDI-TOF MS is primarily utilized for detecting protonated and carbonized species. In the analysis of heterocyclic and unsaturated compounds, because unsaturated compounds undergo more substantial ionization than saturated hydrocarbons, MALDI-TOF MS is limited to detect only a small fraction of unsaturated compounds [104,105]. ^{13}C CP MAS NMR can be employed to detect the composition of carbonaceous precursors, but it cannot provide quantitative results, often resulting in underestimated measurements, especially for aromatic species [104]. Solid-state UV/Vis spectroscopy enables the detection of specific organic groups, including diene and triene functionalities [104]. Confocal Fluorescence Microscopy is commonly used to evaluate structural changes in zeolites at different stages of FCC coking [106]. Space- and time-resolved in situ spectroscopy is mainly employed for detecting carbon species exhibiting fluorescence activity. This technique combines high-temperature in situ spectroscopy with microwave technology, distinguishing graphitic carbon on the external surface of zeolites and determining aromatic components within the zeolite channels [107]. In situ UV/Vis microspectroscopy is employed to determine the types and generation rates of coke species during catalytic reactions [108].

Diaz-Mendoza et al. [109] proposed that coke formation during the isobutane/butene alkylation reaction over zeolites decreases the acidity and acid strength of the catalysts after probing the changes in the acid properties caused by the deactivation processes by microcalorimetric measurements of the differential heat of pyridine adsorption and FTIR spectroscopy of adsorbed pyridine. A fraction of the intermediate strength sites is lost for all deactivated catalysts studied (specially for Beta), but some strong sites are still available for deactivated USY and Beta. This evidence supports a contribution of pore blocking or pore filling (deactivation model II) to the deactivation of these zeolites. The deposits formed have a strong paraffinic character, but evidence for olefin species is also observed. The degree of unsaturation of the surface species formed increases from Beta to USY zeolite, implying that the presence of a high concentration of strong Lewis acid sites promotes the formation of unsaturated compounds.

The deactivation of rare-earth-exchanged zeolite X as catalysts in isobutane/2-butene alkylation between 40°C and 130°C was studied by Feller and his coworkers [110]. The deactivated samples and the isolated deposits were analyzed by a range of techniques, including for the first time (matrix assisted) laser desorption/ionization time-of-flight mass spectroscopy (MALDI-TOF MS, LDI-TOF MS). The compounds found in the deactivated zeolites are, in addition to large alkanes and alkenes, highly unsaturated and highly branched species containing cyclic structures, which are increasingly aromatic as the reaction temperature increases. The deposits in part interact strongly with the acid sites and block the sites for further alkylation reactions. Their structure and route of formation resemble those of conjunct polymers formed in liquid acid-catalyzed alkylation, as listed in Eqs. (5.17) and (5.18).

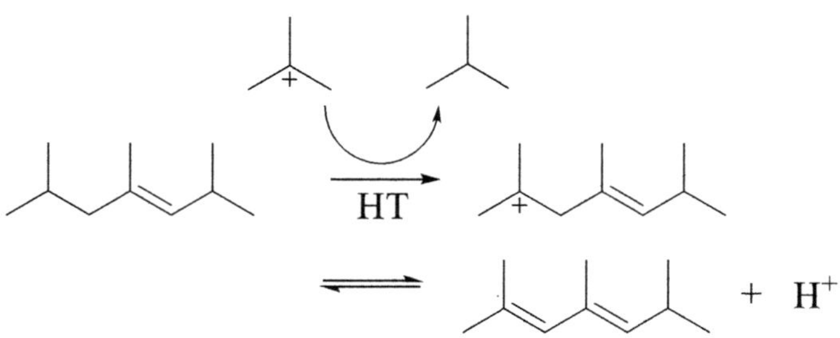

(5.17)

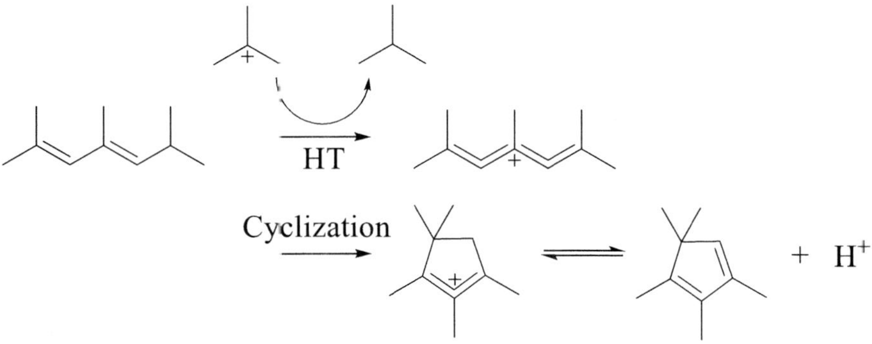

(5.18)

Flego et al. [111] proposed various types of surface species as components and precursors of carbonaceous deposits: alkyl and cycloalkyl carbenium ions, alkenyl and cycloalkenyl carbenium ions, aryl carbocations, radicals, radical ions, simple and condensed aromatics, as shown in Fig. 5.37. Specifically, the authors examined the carbonaceous deposits formed on LaHY-FAU during the semibatch reactor process at 353 K for alkylating isobutane with 1-butene have by in situ UV/Vis, FT-IR, and ^{13}C CP/MAS NMR spectroscopies. The results of UV/ViS (see Fig. 5.38) support unsaturated carbenium ions immediately on adsorption of 1-butene and mixture of 1-butene and isobutane. Due to the presence of monoenyl carbenium ions, the band peak at 315 nm is most intense for the sample taken after the catalytic run at 353 K. The monoenyl carbenium ions are the predominant ions in the residue on the catalyst. In the spectrum of the catalyst used at higher temperature, the band near 370 nm, assigned to dienyl carbenium ions , possesses the same intensity as the monoenyl one. In addition, at higher reaction temperatures, more alkenyl carbenium ions with greater unsaturation are generated.

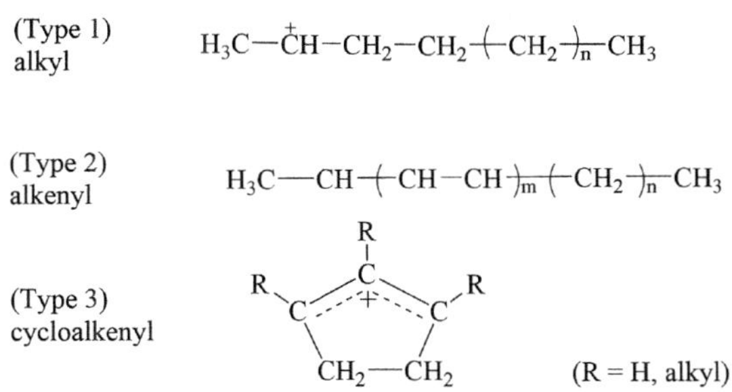

FIG. 5.37 Principal types of carbonium ions formed during parasitic alkene oligomerization reactions at 353 K. *From C. Flego, I. Kiricsi, W.O. Parker, et al., Spectroscopic studies of LaHY-FAU catalyst deactivation in the alkylation of isobutane with 1-butene, Appl. Catal. A-Gen. 124 (1995) 107–119.*

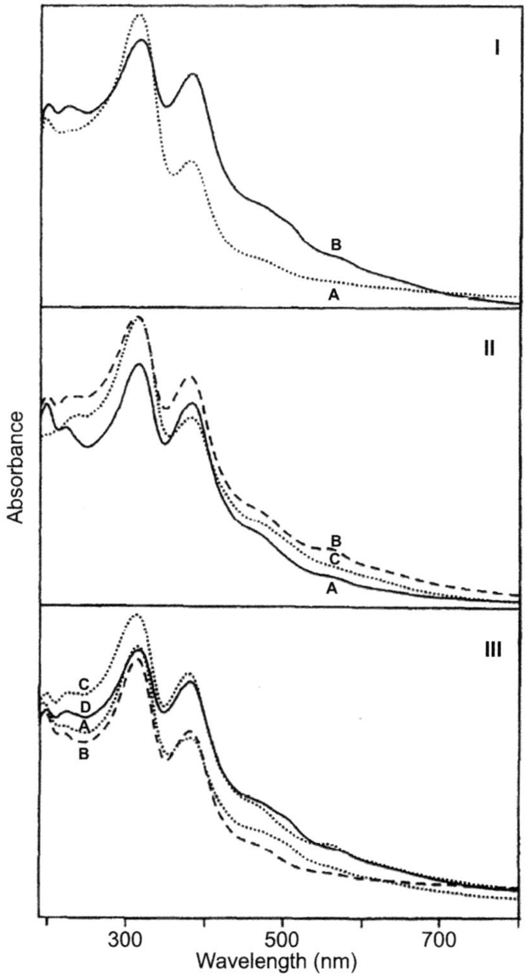

FIG. 5.38 UV-vis spectra of LaHY-FAU used in a semibatch reactor for 7h with (standard conditions) $T_{act}=623\,K$, feed=1-butene/isobutane (1:9), $T_{react}=353\,K$, $P=30\,bar$ and $LSHV=4.07\,h^{-1}$ unless stated otherwise. *From C. Flego, I. Kiricsi, W.O. Parker, et al., Spectroscopic studies of LaHY-FAU catalyst deactivation in the alkylation of isobutane with 1-butene, Appl. Catal. A-Gen. 124 (1995) 107–119.*

The analytical methods mentioned above are capable of analyzing various components, including heterocyclic and unsaturated compounds, aromatic species, specific organic groups (such as diene and triene functional groups), and carbon species exhibiting fluorescence activity. These methods offer new means for quantitatively analyzing the composition of high-carbon-number organic compounds in carbonaceous precursors. Currently, the analysis of the single-component composition and content of high-carbon-number organic compounds in carbonaceous precursors remains a challenge, requiring continued exploration of more precise analytical techniques for carbonaceous precursors.

5.5.3.2 *Deactivation mechanism of solid acid*

The solid acid alkylation catalysts for isobutane/isobutene mainly include zeolite catalysts, solid superacid catalysts, supported heteropolyacid catalysts, and metal halide catalysts. They all exhibit good reactivity in alkylation reactions, but as mentioned earlier, catalyst deactivation occurs, leading to a relatively short lifespan. The deactivation mechanisms vary among different solid acid alkylation catalysts.

For zeolite catalysts, there are primarily two deactivation mechanisms. One involves the deposition of carbonaceous precursors covering the active centers of the zeolite, and the other involves carbonaceous precursors blocking the zeolite channels. The different channel structures result in distinct deactivation modes. In the alkylation reaction between isobutane and 2-butene, LTL zeolites mainly deactivate through channel blocking, whereas ZSM-12 deactivates mainly due to the coverage of active sites [106]. In the case of USY zeolite, accumulated substances (carbonaceous precursors) primarily adsorb in the cage-like structures of the zeolite, leading to a decrease in the catalyst surface area and pore volume, resulting in decreased catalyst activity [112]. In the case of rare-earth-exchanged X zeolite catalyzing isobutane/isobutene alkylation, the covering of acidic centers on the zeolite is the main cause of catalyst deactivation [110]. For LaX zeolite catalysts, catalyst deactivation results from a combination of carbonaceous precursors covering the active centers and blocking the catalyst channels [104]. Shiriyazdanov et al. showed the deactivation mechanism, as listed in Eqs. (5.19) and (5.20) [113]. The main products of conjugate oligomerization are alkanes and compounds with conjugated double bonds, especially with alkylcyclopentadienyl (ACPD) structures. Conjugate oligomerization, as the process of carbon and hydrogen redistribution, is generally represented as a combination of consecutive stages of classical oligomerization and cracking. The formation of ACPD and related compounds from olefins is the main reason for the deactivation of alkylation catalysts: because of the participation in the formation of highly stable cyclopentyl cations, the acid sites at the surface of solid catalysts completely lose their power, which eliminates the possibility of implementing the functions of the hydride transfer from isobutane and disrupts the course of the target alkylation reactions.

$$\text{Alkenes } C_4\text{-}C_8 \xrightarrow{\text{Oligomerization}} \text{Alkenes } C_8\text{-}C_{16} \xrightarrow{\text{Cracking}} \boxed{\begin{array}{l}\text{Alkenes } C_4\text{-}C_8\\ +\\ \text{Alkenes } C_4\text{-}C_8\end{array}} \tag{5.19}$$

$$\text{Dienes } C_4\text{-}C_6 + \text{Alkenes } C_4\text{-}C_8 \xrightarrow{\text{Oligomerization}} \text{Polyenes} \xrightarrow{\text{Cracking}} \boxed{\begin{array}{l}\text{Alkenes } C_4\text{-}C_8\\ +\\ \text{ACPD}\end{array}} \tag{5.20}$$

Solid superacid refers to solid acid catalytic materials with Hammett acidity values (H_0) less than -11.94. Solid superacid alkylation catalysts mainly include SO_4^{2-}/M_xO_y (metal oxide) catalysts and the developed WO_3/M_xO_y and A^{n-} (acidic anion)/M_xO_y types. SO_4^{2-}/M_xO_y catalyst systems are simple to prepare and exhibit high acidity [114]. The results of alkylation reactions indicate that SO_4^{2-}/ZrO_2 and its supported catalysts have high initial catalytic activity but rapidly decline in activity afterward. Due to the high acidity of superacids, the proportion of cracking reactions significantly increases, resulting in a higher content of C5–C7 in the products and poor selectivity [115]. It is found that the primary reason for the deactivation of Brønsted-Lewis conjugate solid superacid in isobutane/isobutene alkylation is the deposition of carbonaceous precursors on the catalyst surface, which covers the active centers [116]. The composition of carbonaceous precursors is mainly long-chain alkanes, and the pore-blocking mainly occurs in the micropores and small pores of the catalyst, rather than the large pores. Additionally, carbonaceous precursor coverage is the main cause of catalyst deactivation, with rapid carbon accumulation and deactivation in the initial stages of the reaction. This study emphasizes that inhibiting the rapid carbon accumulation in the early stages is crucial to preventing catalyst deactivation [117].

Heteropolyacid (HPA) is a type of solid superacid with high acidity, noncorrosiveness, and pseudo-liquid phase properties. For the supported heteropolyacid solid acid catalysts, it was found that under supercritical reaction conditions, catalyst deactivation is caused by the covering of active centers (i.e., strong acid centers), and carbonaceous precursors do not block the catalyst pores [118]. It is found that H-Beta/Al_2O_3 deactivates during isobutane/isobutene alkylation due to the simultaneous blocking of catalyst channels by carbonaceous precursors and the covering of most medium-strength acid centers [119]. For silica-supported heteropoly acid, the specific surface area and pore volume of the deactivated catalyst decrease significantly, and the number and acidity of strong and medium-strong acid centers also decrease, which indirectly affects the activity of the catalyst [120]. It is demonstrated that the carbonaceous precursor blocks the catalyst channels and covers the moderately strong acid sites.

In the field of solid acid catalysts, research on metal halide catalysts mainly involves $AlCl_3$, BF_3, and SbF_3. Additionally, there are a few reports on $KBr/ZrCl_4$, SbF_5/FSO_3H/activated carbon, SbF_5/FSO_3H/graphite, and SbF_5RF/Al_2O_3 [121,122]. Metal halide catalysts exhibit high activity in the presence of a small amount of HCl, but they rapidly deactivate as HCl is lost. The deactivation of metal halide catalysts is mainly due to the loss of active components. Although metal halide catalysts show certain activity and selectivity in isobutane/isobutene alkylation, the active components are prone to loss, causing equipment corrosion and poor performance in continuous operations. Therefore, the application of metal halide catalysts in isobutane/isobutene alkylation is currently limited.

There are three main deactivation methods for commonly used solid acid alkylation catalysts (zeolites, solid super acids, and supported heteropolyacids), including carbonaceous precursors blocking catalyst channels, carbonaceous precursors covering active centers, and the existence of both. The deactivation ways of different types of zeolite catalysts are related to their distinct channel structures. The deactivation modes of channel blocking and active center coverage may act separately or simultaneously in the zeolites. The primary deactivation mode for solid superacid alkylation catalysts is the coverage of active centers by carbonaceous precursors. For supported heteropolyacid catalysts, the main deactivation mode is the simultaneous occurrence of channel blocking and active center coverage. The deactivation mode of metal halide catalysts is mainly the loss of active components.

5.5.4 Controlling of carbon precursors

The acid type and pore structure of zeolite both have a certain impact on catalyst deactivation. Among them, there are mainly two types of acids, namely Brønsted (B acid) and Lewis (L acid), and different types of zeolites possess distinct pore structures. Therefore, the acid properties and pore structures of zeolite are of significant importance in controlling the generation of carbonaceous precursors and delaying catalyst deactivation.

5.5.4.1 Effect of acid properties on the coke precursors

It is generally believed that the alkylation of isobutane/isobutene catalyzed by zeolite occurs predominantly on strong Brønsted acid (B acid) sites, and the activity and catalyst lifespan in the alkylation reaction is related to the strength and concentration of B acid [101,123,124].

The protonation of olefins occurs at B acid centers, forming a relatively stable intermediate through the alkoxyl group [125]. The presence of medium-strength B acid facilitates the hydride transfer reaction between C_8 carbocation and isobutane, thereby enhancing the selectivity of alkylating C_8 target products. Additionally, the distribution of B acid sites on the molecular sieve also affects the catalyst lifespan and alkylation performance. When B acid sites are uniformly distributed both inside and outside the molecular sieve, compared to the distribution solely inside, higher selectivity and yield of alkylating reaction products as well as an extended lifespan of the molecular sieve catalyst [126] are observed when B acid sites are distributed on the external surface of the molecular sieve. Research on USY zeolite reveals that after removing nonframework aluminum, both the B acid concentration and strength of USY zeolite increase, leading to enhanced alkylation reaction activity and molecular sieve lifespan [127,128]. A similar pattern is observed in the catalytic performance of NaX zeolite with different B acid concentrations in the alkylation reaction of butene [129].

5.5.4.2 Effect of structures on the coke precursors

According to the size of the pore diameter, zeolite can be categorized into large-pore, medium-pore, and small-pore zeolite. The different pore structures of zeolite result in variations in their selectivity. The distinct roles of these spatial constraints significantly impact the distribution of alkylate products, the generation of carbonaceous precursors, and the catalyst lifespan in alkylation reactions.

Corma et al. [130] investigated the alkylation performance of zeolite such as USY, β, ZSM-5, MCM-22, and silicalite, finding that the deactivation rate is related to the pore configuration of the zeolites. ZSM-5, with smaller micropores, exhibits a much higher deactivation rate compared to the other four zeolites. It is proposed that the easier blockage of micropores by carbonaceous precursors generated in low-polymerization reactions of olefins within smaller micropores leads to catalyst deactivation. Additionally, Corma et al. [69] suggested that three-dimensional pore structure zeolites are more conducive to product diffusion than one-dimensional pore structure zeolite, resulting in better alkylation performance. However, Yoo et al. [131] compared the alkylation performance of zeolite such as USY, β (three-dimensional pore), mordenite (two-dimensional pore), LTL, and ZSM-12 (one-dimensional pore) in isobutane/2-butene alkylation reactions and found that USY and silicalite deactivate faster than ZSM-12 and β zeolite catalysts under the same conditions. The main reason is that USY has three-dimensional large pores and supercages connecting the large channels, which are blocked by carbonaceous precursors. On the other hand, ZSM-12 and β zeolite have linear channels without supercages, making it difficult for larger hydrocarbons, which are prone to blocking pores. Comparing the alkylation performance of different zeolites such as Y, β, and MCM-41 in isobutane/1-butene alkylation, it is found that β molecular sieve catalysts, with a larger average pore diameter facilitating the diffusion of butene oligomers, have a longer active cycle. Similarly, in the alkylation of isobutane/2-butene using zeolites like USY, β, silicalite, ZSM-5, and MCM-22, zeolites with large pores such as USY, β, and silicalite exhibit higher alkylation product selectivity and longer activity lifetime, while ZSM-5 shows lower activity and rapid deactivation [130]. Some researchers propose that crystal particle size has almost no effect on the selectivity of alkylate products [132]. Others have found that smaller crystal particle size Hβ zeolite exhibits higher selectivity for TMP (2,2,4-trimethylpentane) and a longer catalyst lifetime, which is attributed to the small crystal particle Hβ molecular sieve having a larger specific surface area and more acid active sites [133].

For the isobutane/isobutene alkylation reaction catalyzed by zeolites, the pore configuration, pore size, and crystallite size of the molecular sieve have a significant impact on the generation of carbonaceous precursors and catalyst deactivation. It is generally believed that large-pore zeolites with a three-dimensional pore structure have a higher diffusion rate for high-carbon hydrocarbons, facilitating the diffusion of alkylation reactants and products. Conversely, zeolite with a microporous structure more easily limits the diffusion of butene oligomers, leading to catalyst deactivation.

5.6 Regeneration strategies

5.6.1 Sulfuric acid

5.6.1.1 Freeze crystallization method

In the last century, the United States used freeze crystallization technology for the regeneration treatment of waste H_2SO_4 from the sulfuric acid process. This method utilizes the temperature difference between the crystallization of sulfuric acid and impurities to purify waste H_2SO_4. Under low temperatures, sulfuric acid is crystallized out of the waste H_2SO_4, thereby achieving purification of sulfuric acid.

In this process, cooled isobutane is used for a countercurrent heat exchange with the waste H_2SO_4 stream collected from the reactor in the alkylation process. This causes H_2SO_4 to crystallize from the waste H_2SO_4. Subsequently, a freeze centrifugal filter is employed to separate the crystallized H_2SO_4. The remaining filtrate reaches thermal equilibrium with isobutane at the outlet of the heat exchanger, ceasing sulfuric acid crystallization. The resulting residual liquid comprises approximately 20%–25% of the original waste H_2SO_4, with organic content as high as 50%. The subsequent treatment of the rest of the filtrate can only be done through pyrolysis, where the waste liquid is combusted, generating emissions like CO_2, SO_2, and H_2O, which are released into the atmosphere.

This method allows for the separation of relatively pure sulfuric acid crystals from waste H_2SO_4. However, its drawback lies in the need for low temperatures to achieve sulfuric acid crystallization, which requires a significant amount of cooling in practical production. Furthermore, after crystallization, over half of the waste liquid still needs to be incinerated for treatment, resulting in high energy consumption. Direct emissions from incineration can also be environmentally harmful.

5.6.1.2 SARP chemical process

In 1960, Texco Corporation and Stratford Engineering Limited in the United States developed a chemical method for treating waste H_2SO_4.

This process involves a reaction between propylene and waste H_2SO_4, allowing approximately 90% of the free sulfuric acid molecules in the waste H_2SO_4 to combine with propylene to form propyl sulfate and dipropyl sulfate. Then, at a temperature of 12–18°C, isobutane is used as an extractant to extract dipropyl sulfate in a tray tower. This extract is then returned to the reactor to react with butene, resulting in the formation of isopentane and other compounds, while also releasing H_2SO_4 molecules [134].

This process reduces acid consumption by 65%–75%. The ASO content in the extraction accounts for 31.9%, while the remaining substances are propyl sulfate and a small amount of free sulfuric acid. This method also produces hard-to-treat waste liquids that can only be decomposed at high temperatures. The isopentane produced during the process may affect the quality of alkylate. By now, this method is no longer used in industrial production [134].

5.6.1.3 *Using waste H_2SO_4 to produce other products*

Another approach to treat waste H_2SO_4 is to use the free sulfuric acid in the waste to produce other products. This involves incorporating suitable purification steps in the existing production process to reduce the content of organic impurities in the products.

Geng et al. used waste H_2SO_4 and liquid ammonia as raw materials [135]. They employed a gas flotation method to separate organic impurities from the products, resulting in high-purity ammonium sulfate crystals. This method has a relatively simple technical process, with production costs mainly determined by the price of ammonia and the cost of organic removal. However, using waste acid to produce ammonium sulfate as fertilizer for agricultural production can cause some unremoved organic pollutants to accumulate in crops, posing risks to human health. Additionally, this method can produce a significant amount of waste pollutant, which can only be treated by incineration.

Cheng et al. activated bentonite by using waste H_2SO_4 to produce active white clay applied in the oil refining process [136]. This method is low-cost and can treat a large volume of waste acid, but it also produces a substantial amount of waste water, and the resulting active white clay contains organic impurities.

There have been reports of using waste acid to simultaneously prepare porous white carbon black and process petroleum rust inhibitors. After the waste acid undergoes hydrolysis and separation, diluted sulfuric acid and organic substances are obtained separately. The main raw material for producing white carbon black, sodium silicate, is neutralized with the separated diluted sulfuric acid. After salt precipitation and other operations, the white carbon black product is obtained. The organic oil separated from the waste acid is processed through water washing, multiple saponification, dehydration, and other operations to produce petroleum rust inhibitors. The simplified process flow diagram is shown in Fig. 5.39.

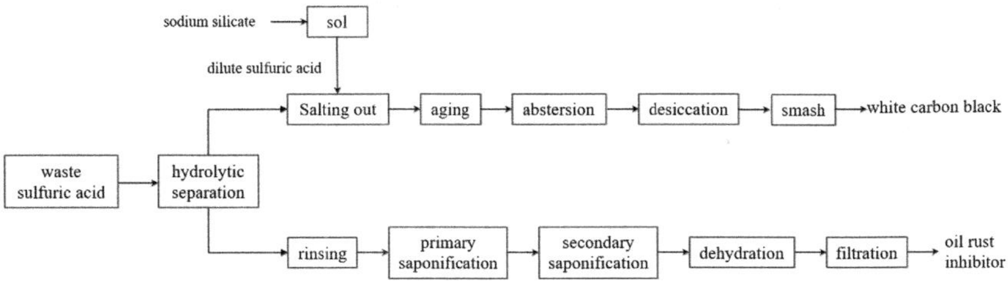

FIG. 5.39 The process for producing activated carbon/black or petroleum rust inhibitor from waste sulfuric acid [135].

Cai used the acidification method based on the two-ore leaching process in mineral extraction [137]. They utilized the MnO_2 in the ore to oxidize and treat the organic impurities in waste acid. Combined with the subsequent adsorption treatment, the content of organic impurities in the product is further reduced. The crystallized $MnSO_4$ product can be applied in industrial production.

$$2FeS_2 + 15MnO_2 + 14H_2SO_4 = Fe_2(SO_4)_3 + 15MnSO_4 + 14H_2O \quad (5.21)$$

Zhou et al. heated the waste acid at a temperature of 200°C, causing ASO to undergo direct polymerization reactions [138]. After subsequent steps involving washing and filtration, the waste sulfuric acid is transformed into dilute sulfuric acid and polymeric carbon materials. These polymeric carbon materials exhibit strong acidity and can be further utilized as solid acid catalysts for the production of biodiesel through the esterification of oleic acid and methanol. The resulting dilute sulfuric acid can be used in the synthesis of sulfate salts. Compared to high-temperature pyrolysis methods, this approach operates at a lower temperature, thus saving processing costs (Fig. 5.40).

This process of separating ASO from sulfuric acid by polymerization is more energy-efficient and generates less carbon emissions than the pyrolysis process. At 453.15 K, the removal of organic in WSAA is higher than 99.5 wt%. The PCMs are amorphous materials with –SH, S=O, $-SO_3H$ functional groups and high total acid densities, which can efficiently catalyze the oleic acid esterification with methanol to produce biodiesel. The conversion rate of oleic acid is 95.03 wt% with the optimized reaction conditions. Higher than 91 wt% of the conversion rate can be obtained over the PCM catalyst after reuse for 5 times, which demonstrates the good stability of the acidic group on PCM. The novel process developed here provides an alternative way to treat waste sulfuric acid from the alkylation process and also a facile and low-cost method to obtain PCM as SAC, which can help to reduce the manufacturing cost of biodiesel. However, during the treatment process, a certain amount of SO_2 is generated, which, if released into the atmosphere, can lead to environmental pollution.

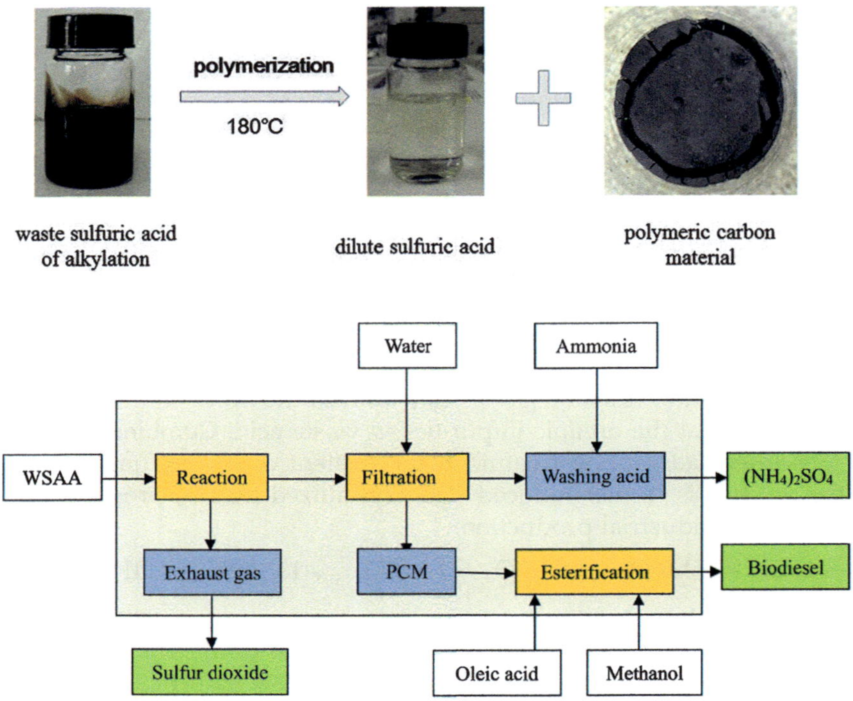

FIG. 5.40 Flow chart of the treatment for waste sulfuric acid of alkylation. *From Z. Zhou, X. Zhang, F. Yang, et al., Polymeric carbon material from waste sulfuric acid of alkylation and its application in biodiesel production, J. Clean. Prod. 215 (2019) 13–21.*

Additionally, considering that biodiesel is still largely in the research stage, the application scale of the produced polymeric carbon material catalysts is relatively limited.

Wang et al. utilized ASO extracted from waste acid as the primary material and employed a self-assembly methodology to fabricate sulfuric acid-catalyzed carbon-based catalysts (RDCMs [139]). The catalytic performance of these RDCMs in the context of coal pyrolysis was scrutinized within a two-stage fixed-bed reactor.

To address the disposal and utilization of hazardous waste H_2SO_4 generated from alkylation in an environmentally sustainable manner, a technology was devised for the simultaneous oxidative breakdown of organics and consumption of H_2SO_4. Pyrolusite, a natural manganese oxide ore, is harnessed for the oxidative degradation and mineralization of organics, with concomitant leaching of Mn from the pyrolusite [140]. The removal efficiency of total organic carbon (TOC) and Mn leaching reaches 46% and 16.44%, respectively. Moreover, pyrites are utilized as a reductant to facilitate Mn extraction from the pyrolusite, achieving an impressive Mn leaching efficiency of 98.31% under optimized conditions:

a liquid/solid ratio of 4/1, stirring speed of 350 rpm, H_2SO_4 concentration of 1.5 mol/L, and temperature of 95°C for 1 h during the initial oxidative degradation phase, followed by a molar ratio of FeS_2/MnO_2 of 0.55/1 for 4 h in the reduction leaching phase. Subsequently, a combined treatment involving MnO_2 and activated carbon is developed to further oxidize and adsorb organics in the solution, resulting in a remarkable TOC removal rate of 84.11%. The purified $MnSO_4$ solution is concentrated to yield high-quality $MnSO_4{\cdot}H_2O$, meeting industrial product standards. This technology demonstrates the promising potential for the efficient removal of hazardous organic contaminants (Fig. 5.41).

This study investigated by Zhang et al. devised a synergistic advanced oxidation process (AOP) to degrade these pollutants using H_2O_2 and sodium persulfate (PS) as oxidants, with apricot shell-derived biochar (OBC) serving as the catalyst [141]. Optimization of critical factors such as reaction temperature, initial oxidant concentration, catalyst dosage, and aeration rate is conducted, with total organic carbon (TOC) and color scale used as indicators.

Under optimal conditions, after 150 min, the removal rates of TOC and color from spent sulfuric acid reach approximately 91% and 96.6%, respectively. Moreover, the developed OBC catalyst demonstrates favorable stability and resistance to acid corrosion, remaining effective for at least four cycles with around 75% TOC removal and 80% color removal. Analysis suggests that the generated $SO_4^{-\bullet}$ and $^{\bullet}OH$ radicals in the process significantly enhance the degradation and elimination of organic pollutants, offering a promising approach for spent sulfuric acid treatment (Fig. 5.42).

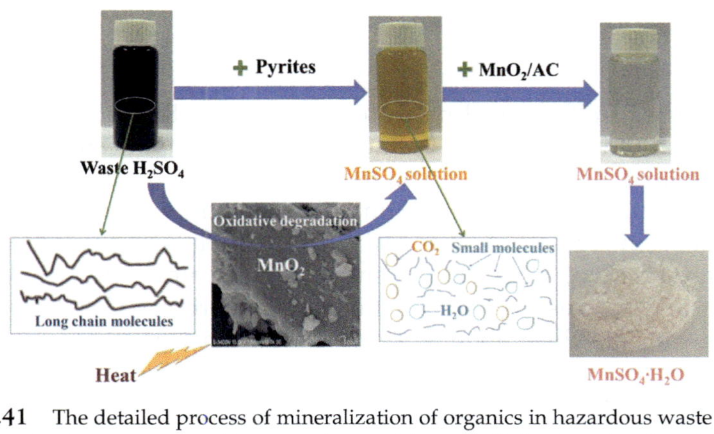

FIG. 5.41 The detailed process of mineralization of organics in hazardous waste sulfuric acid by natural manganese oxide ore and a combined MnO_2/activated carbon treatment. *From X. Cai, F. Shen, Y. Zhang, et al., Mineralization of organics in hazardous waste sulfuric acid by natural manganese oxide ore and a combined MnO_2/activated carbon treatment to produce qualified manganese sulfate, J. Hazard. Mater. 366 (2019) 466–474.*

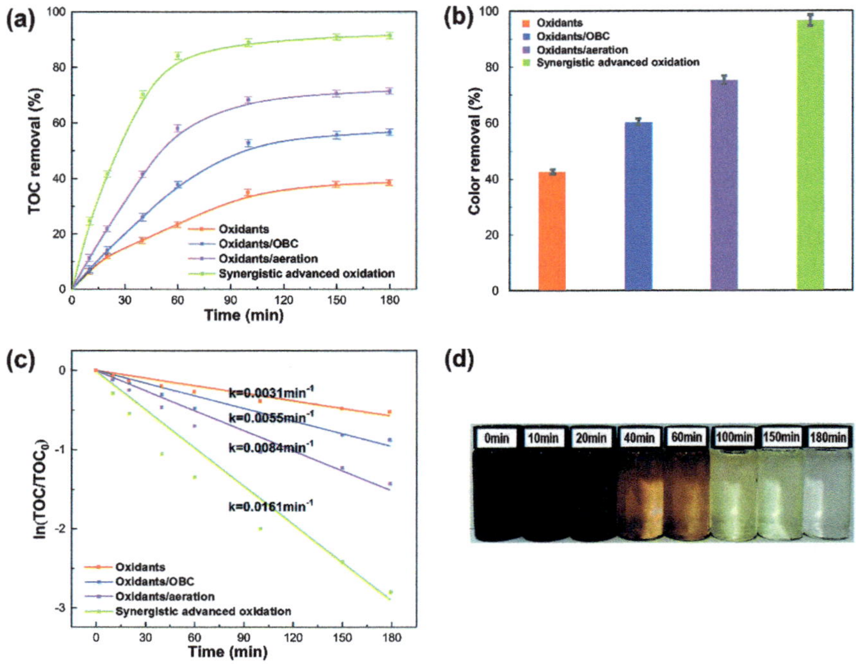

FIG. 5.42 (A) Removal rate of TOC under different reaction systems, (B) removal rate of color under different reaction systems, (C) pseudo-first-order kinetic plot of TOC under different reaction systems, and (D) color change of the sulfuric acid solution with different reaction time in the synergistic advanced oxidation process (initial oxidants concentration = 8 mol (L spent acid)$^{-1}$, catalyst dosage = 1.0 wt% (spent acid weight), aeration rate = 10 L/min, reaction temperature = 100°C). *From J. Zhang, X. Jin, H. Zhao, et al., Synergistic advanced oxidation process for enhanced degradation of organic pollutants in spent sulfuric acid over recoverable apricot shell-derived biochar catalyst, RSC Adv. 12 (2022) 1904–1913.*

The waste sulfuric acid produced during the alkylation process contains a multitude of organic pollutants, posing a significant threat to the environment. Zhang et al. aimed to efficiently utilize waste sulfuric acid for the production of polyaluminum sulfate and the thorough degradation of organic pollutants in the $Al_2(SO_4)_3$ solution using an AOP [142]. A catalyst derived from coconut shells and activated with H_3PO_4 steam, known as CBC, was employed in this oxidation process. H_3PO_4 steam activation can introduce abundant acidic functional groups and enhance the presence of metal ions on the surface of CBC, acting as active sites that facilitated the generation of •OH and •HO_2 radicals from H_2O_2 for effective degradation of organic pollutants [142]. The $Al_2(SO_4)_3$ solution treated under optimal conditions exhibits a remarkable color removal efficiency of 96.2% and total organic carbon removal of 93.8%. This treated solution is then used to produce certified polyaluminum sulfate. This research

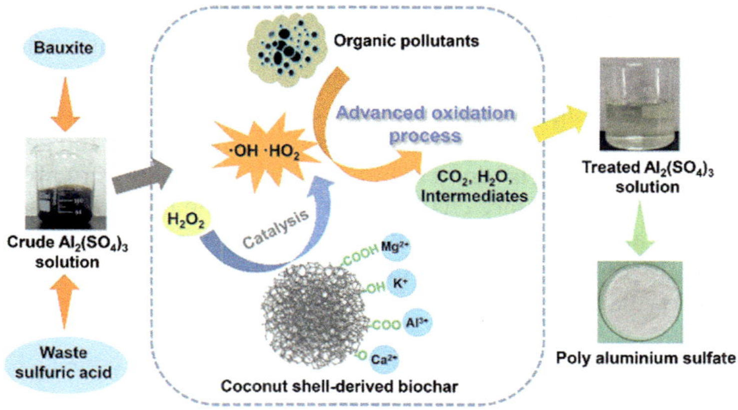

FIG. 5.43 The schematic diagram of the advanced oxidation process. *From J. Zhang, X. Jin, C. Yang, Efficient removal of organic pollutants in waste sulfuric acid by an advanced oxidation process using coconut shell-derived biochar to produce qualified poly aluminium sulfate, Sep. Purif. Technol. 293 (2022) 121057.*

holds significant importance in the valorization of waste sulfuric acid and the utilization of recycled agricultural residues in advanced oxidation processes (Fig. 5.43).

The economic analysis compares the technology with pyrolysis for treating waste sulfuric acid (in Table 5.12). Pyrolysis achieves a slightly higher sulfuric acid recovery rate at 98%. The equipment investments for waste sulfuric acid treatment are \$156,867 for the technology and \$20,392,796 for pyrolysis, showing significantly lower costs for the technology. Total operating costs, including electricity, water, materials, and catalyst costs, are assessed. The total operating cost is \$100 per ton of waste sulfuric acid, a 44% reduction compared to pyrolysis. Revenue from polyaluminum sulfate (PAS) produced using the technology is estimated at \$171 per ton of waste sulfuric acid, higher than that from recovered sulfuric acid via pyrolysis. In China, it is estimated that 8.1×10^5 tons of

TABLE 5.12 Economic analysis of different treatment methods of waste sulfuric acid.

Treatment method	Sulfuric acid recovery (%)	Equipment investment	Total operating cost	Product revenue
This technology	95	156,867\$	100[a]	171[a]
Pyrolysis	98	20,392,796\$	180[a]	103[a]

[a] *\$/ton waste sulfuric acid.*

Data from J. Zhang, X. Jin, C. Yang, Efficient removal of organic pollutants in waste sulfuric acid by an advanced oxidation process using coconut shell-derived biochar to produce qualified poly aluminium sulfate, Sep. Purif. Technol. 293 (2022) 121057.

PAS can be produced with the technology, utilizing 20 tons of coconut shell-derived biochar catalyst that can be reused five times effectively. This method presents economic advantages over conventional pyrolysis, which demands high-temperature and corrosion-resistant equipment.

Currently, approaches aimed at utilizing waste acid for the generation of secondary products demonstrate relatively straightforward procedures. However, they face challenges related to the limited enhancement of product value, restricted effectiveness in organic impurity removal, and the difficulty of managing large quantities of waste byproducts and resistant wastewater.

5.6.1.4 Pyrolysis method

Waste H_2SO_4 recovery often involves methods that directly produce sulfuric acid from raw materials such as SO_2, and these methods can be categorized into the dry contact and wet contact processes. Dry contact refers to the incineration of waste H_2SO_4 into SO_2, followed by a process similar to traditional pyrite acid production. This involves washing, drying, catalytic conversion, absorption, and other steps to regenerate commercial H_2SO_4. The wet process, on the other hand, involves the catalytic conversion of SO_2 into SO_3 in the presence of water vapor and its direct transformation into sulfuric acid. The sulfuric acid-containing gases have already undergone washing during the separation process, eliminating the need for further washing, drying, and purification. The wet process is simpler, allows for easier heat recovery, and is more cost-effective compared to the dry process.

Representative methods in the dry and wet processes include the MECS process, the WSA wet acid process, the CONCAT process, and the low-temperature condensation process in Germany (Fig. 5.44). The WSA wet acid process, developed by the Danish Topsøe company, is a direct

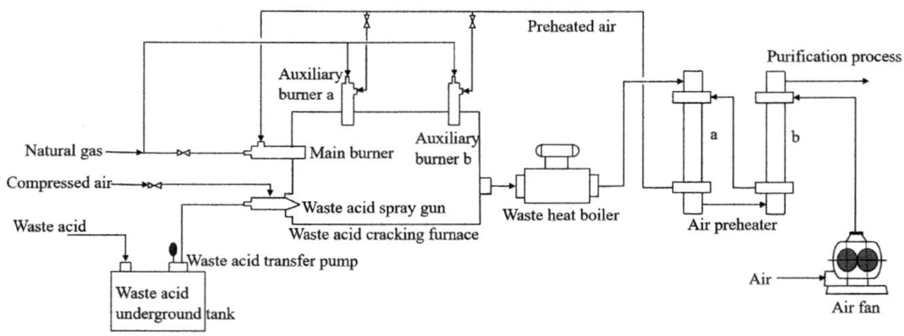

FIG. 5.44 The schematic diagram of the waste acid cracking process. *Modified from J.M. Wang, Process flow and application prospect of WSA wet acid process, Chem. Manag. (2008) 71–75.*

acid production method from sulfur-containing wet gas and is widely used in various industries [143–145]. It achieves a total sulfur recovery rate of up to 99.9%, has a relatively simple process, high energy efficiency, and yields a single product-concentrated sulfuric acid that meets commercial standards. However, the cracking furnace gas temperature exceeds 1000°C, resulting in significant energy consumption and the need for cooling measures. The high content of SO_3 and water vapor in the process gas can also lead to severe equipment corrosion.

The acid cracking process includes a series of complex reactions, such as thermochemical reduction and thermal decomposition. In the sulfuric acid cracking furnace, waste H_2SO_4, fuel gas, and combustion-assisting gas are present. Initially, waste H_2SO_4 decomposes into SO_3 and H_2O, starting at 350°C, gradually reaching equilibrium at temperatures above 500°C. At temperatures exceeding 850°C, SO_3 further decomposes into SO_2 and O_2; concurrently, hydrocarbon organic compounds in the waste acid break down into CO_2 and H_2O. To ensure complete decomposition of SO_3, the temperature inside the cracking furnace typically needs to reach 1100–1150°C. The high-temperature cracking gas generated from these reactions contains SO_2, H_2O, and CO_2. These gases undergo cooling in a waste heat boiler, exchange heat with combustion air, and are then sent to the acid production system to produce industrial sulfuric acid. The obtained H_2SO_4 is then returned to the alkylation unit, achieving resource recycling and reuse of sulfuric acid within the high-temperature cracking unit.

5.6.1.5 Electrolysis method

The principle underlying the electrolysis method for treating waste acid primarily involves the oxidation of pollutants on the electrode surface or under the influence of free radicals generated during electrolysis, leading to their conversion into CO_2 for removal. This method produces relatively pure regenerated acid; however, it consumes a significant amount of electricity during the treatment process, posing challenges for large-scale industrial applications.

Philip developed an electrolytic method for regenerating waste acid. Diluted waste acid undergoes electrolysis in a cell separated by a permeable membrane. The process includes the recovery, reconcentration, and purification of sulfuric acid from the anode chamber. Additionally, enhancements are implemented to address the issue of excessive gas bubble formation during high-temperature electrolysis.

William optimized the electrolytic regeneration method for waste acid. Adjustments were made to the dilution concentration of waste acid, with the water content being 1–2 times the concentration of organic impurities. Organic impurities in the waste sulfuric acid are oxidized to carbon dioxide at the anode, while hydrogen gas is generated at the cathode. Water serves as the apparent oxidant, although the actual oxidant may be persulfuric acid. As the electrolysis process advances, water and hydrocarbons in the waste acid are eliminated, leading to increased purity in the regenerated acid product.

5.6.1.6 Adsorption method

The adsorption method involves the removal of organic impurities from waste acid through the action of solid adsorbents. Adsorbents selected for the regeneration of acid typically need to meet criteria such as acid resistance, strong adsorption capacity, and cost-effectiveness. Materials such as activated clay, activated carbon, and porous adsorption resins are commonly used as adsorbents in industrial applications.

Asof et al. conducted research on the adsorption treatment process for waste acid [146]. They used bentonite obtained from Indonesia as the adsorbent and performed adsorption operations by passing waste sulfuric acid through three serially connected Pyrex high-temperature glass adsorption columns, each with a diameter of 5 cm and a height of 70 cm. The columns were equipped with a liquid distributor made of perforated plates, as shown in Fig. 5.45. The treatment temperature was set at 850°C. Since the interaction mechanism of bentonite with waste acid is primarily physical adsorption, the removal of organic impurities from the waste acid is accompanied by a significant loss of sulfuric acid. If the spent bentonite is not regenerated, it can lead to secondary environmental pollution.

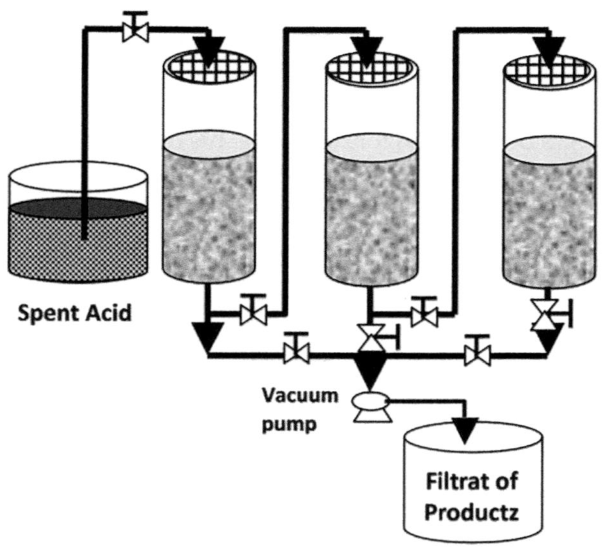

FIG. 5.45 The process of adsorption-based waste acid treatment. *From M. Asof, S.A. Rachman, W.A. Nurmawi, et al., Recovery of H_2SO_4 from spent acid waste using bentonite adsorbent, MATEC Web of Conferences, EDP Sciences, 2017, pp. 02007.*

5.6.1.7 Oxidation method

The oxidation method entails the utilization of potent oxidizing agents to facilitate the degradation of organic pollutants within waste acid, leading to the generation of products such as H_2O and CO_2, ultimately serving the purpose of waste acid regeneration. Commonly employed industrial oxidizing agents in this context encompass H_2O_2, HNO_3, NO_3^-, $HClO_4$, HClO, and O_3.

Li et al. employed concentrated HNO_3 as an oxidizing agent for the treatment of waste acid [147]. Under the conditions of 250°C and a 2% volume fraction of concentrated HNO_3, following a 2 h reaction period, the resulting product exhibits colorless transparency and an impressive COD removal rate of 99%. This method manifests notable efficacy in the oxidation-based removal of organic compounds and yields regenerated acid characterized by elevated acidity. However, it is imperative to note that the utilization of concentrated HNO_3 as an oxidizing agent at elevated temperatures introduces inherent safety risks. Furthermore, it introduces nitrogen-based impurities into the regenerated acid, thereby affecting the purity of the resultant regenerated sulfuric acid.

Li et al. harnessed the Fenton oxidation method to address organic impurities present in waste acid [148]. Initially, the waste sulfuric acid is diluted to achieve an acidity level of 21.18%. Subsequently, Fenton reagents (comprising H_2O_2 and $FeSO_4$) are introduced into the solution. The Fenton oxidation reaction operates on a free radical mechanism basis, entailing the generation of highly oxidative •OH radicals through the interaction between H_2O_2 and Fe_2^+. These radicals further facilitate the oxidation and degradation of organic impurities inherent in the waste sulfuric acid. This method exhibits commendable efficacy in the removal of organic constituents. Nonetheless, it yields regenerated acid with reduced acidity, thereby imposing certain limitations on its practical applicability.

Due to the challenges in waste H_2SO_4 separation technology, current waste H_2SO_4 treatment methods often rely on chemical processes to either generate other products from waste H_2SO_4 or regenerate and reuse the waste H_2SO_4. These approaches have lower added value, higher energy consumption, and have not fully addressed environmental pollution.

5.6.2 Ionic liquid

The IL deactivation due to the loss of Brønsted and Lewis acidity, along with the recovery of activity in CIL, is summarized in Fig. 5.46 [89]. Initially, the outflow of chlorohydrochloride formed via the alkylate leads to a depletion of CIL's Brønsted acidity, which can be restored by introducing substances like HCl or chlorohydrocarbons such as *t*-butyl

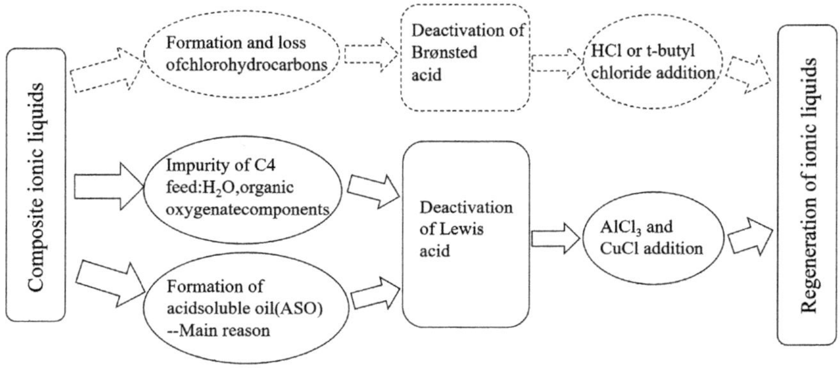

FIG. 5.46 Deactivation and activity recovery process of CIL. *From X. Zhang, R. Zhang, X. Meng, et al., Deactivation mechanism and activity-recovery approach of composite ionic liquids for isobutane alkylation, Appl. Catal. A-Gen. 557 (2018) 64–71.*

chloride or the chlorohydrocarbon-rich boiling fraction of alkylate. However, maintaining Brønsted acidity necessitates the presence of Lewis acidity. Secondly, minute quantities of water or methanol in the feedstock trigger HCl formation, further depleting Lewis acid through the solid formation of $Al_2Cl_7^-$. Consequently, minimizing the presence of these impurities in the feed is essential. Lastly, the complexation of ASO with active compounds primarily causes the deactivation of Lewis acidity and a reduction in C_8 selectivity. Both Lewis acidity (activity) and C_8 selectivity can be restored by adding $AlCl_3$ and CuCl, respectively.

Fig. 5.47 illustrates that the selectivity of alkylate remains consistently high, around 90 wt% C_8, with RON of approximately 95, throughout the 60 day test period, during which both Brønsted and Lewis acid recovery are employed. The chloride content averages around 300 ppm during this duration. Consequently, through the continuous supplementation of active components such as HCl, $AlCl_3$, and CuCl, long-term high activity, selective alkylation, and reduced CIL consumption are achieved.

5.6.3 Solid acids

The regeneration of solid acid alkylation catalysts involves the removal of carbonaceous precursors that cover the active centers or block the pores. This process aims to restore the catalyst active centers, pore structure, and specific surface area, thereby recovering its catalytic activity. Currently, catalyst regeneration methods mainly include hydrogen regeneration, metal hydrogen regeneration, supercritical fluid regeneration, oxidative calcination regeneration, and solvent washing regeneration.

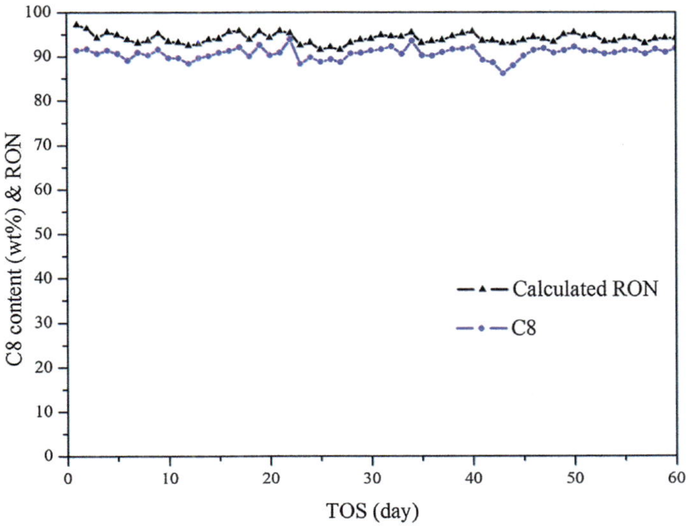

FIG. 5.47 Alkylate RON and C_8 selectivity in a long period pilot alkylation catalyzed by CIL (CIL intake 29.9 kg, CIL to hydrocarbon volume ratio in reactor 1.0, reaction temperature 13°C, C4 flow rate 7.5 L/h with I/O ratio 8:1). *From X. Zhang, R. Zhang, X. Meng, et al., Deactivation mechanism and activity-recovery approach of composite ionic liquids for isobutane alkylation, Appl. Catal. A-Gen. 557 (2018) 64–71.*

5.6.3.1 Hydrogen regeneration

The hydrogen regeneration method regenerates deactivated solid acid catalysts in a hydrogen atmosphere, catalyzing the hydrogenation and cracking of precursor carbon, thereby restoring the catalyst activity. Union Carbide Corporation proposed early on the method of hydrogenation cracking of carbonaceous precursors for catalyst regeneration [149]. Deactivated β zeolite is treated with hydrogen [150], removing the low-hydrogen-containing lean-hydrogen carbon precursors through hydrogenation reactions, thus restoring the catalytic reactivity. It is found that hydrogen treatment requires higher temperatures, and the best regeneration effect (including activity and selectivity) is generally at 550°C. Some scholars [150] have investigated the effect of different regeneration temperatures on the recovery of catalyst activity under atmospheric pressure. Among the selected regeneration temperatures of 350°C, 450°C, and 550°C, the regeneration effect at 550°C is the best. Furthermore, β-zeolite could be regenerated through hydrogen treatment at appropriately low temperatures.

5.6.3.2 Metal hydrogen regeneration

In the metal hydrogen regeneration method, hydrogenation noble metals are loaded onto solid acid catalysts [151], and regeneration is carried out in a hydrogen atmosphere, leading to hydrogenation and removal of precursor carbon on the catalyst, thereby restoring its activity. Noble metals commonly supported on solid acid alkylation catalysts include platinum, palladium, gold, silver, ruthenium, rhodium, and the like. The loading method of precious metals and the use of additives, precursor form, and loading amount will affect the position and dispersion of precious metals on the catalyst, thereby affecting the alkylation reaction and catalyst regeneration. Common methods for loading noble metals onto solid acid alkylation catalysts include impregnation, precipitation, direct synthesis, and ion exchange. Studies on Pt-loaded zeolite A showed that the direct synthesis method can effectively load Pt inside the zeolite A, thus ensuring good metal dispersion. In addition, compounds such as ethylenediaminetetraacetic acid, citric acid, and ammonium hexafluorosilicate are introduced during the loading process, which can serve as competitive adsorption ligands to enhance the dispersion and reducibility of precious metals. Nonnoble metals supported on solid acid alkylation catalysts typically include nickel, molybdenum, cobalt, manganese, chromium, and the like. Bimetallic and multimetallic supports on solid acid alkylation catalysts cover precious metals, such as platinum, palladium, gold, silver, ruthenium, and rhodium, and nonnoble metals such as nickel and cobalt. Currently, there is limited research on solid acid alkylation catalysts supporting nonnoble metals and metal combinations that remove carbonaceous precursors via hydrogen regeneration.

5.6.3.3 Supercritical fluid regeneration

The supercritical fluid regeneration method uses an appropriate supercritical fluid to remove the precursor carbon on the catalyst, thereby restoring its alkylation performance. Supercritical fluid refers to a fluid with a temperature and pressure higher than the critical point. It has powerful solubility and diffusivity, effectively extracting materials from porous matrices, including nonhomogeneous catalysts. Under supercritical reaction conditions, the dissolved and desorbed carbon precursor from porous catalysts can extend the catalyst lifespan and reactivate the deactivated catalysts [152]. He [153] successfully conducted alkylation reactions of isomeric hydrocarbons and butenes using HPW/SiO_2-supported heteropolyacid catalysts under supercritical reaction conditions. The solid acid catalyst maintained a 100% olefin conversion rate even after 1400 h of reaction. Petkovic et al. [112] used different times of stream for alkylation reactions to allow Y-type zeolite catalyst samples to achieve varying degrees of deactivation and then regenerate supercritical fluid isobutane for 60 min. Results from nitrogen physisorption, temperature-programmed oxidation (TPO), diffuse reflectance infrared Fourier transform spectroscopy (DRIFTS), and ultraviolet-visible

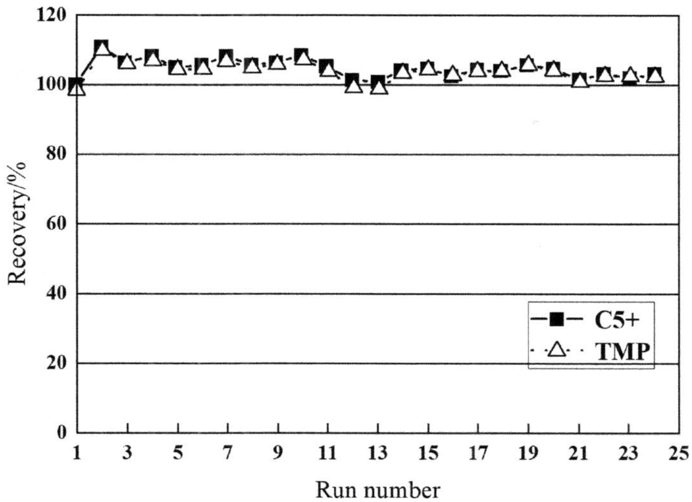

FIG. 5.48 The change in percentage of recovered alkylate (C_5+) and trimethylpentane (TMP) production with each successive run (after being reactivated) (Values are compared with the initial run 100%) [154]

(UV-vis) spectroscopy measurements indicate that the supercritical fluid regeneration method can effectively remove carbonaceous precursors, restore the catalyst surface area and micropore volume, thereby efficiently restoring catalyst activity. Ginosar et al. [154] found that partially deactivated alkylation catalysts can be successfully regenerated by supercritical isobutane. Fig. 5.48 indicates that the catalyst activity and product yield almost do not decrease after 23 reaction/regeneration cycles, maintaining 100% initial activity throughout the entire experimental process. The authors also explored the influence of various fluids, including helium, propane, *n*-butane, isobutane, *n*-pentane, and isopentane, on the regeneration performance of the completely deactivated USY zeolite catalyst during isobutane and 2-butene alkylation. It was found that under supercritical conditions, isobutane exhibits the best regeneration performance, followed by isopentane, while *n*-pentane shows the poorest regeneration performance.

Thompson et al. [155] studied the deactivation and supercritical fluid regeneration of a series of 12-membered ring large-pore zeolites, including acidic Y, β, and ZSM-5 zeolites. The chemical properties of substances on the catalyst surface before and after supercritical fluid regeneration were analyzed using TPO, DRIFTS, and UV. The results show that although most of the coke in the catalyst is removed, hydrocarbon residues are present on all catalysts after supercritical fluid regeneration, and the catalytic activity is not fully restored. For the comparison of the long-term performance of USY and β-zeolite samples, multiple cycles of reaction and regeneration were performed. The experiments revealed that neither catalyst regained their initial performance after supercritical fluid regeneration, and they exhibit a sustained decrease in activity. Nonetheless, the decline in activity for the β-sample is slower. They suggest it might be due to hindrance in the formation of unsaturated cyclic or polycyclic compounds that could poison the catalyst on β-zeolite, which contrasts with what happens in the Y-zeolite catalyst. This suggests that β-zeolite with a high acidity point density is the most suitable zeolite for repeated cycles of reaction and regeneration. Conversely, zeolite catalysts with one-dimensional pore structure, like fibrous zeolite, are unsuitable for regeneration using supercritical fluid due to diffusion limitations.

By selecting a suitable supercritical fluid and utilizing its unique physicochemical properties, the regeneration of deactivated catalysts can be achieved. However, considering that the greater the deactivation, the less complete the regeneration, there may be residual carbon precursor materials, so it is necessary to choose an appropriate regeneration cycle. Isobutane is the most effective in supercritical fluid regeneration, but suitable regeneration conditions must also be considered. Moreover, the operational conditions for supercritical fluid regeneration are stringent and the process is complex. Combining other regeneration processes with supercritical regeneration can improve regeneration performance, but the regeneration process needs further simplification.

5.6.3.4 Oxidative calcination regeneration

Oxidative roasting regeneration, especially under an oxygen atmosphere, stands as a prevalent catalyst regeneration technique in the realms of petroleum refining and Chem Eng—New York. It efficiently eliminates surface coke from the catalyst, widely utilized for the regeneration of catalytic cracking catalysts, catalytic reforming catalysts, and extensively employed in the rejuvenation process of alkylation catalysts. During the regeneration of silk zeolite catalysts using air, RØrvik et al. [156] discovered the necessity of elevated regeneration temperatures (such as 600°C) to achieve optimal regeneration results. Querini [157], in the study on the oxidative roasting regeneration of Y-type molecular sieves post-lanthanum exchange, revealed from TPO results that the extent of coke removal on the catalyst during regeneration primarily relies on the heating rate. Furthermore, it was indicated that even if the coke on the catalyst is not entirely eliminated after helium or air oxidative roasting treatment, the catalyst activity can still partially recover. However, this approach leads to the accumulation of residual coke, resulting in complete deactivation of the catalyst after several cycles. Complete removal of coke from the catalyst requires prolonged treatment at temperatures exceeding 600°C. Querini also examined the Pt/LCHY catalyst with a loaded mass fraction of 0.5% Pt and found that after prolonged ozone treatment, the catalyst can restore its initial activity. This is because ozone treatment not only reduces the quantity of precursor materials for coke deposition on the catalyst but also alters the structure of the precursor, making it more easily combustible. Querini also conducted regeneration experiments using hydrogen peroxide on the deactivated Pt (0.5%)/LCHY catalyst, aiming to utilize the oxygen generated during heating to oxidize coke. At room temperature, hydrogen peroxide is found ineffective in removing coke precursors; when treated at 90°C, most of the coke can be removed with hydrogen peroxide, but it requires a long time. Residual coke after ozone and hydrogen peroxide treatments can be preliminarily removed at temperatures lower than direct air calcination, with ozone as the oxidant showing a more significant effect. Therefore, considering a combined approach of ozone treatment followed by air treatment to lower the temperature of air treatment (from above 500°C to 250°C) is proposed.

Xie et al. [150] employed air treatment, calcining the deactivated β-catalyst at a temperature slightly below that of hydrogen treatment (550°C), followed by regeneration at 520–550°C for 3h before conducting the alkylation reaction. It can be observed from Fig. 5.49 that the activity and selectivity of the catalyst are effectively restored.

Yao and colleagues [158] targeted a nanocomposite Al_2O_3-Hβ molecular sieve catalyst, utilizing an air oxidation calcination regeneration method. They systematically examined the influence of regeneration temperature and airflow rate on the recovery of deactivated catalyst activity.

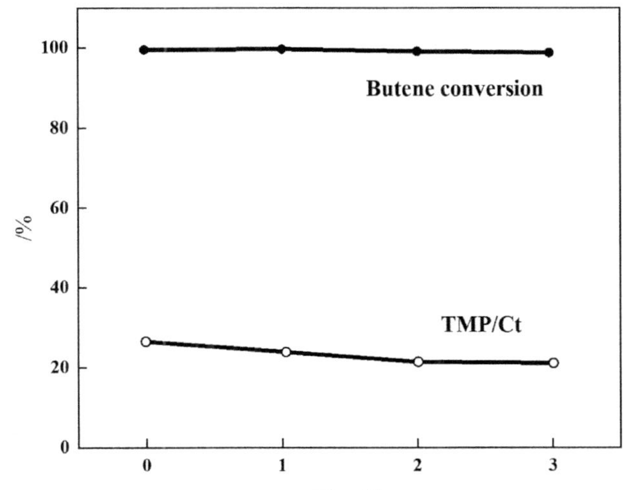

FIG. 5.49 The results of continuous regeneration in the air atmosphere. *From S. Xie, S. He, X. Qian, et al., Regeneration and deactivation of β zeolite catalysts in the alkylation of isobutane/butane, Petrochem. Technol. 33 (2004) 1402–1404.*

The optimal regeneration conditions are identified as 450°C at atmospheric pressure with an airflow rate of 50 mL/min. Qiang et al. [159] selected a zeolite molecular sieve catalyst, an SO_4^{2-}/oxide superacid catalyst, a loaded conjugated solid superacid catalyst, a solid polymer ion exchange resin, and an oxide or molecular sieve catalyst treated with Lewis acid as research objects. The deactivated catalysts after alkylization reaction are brought into contact with a solution containing at least one oxidant for over 10 min. Oxidants include $HClO_4$, $KClO_4$, NH_4ClO_4, $NaClO_4$, high chloric acids and their salts, HClO, KClO, NH_4HClO, NaClO, low chloric acids and their salts, as well as H_2O_2 or HNO_3. Subsequently, regeneration treatment is conducted at 100–250°C in air, oxygen, or a mixture of air and oxygen for 0.5–10 h. Experimental results indicate that this regeneration method is applicable to all the mentioned catalysts, with the most suitability for the supported polyoxometalate catalysts.

5.6.3.5 Washing regeneration

Washing regeneration is typically achieved by selecting an appropriate regenerate and washing the deactivated catalyst in the solvent, thereby removing the accumulated carbon on the deactivated catalyst. Xie et al. [160] studied B-L solid superacid. Considering its thermal instability, which is unsuitable for methods like burning, they chose the washing regeneration method to regenerate the deactivated catalyst. When solvent washing regeneration is performed, the selection of regenerants should consider solvent similarity, miscibility, viscosity, safety and whether it is easily separated from the products. Based on this, Xie et al. [150] chose a mixture of hexadecane and solvent oil No. 120 as the solvent and investigated the performance of intermittent and continuous regeneration. It was found that the regeneration effect of intermittent regeneration was poor and seemed to promote the deactivation of the catalyst. This is because the adsorption of olefins on solid superacid is much higher than that of alkanes. Therefore, during the release of reaction materials and introduction of regenerants for regeneration, the system lacks alkanes, leading to the accumulation of a large amount of adsorbed olefins on the catalyst surface, undergoing low-polymerization reactions. Therefore, the regeneration cycle is shortened to perform regeneration washing before catalyst deactivation, allowing the low-polymer materials formed in the reaction to be washed away before they polymerized into large-molecule alkanes and clogged the pores, thus extending the catalyst lifespan. In addition, continuous regeneration of the reaction and catalyst is effective, as experimental results also indicate that continuous regeneration is effective against the precursor substances of coke formation on the catalyst surface.

Choose what makes the regenerating agent effectively extract and remove the precursor of coke, restore the activity of the deactivated alkylation catalyst, but this method is difficult to completely regenerate the catalyst. It is also found that extracting a large volume of precursor coke from the zeolite pores is challenging, and dichloroethane cannot regenerate β molecular sieves. Therefore, it is necessary to choose the appropriate regeneration agent for different catalysts and select suitable process parameters.

References

[1] L.F. Albright, Present and future alkylation processes in refineries, Ind. Eng. Chem. Res. 48 (2009) 1409–1413.

[2] J.G. Bi, Advances in alkylated oil production technology, Prog. Chem 26 (2007) 934–939.

[3] X.Y. Di, H. Yang, Review of solid acid alkylation process, Contemp. Chem. Ind. (2005). 169-172+187.

[4] Y. Liu, Z.C. Liu, C.P. Huang, et al., Alkylation studies of isobutanewith C4 olefins, Chem. Eng. (2004). New York of Oil and Gas. 12-16+76-77.

[5] Z.T. Cao, Z.W. Qiu, N.N. Zhao, et al., Technical advance of alkylation process, Adv. Fine Petrochem. 17 (2016) 34–37.

[6] M. Tsadkin, A. Badikova, E. Mortikov, The formation of emulsions during the jet mixing of reagent flows as applied to the sulfuric acid alkylation of iso-butane with olefins, Theor. Found. Chem. Eng. 50 (2016) 165–170.

[7] B.S. Wang, J.L. Liu, Y.F. Gao, et al., Numerical simulation of nozzle in sulfuric acid alkylation, Pet. Refin. Eng. 46 (2016) 32–35.

[8] H.B. Cai, G.C. Dai, CFD simulation of flow characteristics in an alkylation reactor by sulfuric acid method—analysis of structural effects, Chem. Eng. Technol. 28 (2012). 391-397+468.

[9] B.A. Cunningham, Sulfuric Acid Alkylation Reactor System With Static Nixers, 1998, US5785933A.

[10] M. Zhang, L. Zhu, Z. Wang, et al., Flow field in a liquid–liquid cyclone reactor for isobutane alkylation catalyzed by ionic liquid, Chem. Eng. Res. Des. 125 (2017) 282–290.

[11] M. Zhang, A. Li, L. Zhu, Z. Wang, et al., A novel liquid-liquid cyclone reactor for ionic liquid catalyzed isobutane alkylation: cold model investigation of the dispersed phase droplet size distribution, Sep. Purif. Technol. 209 (2019) 375–382.

[12] Y. Liu, G. Liu, G. Wu, et al., Alkylation of isobutane and 2-butene in rotating packed-bed reactors: using ionic liquid and solid acid as catalysts, Ind. Eng. Chem. Res. 59 (2020) 14767–14775.

[13] M.H. Dong, B.N. Zong, SINOALKY Sulfuric acid alkylation process technology and its industrial application, Pet. Chem. 50 (2019) 29–32.

[14] G.G. Alexanyan, N.B. Librovitch, Y.A. Prochukhane, Process for the Alkylation of Olefins, 1998, US5777189.

[15] S. Gershuni, Method and Device for the Production of Alkylates, 2008, US7435864.

[16] A. Hassan, E. Bagherzadeh, R.G. Anthony, et al, System and Process for Alkylation, 2012, US8269057.

[17] L. Li, J. Zhang, C. Du, et al., Process intensification of sulfuric acid alkylation using a microstructured chemical system, Ind. Eng. Chem. Res. 57 (2018) 3523–3529.

[18] X. Lin, S. Yan, B. Zhou, et al., Highly efficient synthesis of polyvinyl butyral (PVB) using a membrane dispersion microreactor system and recycling reaction technology, Green Chem. 19 (2017) 2155–2163.

[19] L. Li, J. Zhang, C. Du, et al., Kinetics study of sulfuric acid alkylation of isobutane and butene using a microstructured chemical system, Ind. Eng. Chem. Res. 58 (2018) 1150–1158.
[20] T. Van Gerven, A. Stankiewicz, Structure, energy, synergy, time—the fundamentals of process intensification, Ind. Eng. Chem. Res. 48 (2009) 2465–2474.
[21] K. Guo, Z. Zhang, H. Luo, et al., An innovative approach of the effective mass transfer area in the rotating packed bed, Ind. Eng. Chem. Res. 53 (2014) 4052–4058.
[22] D. Fernandez Rivas, D.C. Boffito, J. Faria-Albanese, Process intensification education contributes to sustainable development goals. Part 2, Educ. Chem. Eng. 32 (2020) 15–24.
[23] P.F. Yang, S. Luo, D.S. Zhang, et al., Extraction of nitrobenzene from aqueous solution in impinging stream-rotating packed bed, Chem. Eng. Process. 124 (2018) 255–260.
[24] J.B. Modak, A. Bhowal, S. Datta, Extraction of dye from aqueous solution in rotating packed bed, J. Hazard. Mater. 304 (2016) 337–342.
[25] H. Liu, T. Hu, D. Wang, et al., Preparation of fluorescent waterborne polyurethane nanodispersion by high-gravity miniemulsion polymerization for multifunctional applications, Chem. Eng. Process. 136 (2019) 36–43.
[26] Y. Tian, S. Mei, L.L. Zhang, et al., Improved H_2SO_4-catalyzed alkylation reaction in a rotating packed bed reactor by adding additives, Can. J. Chem. Eng. 100 (2022) 3395–3407.
[27] Z. Liu, X. Meng, R. Zhang, et al., Reaction performance of isobutane alkylation catalyzed by a composite ionic liquid at a short contact time, AIChE J. 60 (2014) 2244–2253.
[28] Z. Liu, C. Xu, C. Huang, Method for Manufacturing Alkylate Oil With Composite Ionic Liquid Used as Catalyst, US Patent 7,285,698, 2007.
[29] W. Sun, W. Zheng, P. Cao, et al., Probing interfacial behaviors of Brønsted acidic ionic liquids improved isobutane alkylation with C4 olefin catalyzed by sulfuric acid, Chem. Eng. J. 377 (2019) 119744.
[30] M. Zhang, L. Wang, L. Zhu, et al., Phase holdup distribution and dispersion performance in a novel liquid–liquid cyclone reactor of isobutane alkylation catalyzed by ionic liquid, Chem. Eng. Res. Des. 125 (2017) 257–264.
[31] S. Duan, M. Chen, R. Zhang, et al., Effect of light phase inlet structure on liquid-liquid cyclone reactor for isobutane alkylation catalyzed by ionic liquid, Chem. Eng. Process. 140 (2019) 22–28.
[32] J. Bi, P. Yang, J. Duan, et al., Numerical investigation of secondary flow in a cyclone reactor for alkylation catalyzed by ionic liquid, Ind. Eng. Chem. Res. 61 (2022) 16461–16469.
[33] R.A. Strauss, R.R. Hemrajani, HF Alkylation Reactor, US7947232B2, 2011.
[34] L. Qi, X. Meng, R. Zhang, et al., Droplet size distribution and droplet size correlation of chloroaluminate ionic liquid–heptane dispersion in a stirred vessel, Chem. Eng. J. 268 (2015) 116–124.
[35] H.D. Velázquez, N. Likhanova, N. Aljammal, et al., New insights into the progress on the isobutane/butene alkylation reaction and related processes for high-quality fuel production. A critical review, Energ. Fuels 34 (2020) 15525–15556.
[36] R. Kore, A.M. Scurto, M.B. Shiflett, Review of isobutane alkylation technology using ionic liquid-based catalysts—where do we stand? Ind. Eng. Chem. Res. 59 (2020) 15811–15838.
[37] W. Zheng, P. Cao, W. Sun, et al., Modeling of interfacial behaviors of isobutane alkylation with 2-butene catalyzed by sulfuric acid/Brønsted acidic ionic liquid, in: 2018 AIChE Annual Meeting, AIChE J, 2018.
[38] M. Narasimha, R. Sripriya, P. Banerjee, CFD modelling of hydrocyclone Äîprediction of cut size, Int. J. Miner. Process. 75 (2005) 53–68.
[39] K. Neumann, S. Hunold, K. Groß, A. Górak, Experimental investigations on the upper operating limit in rotating packed beds, Chem. Eng. Process. 121 (2017) 240–247.
[40] Y. Tian, Z. Li, S. Mei, et al., Alkylation of isobutane and 2-butene by concentrated sulfuric acid in a rotating packed bed reactor, Ind. Eng. Chem. Res. 57 (2018) 13362–13371.

[41] D.J. Ende, R.E. Eckert, L.F. Albright, Interfacial area of dispersions of sulfuric acid and hydrocarbons, Ind. Eng. Chem. Res. 34 (1995) 4343–4350.
[42] Y. Li, R. Liu, G. Zhao, et al., Simulation and optimization of fixed bed solid acid catalyzed isobutane/2-butene alkylation process, Fuel 216 (2018) 686–696.
[43] B. Bourgeois, Focus on catalysts, Albemarle's alky Star catalyst successfully employed in the world's first solid acid alkylation unit in shandong, China, facility, Focus. Catal. 2016 (2016) 4.
[44] L.F. Albright, Alkylation of isobutane with C3–C5 olefins: feedstock consumption, acid usage, and alkylate quality for different processes, Ind. Eng. Chem. Res. 41 (2002) 5627–5631.
[45] J.H. Morgenthaler, Process for the Regeneration of Spent Sulfuric Acid, US 4,376,107, 1983.
[46] S. Miron, R.J. Lee, Molecular structure of conjunct polymers, J. Chem. Eng. Data 8 (2002) 150–160.
[47] L.F. Albright, M.A. Spalding, C.G. Kopser, et al., Alkylation of isobutane with C4 olefins. 2. Production and characterization of conjunct polymers, Ind. Eng. Chem. Res. 27 (1988) 386–391.
[48] Y. Xin, Y. Hu, M. Li, et al., Isobutane alkylation catalyzed by H_2SO_4: effect of H_2SO_4 acid impurities on alkylate distribution, Energy Fuel 35 (2020) 1664–1676.
[49] L.F. Albright, Alkylation of isobutane with C3–C5 olefins to produce high-quality gasolines: physicochemical sequence of events, Ind. Eng. Chem. Res. 42 (2003) 4283–4289.
[50] S.I. Hommeltoft, Isobutane alkylation: recent developments and future perspectives, Appl. Catal. A-Gen. 221 (2001) 421–428.
[51] J.G. Bi, Advances in alkylated oil production technology, Chem. Ind. Eng. Prog. (2007) 934–939.
[52] V. Ipatieff, A. Grosse, Alkylation of aromatics with olefins in the presence of boron fluoride, J. Am. Chem. Soc. 58 (1936) 2339.
[53] H. Bloch, Chemistry of catalyst degradation, in: ACS-AIChE Joint Meeting, 1958.
[54] S. Miron, R.J. Lee, Molecular structure of conjunct polymers, J. Chem. Eng. Data 8 (1963) 150–160.
[55] A.W. Francis, Estimation of the unsaturated content of petroleum products, Ind. Eng. Chem. Res. 18 (1926) 821–822.
[56] H.D. Hoffman, C.E. Green, A rapid method for the determination of iodine number, Oil Soap 16 (1939) 236–238.
[57] W. Hanson, Directions for the use ofthe modified Menzies-Wright molecular weight apparatus, Mellon Institute Report, Pittsburgh, PA, 1943.
[58] W.E. Hanson, J.R. Bowman, Determination of the molecular weights of oils, Ind. Eng. Chem. Anal. Ed. 11 (1939) 440–442.
[59] A.W.C. Menzies, S.L. Wright Jr., The application of a differential thermometer in ebullioscopy, J. Am. Chem. Soc. 43 (1921) 2314–2323.
[60] L.F. Albright, Mechanism for Alkylation of Isobutane With Light Olefins, ACS Publications, 1977.
[61] L.F. Albright, Degradation and Isomerization Reactions Occurring During Alkylation of Isobutane With Light Olefins, 1975.
[62] L.F. Albright, A.R. Goldsby, Industrial and Laboratory Alkylations, ACS Publications, 1977.
[63] Y.Y. Xin, Y.F. Hu, L. Mengchen, et al., Isobutane alkylation catalyzed by H_2SO_4: Effect of H_2SO_4 acid impurities on alkylate distribution, Energ. Fuel 35 (2021) 1664–1676.
[64] L. Fieser, M. Fieser, Natural Products Related to Phenanthrene, Reinhold Pub, Co., 1949, p. 192.
[65] R.B. Woodward, Structure and absorption spectra. III. Normal conjugated dienes, J. Am. Chem. Soc. 64 (1942) 72–75.

[66] L. Wang, G. Zhao, X. Yao, et al., Adamantane-based cation and $[MF_n]^-$ anion synergistically enhanced catalytic performance of sulfuric acid for isobutane alkylation, Ind. Eng. Chem. Res. 56 (2017) 7920–7929.

[67] L.F. Albright, K.V. Wood, Alkylation of Isobutane with C3–C4 Olefins: identification and chemistry of heavy-end production, Ind. Eng. Chem. Res. 36 (1997) 2110–2120.

[68] H. Zhang, H. Tang, J. Xu, et al., Enhancement of isobutane/butene alkylation by aromatic-compound additives in strong Brønsted acid, Fuel 231 (2018) 224–233.

[69] A. Corma, A. Martinez, Chemistry, catalysts, and processes for isoparaffin–olefin alkylation: actual situation and future trends, Catal. Rev. 35 (1993) 483–570.

[70] A.S. Berenblyum, E.A. Katsman, R. Berenblyum, et al., Modeling of side reactions of isobutane alkylation with butenes catalyzed by trifluoromethane sulfonic acid, Appl. Catal. A-Gen. 284 (2005) 207–214.

[71] Z.H. Ma, W.Z. Zheng, W.Z. Sun, et al., Enhanced catalytic performance of H_2SO_4-catalyzed C4 alkylation by formyl functional $[N_{1,1,1,1}][C_{10}SO_4]$ additive, AIChE J. 69 (2023) e18179.

[72] L.F. Albright, M.A. Spalding, J.A. Nowinski, et al., Alkylation of isobutane with C4 olefins. 1. First-step reactions using sulfuric acid catalyst, Ind. Eng. Chem. Res. 27 (1988) 381–386.

[73] L.F. Albright, K. Li, Alkylation of isobutane with light olefins using sulfuric acid. Reaction mechanism and comparison with HF alkylation, Ind. Eng. Chem. Process 9 (1970) 447–454.

[74] A. Lutsyk, E. Rudakov, V. Sumanov, et al., Influence of alkylation catalyst composition on solubility of isobutane in sulfuric acid, Chem. Technol. Fuels Oils 27 (1991) 388–389.

[75] D. Wang, T. Zhang, Y. Yang, et al., Simulation and design microreactor configured with micromixers to intensify the isobutane/1-butene alkylation process, J. Taiwan Inst. Chem. Eng. 98 (2019) 53–62.

[76] L.F. Albright, Two-Step Alkylation of Isobutane with C4 Olefins: Reaction of Isobutane with Initial Reaction Products, ACS Publications, 1977.

[77] L.M. Lee, P. Harriott, The kinetics of isobutane alkylation in sulfuric acid, Ind. Eng. Chem. Process 16 (1977) 282–287.

[78] L.F. Albright, L. Houle, A.M. Sumutka, R. Eckert, Alkylation of isobutane with butenes: effect of sulfuric acid compositions, Ind. Eng. Chem. Process 11 (1972) 446–450.

[79] H. Wang, X. Meng, G. Zhao, et al., Isobutane/butene alkylation catalyzed by ionic liquids: a more sustainable process for clean oil production, Green Chem. 19 (2017) 1462–1489.

[80] L.F. Albright, M.A. Spalding, J. Faunce, et al., Alkylation of isobutane with C4 olefins. 3. two-step process using sulfuric acid as catalyst, Ind. Eng. Chem. Res. 27 (1988) 391–397.

[81] S.I. Hommeltoft, O. Ekelund, J. Zavilla, Role of ester intermediates in isobutane alkylation and its consequence for the choice of catalyst system, Ind. Eng. Chem. Res. 36 (1997) 3491–3497.

[82] L.F. Albright, Two-Step Alkylation of Isobutane With C4 Olefins: Reactions of C4 Olefins With Sulfuric Acid, ACS Publications, 1977.

[83] R. Shlegeris, L. Albright, Alkylation of isobutane with various olefins in the presence of sulfuric acid, Ind. Eng. Chem. Process 8 (1969) 92–98.

[84] A. Berenblyum, L. Ovsyannikova, E. Katsman, et al., Acid soluble oil, by-product formed in isobutane alkylation with alkene in the presence of trifluoro methane sulfonic acid: Part I acid soluble oil composition and its poisoning effect, Appl. Catal. A-Gen. 232 (2002) 51–58.

[85] J. Zhang, C. Huang, B. Chen, et al., Isobutane/2-butene alkylation catalyzed by chloroaluminate ionic liquids in the presence of aromatic additives, J. Catal. 249 (2007) 261–268.

[86] C.P. Huang, Z.C. Liu, C.M. Xu, et al., Effects of additives on the properties of chloroaluminate ionic liquids catalyst for alkylation of isobutane and butene, Appl. Catal. A-Gen. 277 (2004) 41–43.

[87] Y. Liu, R. Hu, C. Xu, et al., Alkylation of isobutene with 2-butene using composite ionic liquid catalysts, Appl. Catal. A-Gen. 346 (2008) 189–193.
[88] H. Ma, R. Zhang, X. Meng, et al., Solid formation during composite-ionic-liquid-catalyzed isobutane alkylation, Energ. Fuel 28 (2014) 5389–5395.
[89] X. Zhang, R. Zhang, X. Meng, et al., Deactivation mechanism and activity-recovery approach of composite ionic liquids for isobutane alkylation, Appl. Catal. A-Gen. 557 (2018) 64–71.
[90] Z. Liu, C. Xu, R. Zhang, et al., Methods for Regenerating and Maintaining Activity of Ionic Liquid Catalyst and Producing Alkylate, U.S. Patent Application 13/383,946, 2012.
[91] S. Aschauer, L. Schilder, W. Korth, et al., Liquid-phase isobutane/butene-alkylation using promoted Lewis-acidic IL-catalysts, Catal. Lett. 141 (2011) 1405–1419.
[92] Z. Liu, C. Xu, R. Zhang, et al., Alkylation method using ionic liquid as catalyst, US Patent 9,096,487, 2015.
[93] J. Cui, J. de With, P.A.A. Klusener, et al., Identification of acidic species in chloroaluminate ionic liquid catalysts, J. Catal. 320 (2014) 26–32.
[94] S. Elomari, T.V. Harris, Regeneration of Acidic Catalysts, Google Patents, 2010.
[95] S.A. Elomari, H.-K.C. Timken, Recovery and Use of Conjunct Polymers from Ionic Liquid Catalysts, Google Patents, 2010.
[96] K. Ren, J. Long, Q. Ren, et al., A multi-scale computational study on the mechanism of protonation of isobutene and 2-butene over HY zeolite, Mol. Simul. 43 (2017) 1348–1355.
[97] Y.Y. Lu, Y.X. Li, J. Long, et al., Molecular simulation of isobutylene protonation reaction in HY-catalyzed C4 alkylation, J. Pet. (Pet. Process.) 30 (2014) 765–771.
[98] K. Ren, J. Long, Q. Ren, et al., Simulation of addition reactions in the HY molecular sieve-catalyzed isobutane/butene alkylation reaction, J. Pet. (Pet. Process.) 34 (2018) 24–33.
[99] K. Ren, J. Long, Q. Ren, et al., Simulation of hydrogen-negative ion transfer processes in the HY molecular sieve-catalyzed isobutane/butene alkylation reaction, J. Pet. (Pet. Process.) 33 (2017) 1061–1071.
[100] Y.Y. Lu, Y.X. Li, J. Long, et al., Molecular simulation of C4 alkylation reaction mechanism, J. Pet. (Pet. Process.) 32 (2016) 14–20.
[101] Y.C. Hou, C.X. Zhang, L.Y. Xiang, et al., Progress on the conformational relationship of molecular sieve catalysts in isobutane-butene alkylation reaction, J. Pet. (Pet. Process.) 34 (2018) 410–418.
[102] N. Gao, C.X. Zhang, Y.X. Li, Advances in Pt-loaded molecular sieve catalysts, Prog. Chem 35 (2016) 506–512.
[103] K. Peng, C.X. Zhang, Y.X. Li, Advances in regeneration methods for solid acid catalysts for isobutane/butene alkylation, Prog. Chem. 34 (2015) 3296–3302.
[104] C. Sievers, I. Zuazo, A. Guzman, et al., Stages of aging and deactivation of zeolite LaX in isobutane/2-butene alkylation, J. Catal. 246 (2007) 315–324.
[105] J. Wang, P. Sporns, MALDI-TOF MS analysis of isoflavones in soy products, J. Agric. Food Chem. 48 (2000) 5887–5892.
[106] I.L. Buurmans, J. Ruiz-Martínez, W.V. Knowles, et al., Catalytic activity in individual cracking catalyst particles imaged throughout different life stages by selective staining, Nat. Chem. 3 (2011) 862–867.
[107] D. Mores, E. Stavitski, M.H. Kox, et al., Space-and time-resolved in-situ spectroscopy on the coke formation in molecular sieves: methanol-to-olefin conversion over H-ZSM-5 and H-SAPO-34, Chem. Eur. J. 14 (2008) 11320–11327.
[108] D. Mores, J. Kornatowski, U. Olsbye, et al., Coke formation during the methanol-to-olefin conversion: in situ microspectroscopy on individual H-ZSM-5 crystals with different Brønsted acidity, Chem. Eur. J. 17 (2011) 2874–2884.
[109] F.A. Diaz-Mendoza, L. Pernett-Bolaño, N. Cardona-Martínez, Effect of catalyst deactivation on the acid properties of zeolites used for isobutane/butene alkylation, Thermochim. Acta 312 (1998) 47–61.

[110] A. Feller, J.O. Barth, A. Guzman, et al., Deactivation pathways in zeolite-catalyzed isobutane/butene alkylation, J. Catal. 220 (2003) 192–206.
[111] C. Flego, I. Kiricsi, W.O. Parker, et al., Spectroscopic studies of LaHY-FAU catalyst deactivation in the alkylation of isobutane with 1-butene, Appl. Catal. A-Gen. 124 (1995) 107–119.
[112] L.M. Petkovic, D.M. Ginosar, The effect of supercritical isobutane regeneration on the nature of hydrocarbons deposited on a USY zeolite catalyst utilized for isobutane/butene alkylation, Appl. Catal. A-Gen. 275 (2004) 235–245.
[113] R. Shiriyazdanov, Regeneration of zeolite-containing catalysts of alkylation of isobutane with the butane-butene faction in supercritical carbon dioxide, Russ. J. Phys. Chem. B 5 (2011) 1080–1083.
[114] W.L. Wang, B.J. Liu, Y.T. Xu, Advances in solid acid catalysts for isobutane/butene alkylation, Oil Gas Chem. 37 (2008) 110–114.
[115] K. Arata, I. Toyoshima, High catalytic activity of Iron sulfates for Friedel-Crafts type benzylation of toluene with benzyl chloride, Chem. Lett. (1974) 929–930.
[116] F.Q. Xie Wenhua, Regeneration studies of solid super acids in isobutane/butene alkylation reactions, Mol. Catal. (2004) 326–331.
[117] Z. Qiang, J.Q. Ze, M.E. Ze, Performance and deactivation studies of isobutane/1-butene alkylation reaction with solid superacid (SO_4^{2-}/ZrO_2), J. Chem. Educ. (2005) 1130–1132.
[118] Y.G. He, Y.F. He, Structural and physicochemical characterization of coke precursors on the surface of solid acid alkylation catalysts, J. Catal. (2000) 175–178.
[119] J.H. Yao, H.T. Wang, K.H. Hou, Deactivation and regeneration of Hβ-Al_2O_3 alkylation catalysts, Pet. Chem. 44 (2013) 71–76.
[120] Y.H. Ruan, Y.F. Liu, Z.C. Liu, Catalytic effect of silica-loaded heteropolyacids on the alkylation of isobutane with butene I. catalyst preparation, characterization and deactivation, J. Catal. (2004) 948–954.
[121] A. Schneider, R.M. Kennedy, Reactions of hydrocarbons induced by alkyl fluoride—boron trifluoride. II. Self-alkylation and disproportionation of isoparaffins1, J. Am. Chem. Soc. 73 (1951) 5017–5023.
[122] M. Misono, Heterogeneous catalysis by heteropoly compounds of molybdenum and tungsten, Catal. Rev. 29 (1987) 269–321.
[123] C. Sievers, J.S. Liebert, M.M. Stratmann, et al., Comparison of zeolites LaX and LaY as catalysts for isobutane/2-butene alkylation, Appl. Catal. A-Gen. 336 (2008) 89–100.
[124] A. Zheng, S. Li, S.B. Liu, et al., Acidic properties and structure–activity correlations of solid acid catalysts revealed by solid-state NMR spectroscopy, Acc. Chem. Res. 49 (2016) 655–663.
[125] M. Boronat, P. Viruela, A. Corma, Reaction intermediates in acid catalysis by zeolites: prediction of the relative tendency to form alkoxides or carbocations as a function of hydrocarbon nature and active site structure, J. Am. Chem. Soc. 126 (2004) 3300–3309.
[126] M.F. Simpson, J. Wei, S. Sundaresan, Kinetic analysis of isobutane/butene alkylation over ultrastable H−Y zeolite, Ind. Eng. Chem. Res. 35 (1996) 3861–3873.
[127] A. Corma, A. Martinez, C. Martinez, The role of extraframework aluminum species in USY catalysts during isobutane/2-butene alkylation, Appl. Catal. A-Gen. 134 (1996) 169–182.
[128] A. Corma, A. Martinez, P. Arroyo, et al., Isobutane/2-butene alkylation on zeolite beta: influence of post-synthesis treatments, Appl. Catal. A-Gen. 142 (1996) 139–150.
[129] A. Feller, A. Guzman, I. Zuazo, et al., On the mechanism of catalyzed isobutane/butene alkylation by zeolites, J. Catal. 224 (2004) 80–93.
[130] A. Corma, A. Martinez, C. Martinez, Isobutane/2-butene alkylation on MCM-22 catalyst. Influence of zeolite structure and acidity on activity and selectivity, Catal. Lett. 28 (2) (1994) 187–201.

[131] K. Yoo, E.C. Burckle, P.G. Smirniotis, Comparison of protonated zeolites with various dimensionalities for the liquid phase alkylation of i-butane with 2-butene, Catal. Lett. 74 (2001) 85–90.

[132] M. Stöcker, H. Mostad, T. Rørvik, Isobutane/2-butene alkylation on faujasite-type zeolites (H EMT and H FAU), Catal. Lett. 28 (1994) 203–209.

[133] R.W. Wang, S. Lin, X.J. Meng, et al., Influence of catalysts particle size alkylation of isobutane with butene over Hβ zeolites, Chem. J. Chin. Univ. 24 (2003) 205–207.

[134] A. Goldsby, Chemical Recovery of Used Sulfuric Acid Alkylation Catalyst, 1977.

[135] Y. Geng, W.Z. Qi, H. Liu, et al., A new process for preparing ammonium sulfate by alkylating waste acid, Chem. Eng. New York 43 (2015). 1-4+24.

[136] H.J. Cheng, C.S. Hu, H.W. Lv, et al., Activated clay was produced by alkylating waste sulfuric acid, Environ. Prot. Petrochem. Ind. 27 (2004) 57–61.

[137] X.N. Cai, Oxidative degradation—Adsorption Treatment of Alkylated Waste Sulfuric Acid to Prepare Manganese Sulfate and Its Waste Residue Resource Utilization, Guangxi University, 2018.

[138] Z. Zhou, X. Zhang, F. Yang, et al., Polymeric carbon material from waste sulfuric acid of alkylation and its application in biodiesel production, J. Clean. Prod. 215 (2019) 13–21.

[139] D. Wang, Z. Chen, Z. Zhou, et al., Catalytic upgrading of volatiles from coal pyrolysis over sulfated carbon-based catalysts derived from waste red oil, Fuel Process. Technol. 189 (2019) 98–109.

[140] X. Cai, F. Shen, Y. Zhang, et al., Mineralization of organics in hazardous waste sulfuric acid by natural manganese oxide ore and a combined MnO_2/activated carbon treatment to produce qualified manganese sulfate, J. Hazard. Mater. 366 (2019) 466–474.

[141] J. Zhang, X. Jin, H. Zhao, et al., Synergistic advanced oxidation process for enhanced degradation of organic pollutants in spent sulfuric acid over recoverable apricot shell-derived biochar catalyst, RSC Adv. 12 (2022) 1904–1913.

[142] J. Zhang, X. Jin, C. Yang, Efficient removal of organic pollutants in waste sulfuric acid by an advanced oxidation process using coconut shell-derived biochar to produce qualified poly aluminium sulfate, Sep. Purif. Technol. 293 (2022) 121057.

[143] J.M. Wang, Process flow and application prospect of WSA wet acid process, Chem. Manag. (2008) 71–75.

[144] F. Zhang, WSA wet acid process and its application in China, in: Sulfur and Phosphorus Design and Powder Engineering, 2011, pp. 1–5.

[145] S.L. Xuan, Application of WSA wet sulfuric acid process, Anhui Chem. Ind. B (2014) 53–55.

[146] M. Asof, S.A. Rachman, W.A. Nurmawi, et al., Recovery of H_2SO_4 from spent acid waste using bentonite adsorbent, in: MATEC Web of Conferences, EDP Sciences, 2017, p. 02007.

[147] Z. Li, J.F. Zhou, L.H. Zhang, et al., Study on regeneration of alkylated waste sulfuric acid by oxidation with concentrated nitric acid, Chem. Ind. 46 (2017) 884–886. 890.

[148] C.X. Li, Y.B. Zhang, Application of Fenton reaction in comprehensive utilization of alkylated waste sulfuric acid, Guangdong Chem. Ind. 44 (2017) 208–209.

[149] C.L. Yang, Isoparaffin Alkylation Process With Periodic Catalyst Regeneration, US Patent 3,893,942, 1975.

[150] S. Xie, S. He, X. Qian, et al., Regeneration and deactivation of β zeolite catalysts in the alkylation of isobutane/butane, Petrochem. Technol. 33 (2004) 1402–1404.

[151] J. Li, Y.X. Song, Advances in the application of zeolites in the alkylation of isobutane/butene, Prog. Chem. (2013) 122–126.

[152] D.M. Ginosar, D.N. Thompson, K.C. Burch, Recovery of alkylation activity in deactivated USY catalyst using supercritical fluids: a comparison of light hydrocarbons, Appl. Catal. A-Gen. 262 (2004) 223–231.

[153] Y. He, Alkylation of isoparaffin with olefin under supercritical fluid state, Chin. J. Catal. 20 (1999) 408–413.

[154] D.M. Ginosar, D.N. Thompson, K. Coates, et al., Method for Reactivating Solid Catalysts Used in Alkylation Reactions, US Patent 6,579,821, 2003.
[155] D.N. Thompson, D.M. Ginosar, K.C. Burch, Regeneration of a deactivated USY alkylation catalyst using supercritical isobutane, Appl. Catal. A-Gen. 279 (2005) 109–116.
[156] T. Rørvik, H. Mostad, O.H. Ellestad, et al., Isobutane/2-butene alkylation over faujasite type zeolites in a slurry reactor. Effect of operating conditions and catalyst regeneration, Appl. Catal. A-Gen. 137 (1996) 235–253.
[157] C.A. Querini, Isobutane/butene alkylation: regeneration of solid acid catalysts, Catal. Today 62 (2000) 135–143.
[158] J.H. Yao, H.T. Wang, H.K. Hu, Deactivation and regeneration of Hβ-Al_2O_3 alkylation catalysts, Pet. Chem. 44 (2013) 73–78.
[159] F. Qiang, H.Y. Kung, X. Wenhua, et al., A Low-Temperature Regeneration Method for Solid Acid Alkylation, Catalysts (2021) CN1102438C.
[160] W.H. Xie, F. Qiang, Regeneration studies of solid super acids in isobutane/butene alkylation reactions, Mol. Catal. 18 (2004) 326–331.

CHAPTER

6

Alkylation process intensification

Through the investigation of reaction mechanisms, kinetics, and industrial reactors, a comprehensive understanding of the alkylation reaction from the micro- to macro-level has been achieved, which is the foundation for process intensification. From the perspective of process design, process optimization and intensification represent a vital part in terms of economics and environment protection, as well as in guiding decision-making within the chemical industry market. This chapter focuses on summarizing novel methods for process intensification and optimization, presenting the intensification mechanism of interfacial behaviors, and outlining the principles of novel additive design.

6.1 Multiobjective (MO) optimization

In the chemical industries, the production of a particular product can rapidly shift from being cost-competitive to unsellable. Competition is often intense, and even slight changes in market prices can be critical. Product price is influenced by both raw material expenses and production costs. While raw material costs cannot be controlled, production costs can be. Analyzing the behavior of various alternative process configurations is essential for achieving the most efficient process with the lowest production costs. Additionally, it is important to identify the parameters that have the greatest impact on the operability of the unit. This multidisciplinary study concerns the optimal design of processes aimed at maximizing profit and reducing environmental impacts. This can be accomplished through a combination of conventional chemical process design practices, environmental impact assessments, and advanced mathematical optimization methods.

Aliphatic Alkylation
https://doi.org/10.1016/B978-0-443-21984-9.00003-9

Specifically, García et al. introduced a hybrid simulation-MO optimization approach that simultaneously optimizes production costs and minimizes the environmental impacts associated with C4 alkylation [1]. This approach has also facilitated the determination of flowsheet configurations and process variables required to produce isooctane in a manner that fulfills the aforementioned dual objective.

The MO optimization approach, initially applied by Azapagic [2], has proven effective in providing an optimal set of design parameters that address both economic and environmental concerns. This methodology has been applied in various case studies, including those by Erol and Thöming [3], emphasizing the eco-design of reuse and recycling networks. The outcomes of MO optimization are depicted as a Pareto curve, which illustrates the relationship between an environmental impact variable and economic potential (EP) (see Fig. 6.1). Within the Pareto curve, the best set of alternatives is presented, simultaneously achieving both economic and environmental objectives. An inferior solution occurs when a superior solution can be attained without compromising the remaining objective functions. Such a solution is also referred to as a dominated solution. Hence, the Pareto curve encompasses all noninferior solutions to the optimization problem, also known as nondominated alternatives [4,5].

There are two commercial alkylation processes including (a) the external refrigeration process, and (b) the auto-refrigeration process, as shown in Figs. 6.2 and 6.3, respectively.

For economic analysis, the EP as defined in Smith [6] was utilized, as this criterion is favored during the initial design phases. It is especially valuable for comparing the relative benefits of various structural options

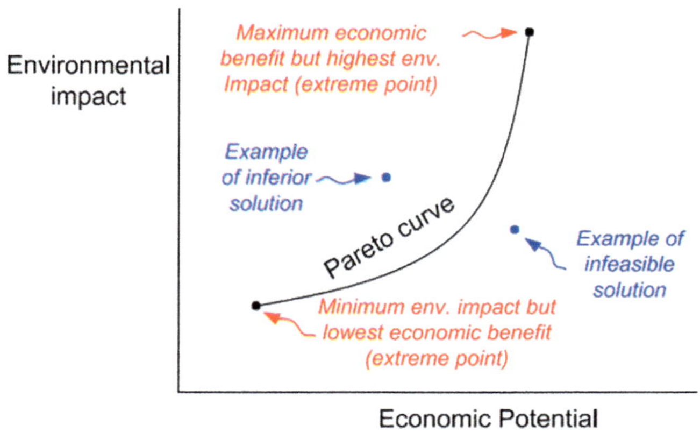

FIG. 6.1 Schematic representation of the Pareto curve obtainable by the multiobjective (MO) optimization method. *From reference N. García, M.J. Fernández-Torres, J.A. Caballero, Simultaneous environmental and economic process synthesis of isobutane alkylation, J. Clean. Prod. 81 (2014) 270–280.*

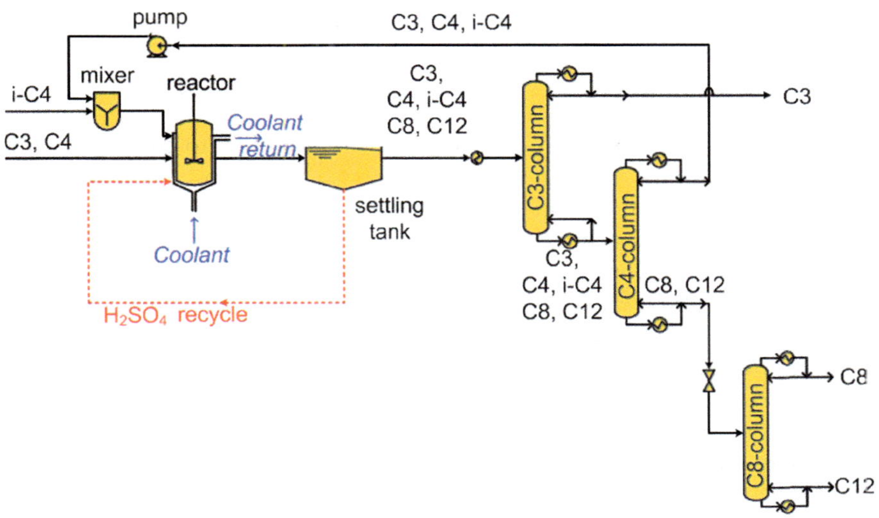

FIG. 6.2 Flowsheet diagram with external refrigeration taken from the initial configuration. *From reference N. García, M.J. Fernández-Torres, J.A. Caballero, Simultaneous environmental and economic process synthesis of isobutane alkylation, J. Clean. Prod. 81 (2014) 270–280.*

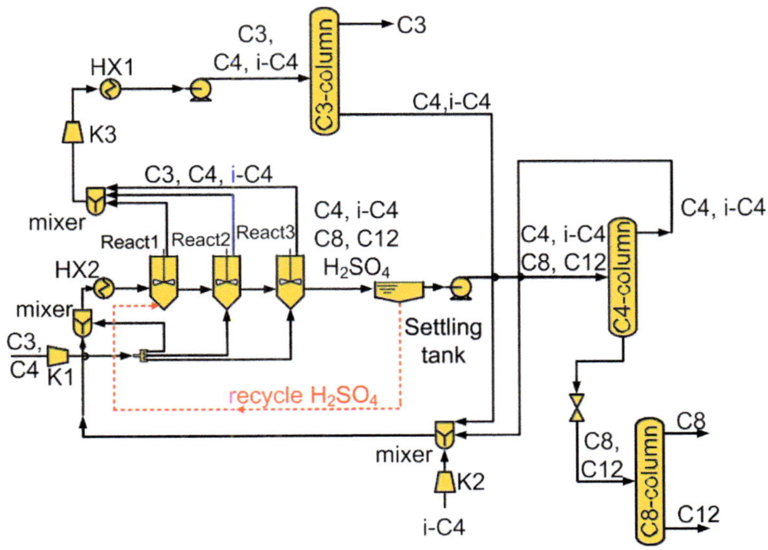

FIG. 6.3 Flowsheet diagram with auto-refrigerated reactors. K_i and HX_i refer to compressors and heat exchanger, respectively. *From reference N. García, M.J. Fernández-Torres, J.A. Caballero, Simultaneous environmental and economic process synthesis of isobutane alkylation, J. Clean. Prod. 81 (2014) 270–280.*

in the flowsheet and different configurations of operating parameters. This term may be represented as

$$EP = \text{value of products} - \text{annualized fixed costs} - \text{variable costs} \quad (6.1)$$

The MO optimization results, depicted by two Pareto curves, are integrated with three configurations labeled "1st Solution," "2nd Solution," and "3rd Solution" in Fig. 6.4. No single optimal configuration within the superstructure achieves the best results while simultaneously meets both criteria (maximizing EP and minimizing either of the two environmental indicators: EI-99 or GWP). These composite Pareto curves illustrate the optimal configurations for conducting C4 alkylation. Each solution or configuration is concisely presented and described in Fig. 6.5.

Fig. 6.4 demonstrates that the 1st solution, detailed in Table 6.3 for the corresponding operating conditions, exhibits the lowest environmental impacts with regard to both environmental indicators, EI-99 and GWP. However, its EP is the weakest when compared to the 2nd and 3rd solutions or configurations. Conversely, the auto-refrigerated alternative represents the best solution based solely on economic criteria but incurs the highest environmental impact near its right end. Therefore, decision-makers must assess these results to determine which configuration (along with the corresponding optimal operating conditions) to ultimately employ.

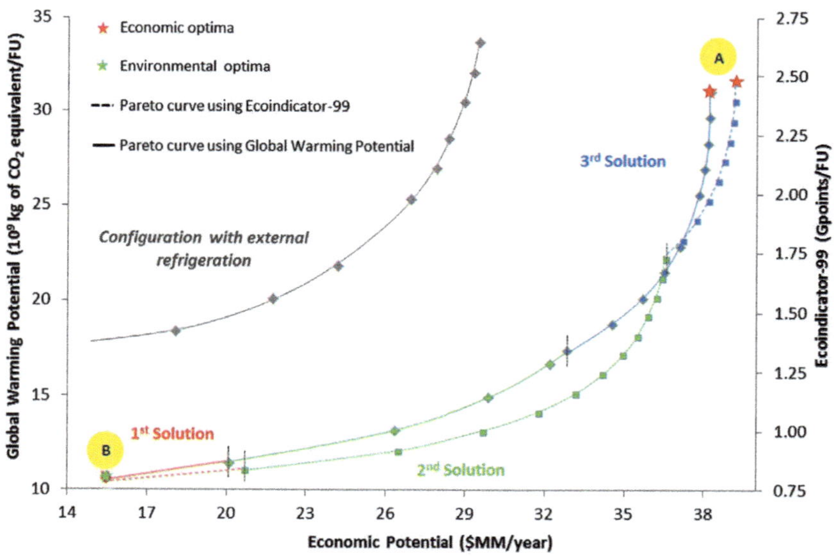

FIG. 6.4 Pareto curves for both environmental indicators (EI-99 and GWP). *From reference N. García, M.J. Fernández-Torres, J.A. Caballero, Simultaneous environmental and economic process synthesis of isobutane alkylation, J. Clean. Prod. 81 (2014) 270–280.*

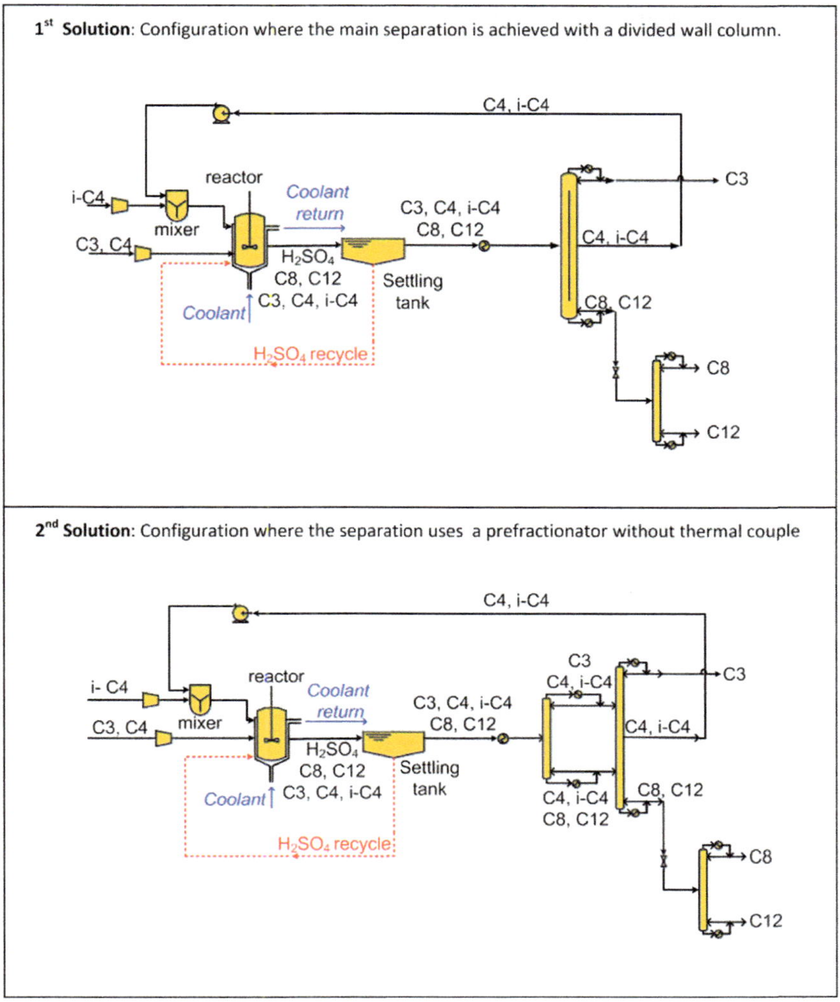

FIG. 6.5 Configurations that fall on the Pareto curve (set of noninferior solutions) in Fig. 6.4. *From reference N. García, M.J. Fernández-Torres, J.A. Caballero, Simultaneous environmental and economic process synthesis of isobutane alkylation, J. Clean. Prod. 81 (2014) 270–280.*

Table 6.1 furnishes details about the extreme points on the two Pareto curves. In the case involving the GWP indicator, the breakeven selling price of isooctane was calculated by setting the value of EP to zero (no profit) in Eq. (6.1) and using the information from Tables 6.2 and 6.3. The calculated breakeven selling price of isooctane was \$0.42/kg for the economic optimum and \$0.86/kg for the environmental optimum. This parameter should be considered, as market prices for isooctane can

TABLE 6.1 Information about the p-anchor points.

	Maximizing economic potential (EP)			
	Minimizing Ecoindicator-99		Minimizing GWP	
	Economic optimum	Environmental optimum	Economic optimum	Environmental optimum
EP (economic potential), Eq. (6.1) for a selling price of C8 of \$1.9/kg	\$39.2MM/yr	\$15.5MM/yr	\$38.2MM/yr	\$15.5MM/yr
Environmental indicator	2.5×10^9 Point/FU	0.8×10^9 Point/FU	31.0×10^9kg CO_2eq/FU	10.6×10^9 kg CO_2eq/FU

Data from reference N. García, M.J. Fernández-Torres, J.A. Caballero, Simultaneous environmental and economic process synthesis of isobutane alkylation, J. Clean. Prod. 81 (2014) 270–280.

TABLE 6.2 Details of the configuration of the superstructure which drives to the maximum EP—3rd solution in Fig. 6.4 (GWP curve—Point A) and Fig. 6.5.

Main equipment characteristics	
Reactor (1, 2, or 3)	Heat exchanger 1 (HX_2)
Volume = 13.89 m^3	Heat flow = −2666.7 kW
Heat flow = 0 kW	
	C4 column
	Number of trays: (r) and 5 (s)
C3 column	
Number of trays: 8 (rectifying) 24 (stripping)	Reboiler heat = 10,373.1 kW
Reboiler heat: 1348.4 kW	Condenser heat = 4935.8 kW
Condenser heat: 1032.3 kW	Distillate temperature = 47.4°C
Distillate temperature = 48.3°C	Bottoms temperature = 182.7°C
Bottoms temperature = 85.5°C	Lights ump:
	Shaft work = 5.6 kW
Compressor 1 (K_1):	
Shaft work = 14.5 kW	Heavies pump:
	Shaft work = 18.5 kW
Compressor 1 (K_2);	
Shaft work = 11.3 kW	C8 column:

TABLE 6.2 Details of the configuration of the superstructure which drives to the maximum EP—3rd solution in Fig. 6.4 (GWP curve—Point A) and Fig. 6.5—cont'd

Main equipment characteristics	
Compressor 1(K_3):	Number of trays: 11 (r) and 4 (s)
Shaft work = 166.1 kW	Reboiler beat = 64.1 kW
	Condenser heat = 202.1 kW
Heat exchanger 1 (HX_1):	Distillate temperature = 95.6°C
Heat flow = −785.2 kW	Bottoms temperature = 165.5°C
Total molar now rates and molar fractions of selected streams:	
• C3 + C4 feed stream:	• C12 product stream
Flow rate = 40.0 kmol/h	Flow rate = 1.6 kmol/h
$X_{C4} = 0.7854$	$X_{C12} = 0.8214$
• *i*-C4 feed stream:	• C3 product stream
Flow rate = 31.2 kmol/h	Flow rate = 10.0 kmol/h
$X_{i\text{-}C4} = 1$	$X_{C3} = 0.8611$
• C8 product stream:	• Recycle stream (−1.6°C)
Flow rate = 28.5 kmol/h	Flow rate = 1123.1 kmol/h
$X_{C8} = 0.9915$	$X_{i\text{-}C4} = 0.9583$

Data from reference N. García, M.J. Fernández-Torres, J.A. Caballero, Simultaneous environmental and economic process synthesis of isobutane alkylation, J. Clean. Prod. 81 (2014) 270–280.

influence decision-makers regarding the most suitable configuration in terms of competitiveness with other isooctane suppliers.

This process leads to the generation of three different structural alternatives. These alternatives are more sustainable by design compared to those derived from other commercial strategies, as they feature lower energy consumption and cleaner production.

Additionally, a simple reactor-separation-recycle model of the plant is established by Dragnea et al., who calculate the two control structures with the effect of reactor volume and reaction temperature on conversion, selectivity, and recycle flowrate [7]. The established control structures are different based on the following specifications:

Control structure 1: Both reactants flowrates at the process inlet are fixed, as presented in Fig. 6.6.

$$F_{A0} = SP_A \tag{6.2}$$

$$F_{B0} = SP_B \tag{6.3}$$

TABLE 6.3 Details of the configuration of the superstructure which drives to the minimum Global Warming Potential—1st solution in Fig. 6.4 (GWP curve—Point A) and Fig. 6.5.

Main equipment characteristics:	
C3 + C4 compressor:	Divided wall column:
Shaft work = 78.0 kW	Number of trays in each section;
i-C4 compressor:	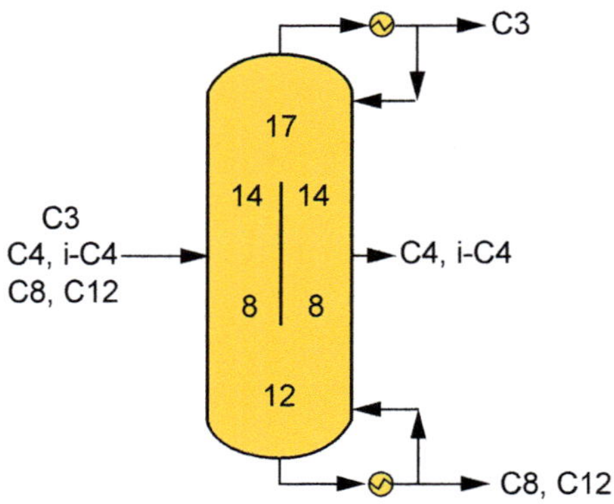
Shaft work = 45.6 kW	
Reactor:	
Volume = 87.19 m^3	
Heat flow = −1475.2 kW	
C8 column:	Reboiler heat = 1493.7 kW
Number of trays: 5 (rectifying) 4 (stripping)	Condenser heat = 533.2 kW
Reboiler heat: 22.5 kW	Distillate temperature = 16.9°C
Condenser heat: 173.9 kW	Bottoms temperature = 215.6°C
Distillate temperature = 100.0°C	Recycle pump:
Bottoms temperature-207.7°C	Shaft work = 0.2 kW
Heat exchanger:	
Heat flow = +404.6 kW	
Total molar flowrates and molar fractions of selected streams:	
• C3 + C4 feed stream:	• C12 product stream
Row rate-40.0 kmol/h	Flowrate = 7.6 kmol/h
X_{C4} = 0.7854	X_{C12} = 0.9783
• *i*-C4 feed stream:	• C3 product stream:
Flowrate = 24.0 kmol/h	Flowrate = 8.6 kmol/h
$X_{i\text{-}C4=}1$	X_{C3} = 0.9943
C8 product stream:	• Recycle stream (55.0°C)
Flowrate = 16.4 kmol/h	Flowrate = 107.3 kmol/h
X_{C8} = 0.9953	$X_{i\text{-}C4}$ = 0.9703

Data from reference N. García, M.J. Fernández-Torres, J.A. Caballero, Simultaneous environmental and economic process synthesis of isobutane alkylation, J. Clean. Prod. 81 (2014) 270–280.

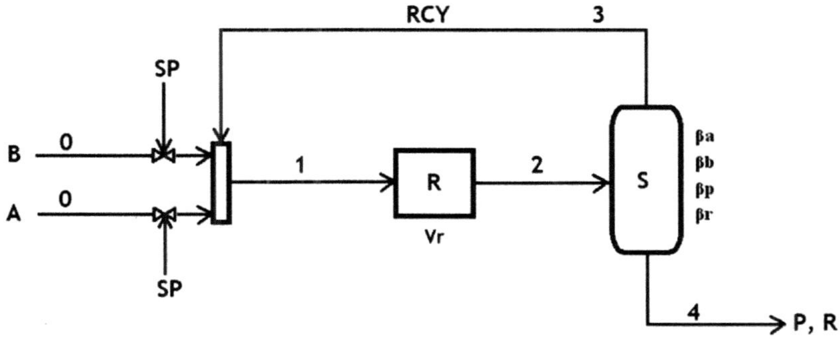

FIG. 6.6 Reaction-separation-recycle system, fixed fresh reactants. *From reference P. Dragnea, C.S. Bîldea, Process optimization of a butane-butene alkylation plant, UPB Sci. Bull. 76 (2014) 71–84.*

The mass balance of the process, calculated for total reactor volume of 30 m^3 and reaction temperature of 263 K, is presented in Table 6.4.

It should be remarked that the product flowrates, $F_{P,4}$ and $F_{R,4}$, are solely determined by the process-inlet flowrates $F_{A,0}$ and $F_{B,0}$. Both production rate and selectivity can be directly set in this control structure.

The influence of the reactor volume on butene conversion and recycle flowrate, at different reactor temperatures, is presented in Fig. 6.7.

Control structure 2: The butene flow rate at process inlet and combined fresh isobutane are fixed (Fig. 6.8).

$$F_{A0} = SP_A \tag{6.4}$$

$$F_{B0} + F_{B3} + F_{A3} + F_{B0} + F_{B0} = SP_B \tag{6.5}$$

TABLE 6.4 Material balance of control structure 1.

	Stream				
Component	**F_0 (kmol/h)**	**F_1 (kmol/h)**	**F_2 (kmol/h)**	**F_3 (kmol/h)**	**F_4 (kmol/h)**
A	52.5	94.7	42.2	42.2	0
B	50	1663.9	1613.9	1613.9	0
P	0	0	47.5	0	47.5
R	0	0	2.5	0	2.5

Data from reference P. Dragnea, C.S. Bîldea, Process optimization of a butane-butene alkylation plant, UPB Sci. Bull. 76 (2014) 71–84.

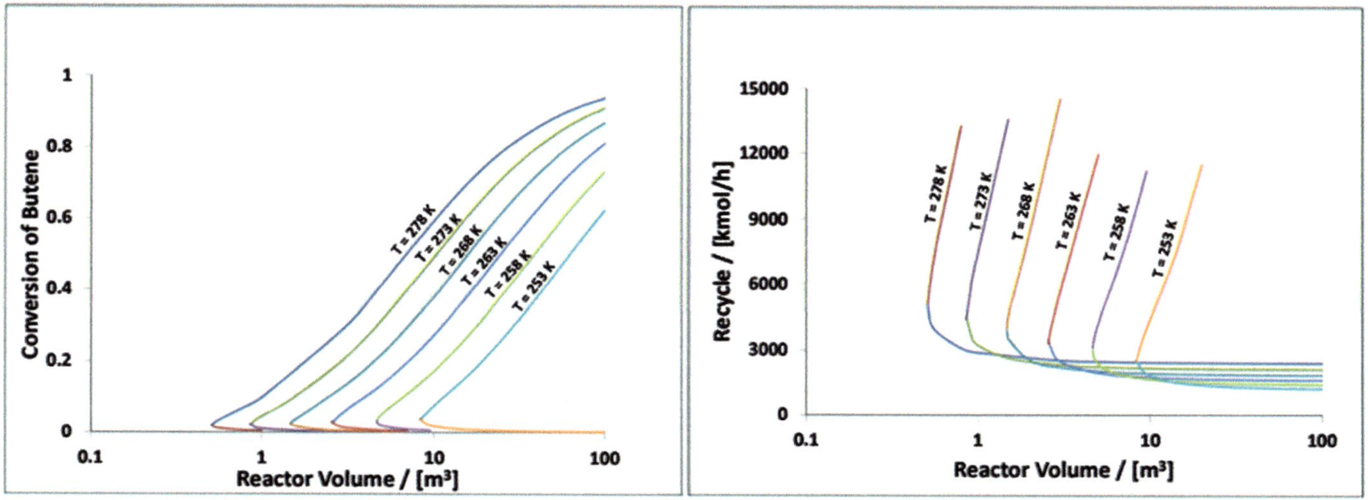

FIG. 6.7 Influence of reactor volume on conversion and recycle flowrate at different temperatures. *From reference P. Dragnea, C.S. Bîldea, Process optimization of a butane-butene alkylation plant, UPB Sci. Bull. 76 (2014) 71–84.*

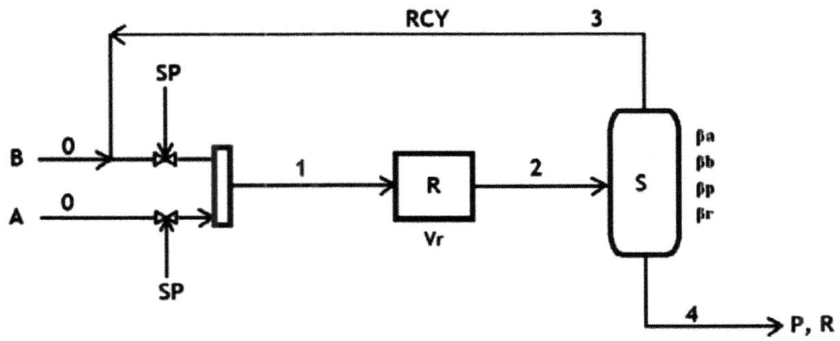

FIG. 6.8 Reaction-separation-recycle system, fixed flowrates at reactor inlet. *From reference P. Dragnea, C.S. Bîldea, Process optimization of a butane-butene alkylation plant, UPB Sci. Bull. 76 (2014) 71–84.*

In structure 2, the reactant streams at the reactor inlet are fixed. The fresh butene is introduced under flow control, and the fresh isobutane balances the recycle flow to maintain constant the combined flow at the reactor inlet. The mathematical model consists of the same equations as the previous control structure, with the only difference being the specified variables.

Fig. 6.9 presents the recycle flow rate versus the reactor volume at different temperatures. For each temperature, there is a minimum reactor volume for which two steady states exist. For constant reaction temperature, large reactor leads to low recycle flow (Fig. 6.9) and high conversion (Fig. 6.10). In contrast to the previous control structure, the recycle flow rate is less sensitive to changes of the reactor volume or reaction temperature.

This operating scheme offers the advantage of maintaining nearly constant flows at the reactor and separation section inlets. This affords greater process flexibility, as the separation section is not subjected to significant disturbances in the inlet flow. However, it does not allow for direct control of selectivity, potentially resulting in undesired products.

Except for the abovementioned methods, Bird et al. utilized the response surface methods to optimize the operations of the alkylation unit [8]. The initial step in process improvement involves identifying the general direction using a first-order profit response surface constructed from available operating data. Subsequently, the performance of the unit is assessed at the refinery by systematically adjusting the key factors in the anticipated direction of process enhancement. The operating outcomes are then utilized to create a localized second-order profit response surface, yielding a revised set of optimal targets. Multiple linear regression models

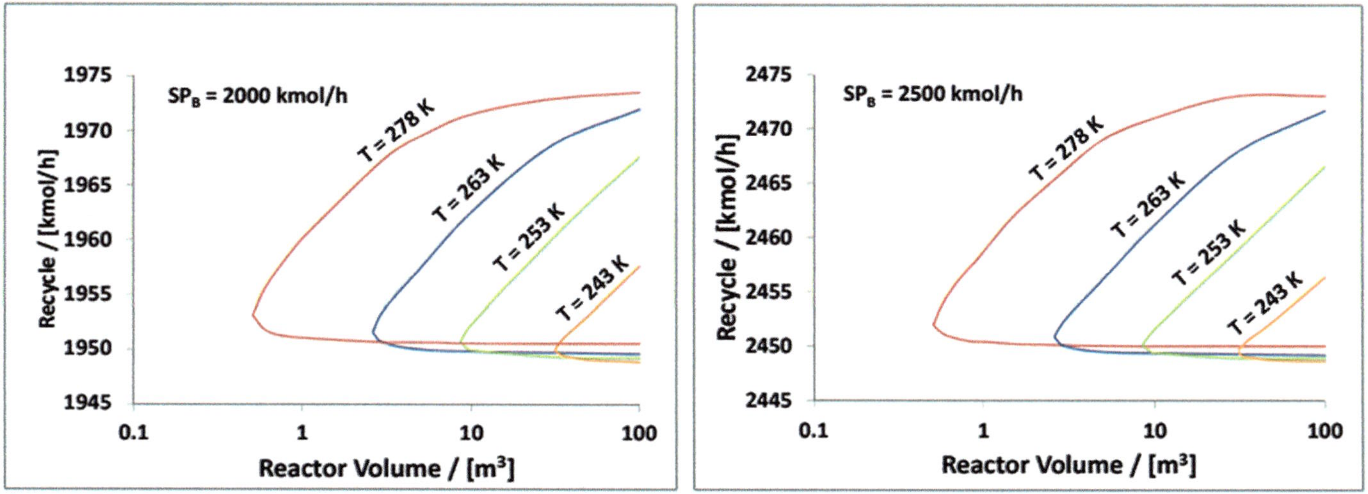

FIG. 6.9 Influence of reactor volume on recycle flow. *From reference P. Dragnea, C.S. Bîldea, Process optimization of a butane-butene alkylation plant, UPB Sci. Bull. 76 (2014) 71–84.*

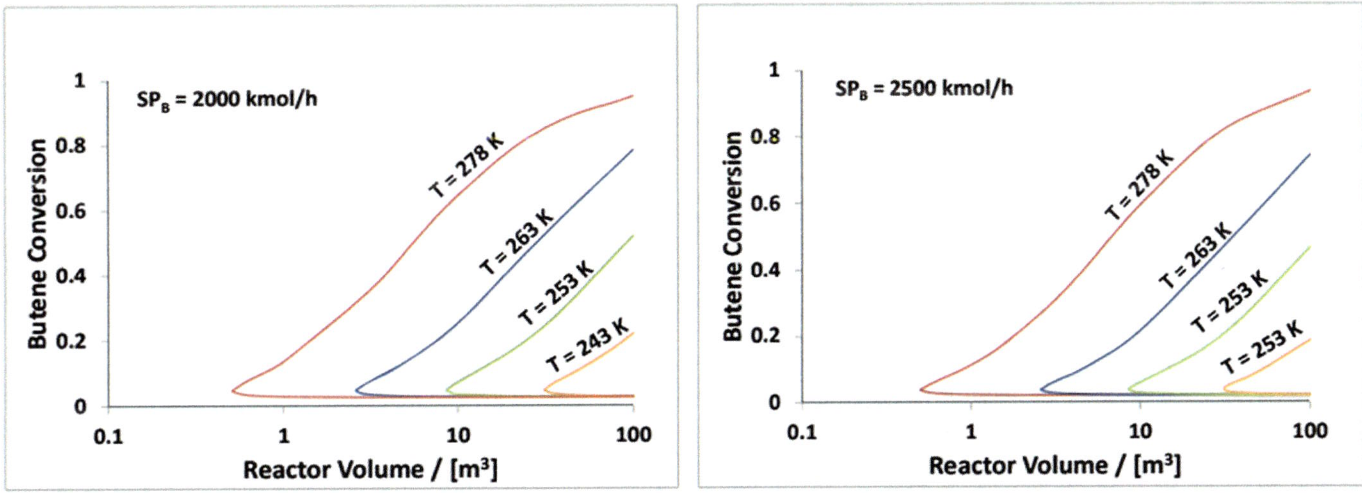

FIG. 6.10 Influence of reactor volume on conversion. *From reference P. Dragnea, C.S. Bîldea, Process optimization of a butane-butene alkylation plant, UPB Sci. Bull. 76 (2014) 71–84.*

are employed to forecast alkylate yield, octane number, and isostripper or de-isobutanizer reboiler duty based on key process variables. Figs. 6.11–6.13 depict the alkylate yield, octane number, and the isostripper reboiler duty response surfaces.

Fig. 6.14 indicates that the RON is maximized at higher I/O ratios and within the 90%–92% acid strength range. The decline in RON at higher reactor temperatures is depicted in Fig. 6.15, which illustrates a contour map of the RON as a function of I/O ratio and reactor temperature. The reduction in RON appears to be more pronounced at lower I/O ratios. The analysis estimated an optimal I/O ratio target of 12 and an optimum acid strength in the range of 92%–94%.

Numerous research studies have demonstrated that the RON and product components are significantly influenced by factors such as feedstock composition, reactor structure, and operating conditions. These findings are based on mathematical and correlation models applied in specific reactors. For instance, Nurmakanova et al. utilized density functional theory (DFT) to calculate thermodynamic characteristics and reaction kinetics factors in order to develop a mathematical model of isobutane

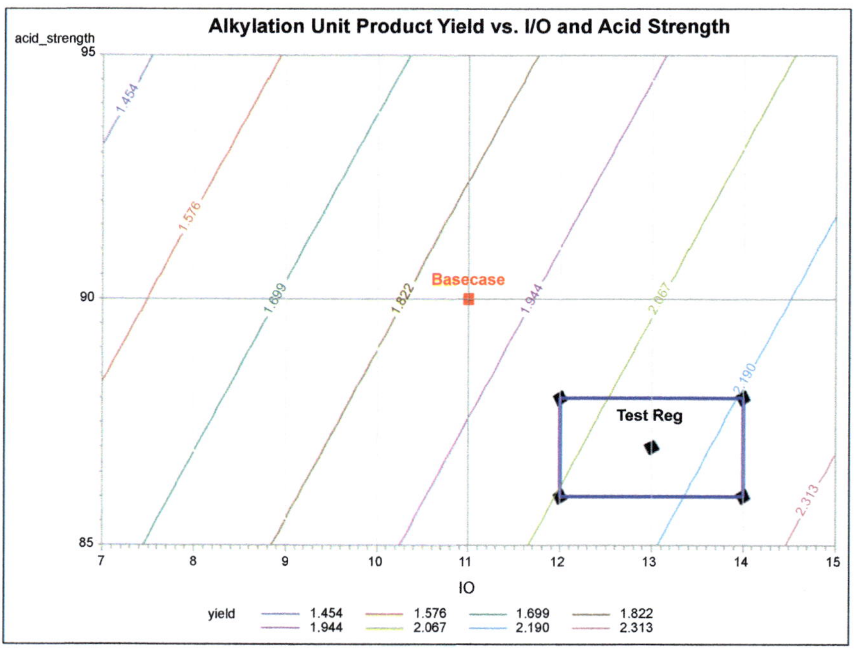

FIG. 6.11 Yield vs I/O and acid strength contour map. *From reference J. Bird, D. Seillier, M. Teders, Optimization of Refining Alkylation Process Unit Using Response Surface Methods.*

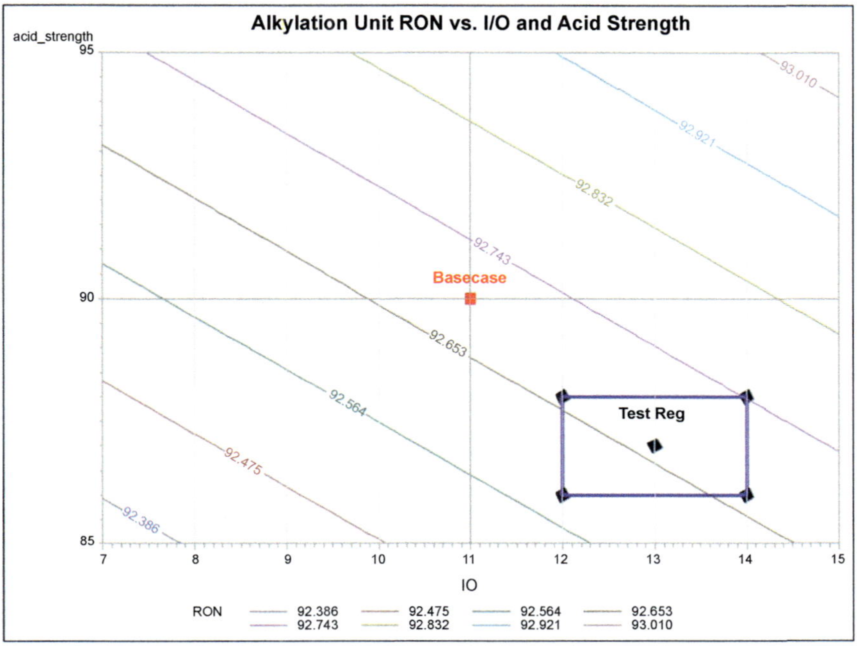

FIG. 6.12 Octane number vs I/O and acid strength contour map. *From reference J. Bird, D. Seillier, M. Teders, Optimization of Refining Alkylation Process Unit Using Response Surface Methods.*

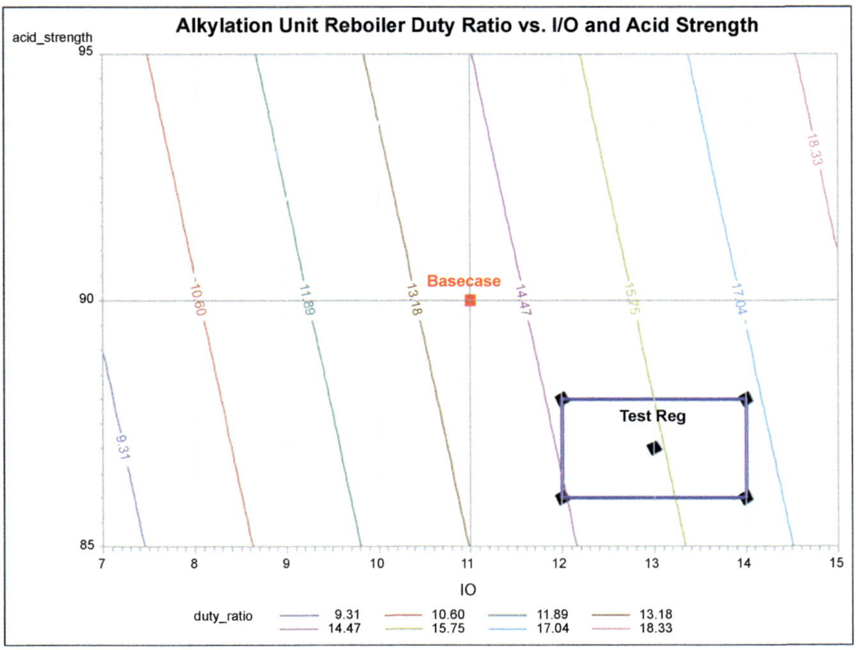

FIG. 6.13 Reboiler duty ratio vs I/O and acid strength contour map. *From reference J. Bird, D. Seillier, M. Teders, Optimization of Refining Alkylation Process Unit Using Response Surface Methods.*

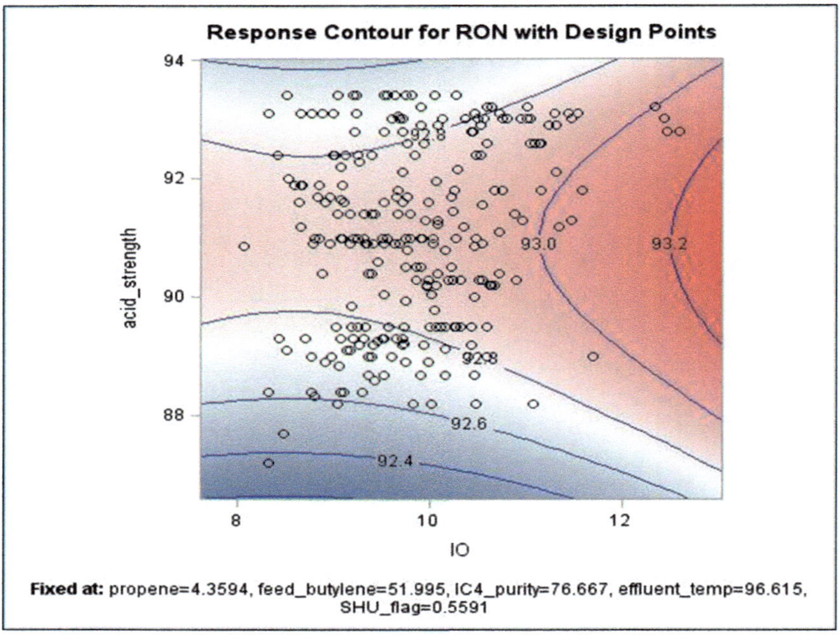

FIG. 6.14 RON per barrel vs I/O and acid strength contour map. *From reference J. Bird, D. Seillier, M. Teders, Optimization of Refining Alkylation Process Unit Using Response Surface Methods.*

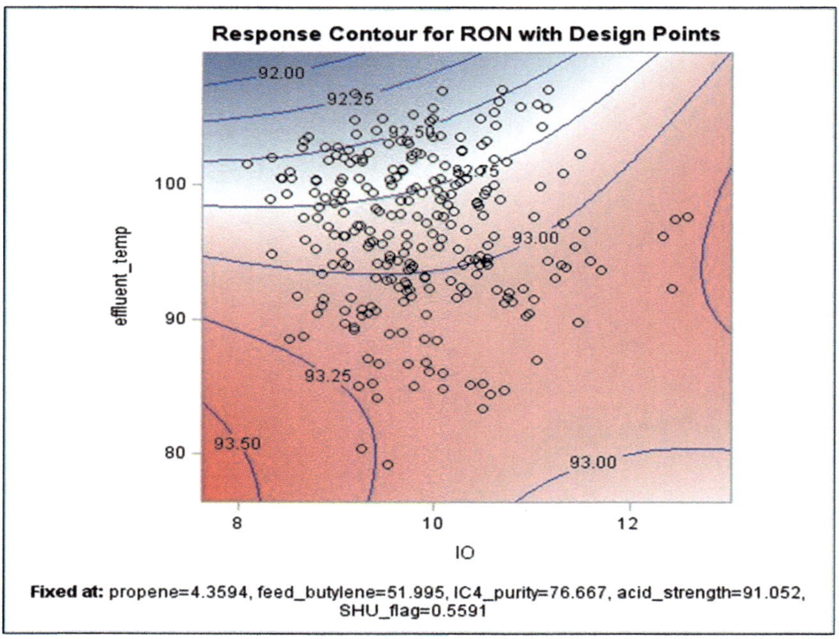

FIG. 6.15 RON vs I/O and reactor temperature contour map. *From reference J. Bird, D. Seillier, M. Teders, Optimization of Refining Alkylation Process Unit Using Response Surface Methods.*

alkylation with olefins catalyzed by H_2SO_4 in a hollow, horizontal cylinder [9]. This model was employed to predict the product distribution resulting from changes in the feedstock compositions. Similarly, Ivashkina et al. utilized DFT calculations to develop a mathematical model in a STRATCO reactor, aiming to elucidate the influence of feedstock compositions on product distribution and RON of alkylate [10]. Furthermore, Liu et al. established a correlation model for the IL-catalyzed alkylation process in STR, effectively linking different operating conditions to the product distribution and RON [11]. Tian et al. were the first to employ an artificial neural network (ANN) model, which was optimized to assess the C4 alkylation process catalyzed by H_2SO_4 in both the stirred tank reactor (STR) and rotating packed bed (RPB) [12]. The ANN model encompassed 10 independent input parameters, including feedstock compositions, operating conditions, and reactor types. Three key elements of the ANN model—algorithm, transfer function, and framework—were scrutinized to determine the most suitable model. Subsequently, predictions were made for the product distribution and RON of the H_2SO_4-catalyzed isobutane/butene alkylation process in both the STR and RPB, followed by parameter analysis.

To derive the optimal ANN model: a network topology of 10–20–30–5 employing the BR main algorithm and tan-sigmoid transfer function. Notably, the ANN model exhibited a notably higher correlation with a deviation of ±2% between the experimental and predicted values for RON. The comparison of weights in parameter analysis for RON reveals that $\mathrm{A/HC} > \mathrm{t} > \mathrm{NRPB} > n_1 \approx \mathrm{P} \approx N_{\mathrm{STR}}$ (positive effects), while $n_3 > T > n_2$ (negative effects). It is evident that A/HC and t have substantial effects on RON. Furthermore, the parameter analysis of the ANN model is utilized to assess the impact of operating conditions on products in both STR and RPB, highlighting the significant influence of rotational speed on the alkylation process.

6.2 Interfacial behaviors

Isobutane and butene exhibit closely aligned physical properties, with notably high-saturated vapor pressures that make them easy to vaporization. In their liquid states, these compounds have lower densities and viscosities than water under similar conditions. From a theoretical standpoint, these physical characteristics in the context of H_2SO_4-catalyzed alkylation introduce conditions conducive to phase separation phenomena among catalysts, reactants, and products. This property proves advantageous for postreaction separation processes from the perspective of product-catalyst segregation, but it poses a significant challenge from the reaction standpoint due to the need for thorough reactant-catalyst contact to facilitate optimal conversion into the target products. For example, the industrial Stratco process relies on high-power agitators equipped with impellers and nozzles as a core technological feature. This setup enables forced mechanical mixing, leading to emulsification between concentrated H_2SO_4 and C4 hydrocarbons, thereby enhancing the progression of the alkylation reaction. The intensified nature of both reaction and mass transfer processes in the industrial setting necessitates subsequent separation procedures, including settling, acid washing, alkali washing, and other methods, to ensure the quality of the alkylate after production.

In the context of H_2SO_4-catalyzed alkylation reactions, the alkylation of isobutane and butene constitutes a biphasic, rapid reaction, where the interphase mass transfer rates plays a pivotal role in determining the overall macroscopic reaction rate. Extensive contact between the concentrated sulfuric acid catalyst and the C4 reactants, namely isobutane and butene, is crucial for efficient alkylation and conversion into the desired target products. The state of mass transfer dispersion between concentrated H_2SO_4 and C4 hydrocarbons holds paramount significance in the context of the alkylation reaction.

Moreover, transferring isobutane into the acid phase presents considerable challenges relative to butene. The concentration of isobutane within the reactant system serves as both the driving force for propelling the alkylation reaction forward and a pivotal factor influencing the reaction kinetics. A higher I/O ratio is advantageous for the progression of the primary reaction, which is typically achieved in existing H_2SO_4-catalyzed alkylation processes through specialized reactors or agitation equipment to increase the interfacial contact area required for the reaction. To further understand the interfacial characteristic, there are many detailed investigations about the interfacial features of C4 alkylation catalyzed by H_2SO_4 or ILs using experiments and MD simulation [13,14].

As suggested earlier by Albright et al., these variables including agitation speed, feed ratio of isobutane to olefin, and residence time affect the dissolved ratio of isobutane to olefin in the acid phase or at least in the boundary layer of the acid phase [15]. Other variables, which are also likely interchangeable with the ones already mentioned, include acid-to-hydrocarbon ratio, type of emulsion (acid-continuous or hydrocarbon-continuous), and reactor modification. Moreover, in 1970, Li et al. have reported extensive data, indicating that this transfer step is the most controlling step of the entire process, which will affect the character of the emulsion in the reactor, and the transfer rate of the isobutane to the acid phase. At low dissolved ratios, most of the olefin reacts in the absence of isobutane. As the dissolved ratio is increased, an increased fraction of the olefin reacts with isobutane to produce high-quality alkylate.

The increased dissolved ratio of reactants also produces higher yields and lower acid consumption. These results confirm the importance of the physical transfer steps in controlling the alkylation reactions.

Zhao et al. successfully measured the interfacial properties in surfactant/H_2SO_4 and H_2SO_4 system, such as interfacial tension, dispersion of isobutane in acid phase, and isobutane solubility. The addition of the surfactant in H_2SO_4 could significantly influence the interfacial properties. These is a reduction in interfacial tension with the presence of the $[N_{12,1,1,1}]Cl$ surfactant, as depicted in Fig. 6.16. This suggests that the interfacial tension between H_2SO_4 and isobutane also decreases upon the addition of the ammonium surfactant. The decrease in interfacial tension contributes to enhanced mixing of the acid and hydrocarbon phases, thereby improving mass transfer within the surfactant/H_2SO_4 system. Improved mixing can lead to larger interfacial areas, promoting the C4 alkylation reaction at the interfacial regions while reducing the occurrence of side reactions.

The dispersion of isobutane in the catalytic systems with and without the ammonium surfactant was observed using a digital microscope as presented in Fig. 6.17. It is evident that the droplet size of the $[N_{12,1,1,1}]Cl/H_2SO_4$ system was notably smaller than that of the H_2SO_4 system. This suggests that the addition of the ammonium surfactant can significantly enhance the dispersion of isobutane in H_2SO_4, thereby increasing the hydrocarbon-acid interfacial area. This, in turn, leads to the improved

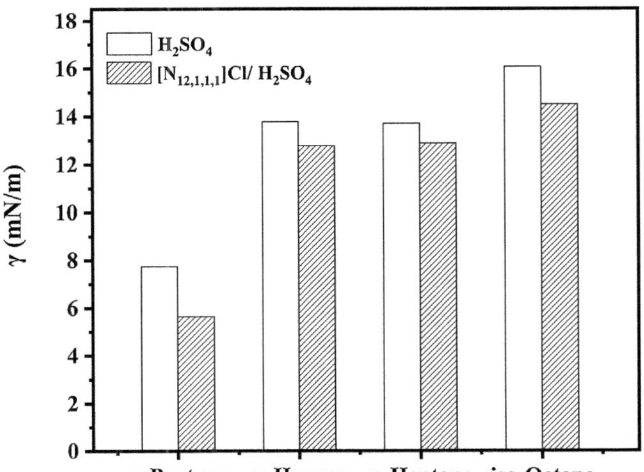

FIG. 6.16 Interfacial tension of alkane/H_2SO_4 and alkane/$[N_{12,1,1,1}]Cl/H_2SO_4$ systems. *From reference Y. Zhao, T. Li, X. Meng, H. Wang, Y. Zhang, H. Wang, S. Zhang, Improvement of product distribution through enhanced mass transfer in isobutane/butene alkylation, Chem. Eng. Res. Des. 143 (2019) 190–200.*

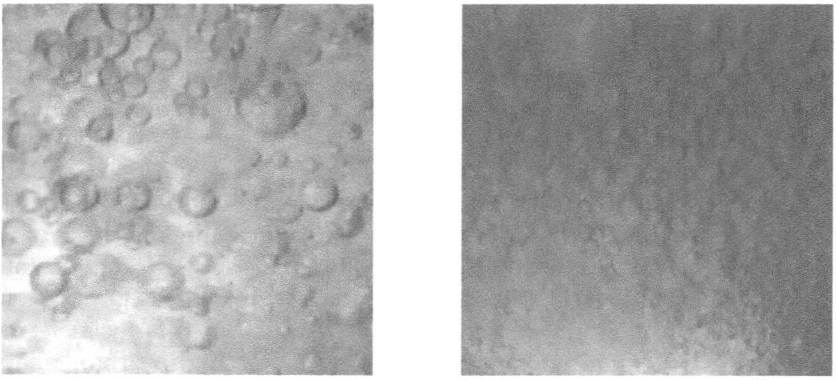

FIG. 6.17 Photos of isobutane mixing with the catalytic systems under electron microscope: isobutane in H_2SO_4 (left) and isobutane in surfactant/H_2SO_4 (right). *From reference Y. Zhao, T. Li, X. Meng, H. Wang, Y. Zhang, H. Wang, S. Zhang, Improvement of product distribution through enhanced mass transfer in isobutane/butene alkylation, Chem. Eng. Res. Des. 143 (2019) 190–200.*

mass transfer, reduced isomerization rate, and consequently better product distribution, consistent with the findings on interfacial tension reported by Ende et al. [16].

The occurrence of side reactions can be attributed to the limited solubility of isobutane in H_2SO_4, as the solubility of butene in the acid is higher than that of isobutane, leading to rapid olefin oligomerization. In addition, the solubility of isobutane in the catalytic systems, specifically H_2SO_4 and $[N_{12,1,1,1}]Cl/H_2SO_4$, was estimated by measuring pressure changes before and after the dissolution of isobutane in the acid. Under the current experimental conditions, the pressure drop before and after isobutane dissolution in the H_2SO_4 system is 0.003 MPa, while the pressure drop increases to 0.006 MPa when the ammonium surfactant (40 mmol/L) is present. This observation illustrates that the presence of the surfactant in the catalytic system can enhance the solubility of isobutane in H_2SO_4. Chen et al. noted that the addition of minor amounts of surfactants can increase the solubility ratio of isobutane/alkenes in sulfuric acid, and the yield of TMPs is found to increase linearly with the increased solubility ratio of isobutane/alkenes in the acid [17]. According to previous literature, the high concentration of isobutane can accelerate the rate of hydrogen transfer from isobutane to the C_8^+, which promotes the generation of the desired C_8 component, particularly TMPs, and inhibits the occurrence of side reactions such as cracking and oligomerization [18].

To gain a deep molecular-level understanding about the effects of surfactants on the acid/C4 hydrocarbon miscibility, the interfacial behaviors of cationic and anionic surfactants with long alkyl chains in H_2SO_4 have been successfully investigated in detail using MD simulations [13]. MD simulations gives the molecular-level understanding about the interfacial thickness and dispersion, orientation of the long alkyl chains of surfactants, interfacial tension, and diffusion coefficients of C4 hydrocarbons as shown in Fig. 6.18–6.20.

The MD results indicate that at higher concentrations, surfactants mainly concentrate in the acid/hydrocarbons interfacial regions, leading to a broader occupation of these zones. Specifically, for hexadecyltrimethylammonium bromide (CTAB) and dihexadecyldimethylammoniumchloride (DDAB) cationic surfactants, better dispersion in interfacial regions is observed at lower concentrations (<1.0 wt%), with obvious clustering at higher concentrations. In contrast, the sodium dodecyl benzene sulfonate (SDBS) anionic surfactant tends to self-aggregate at larger concentrations (>2.0 wt%). Orientation analysis reveals that the long alkyl chains of surfactants tend to be perpendicular to the acid/hydrocarbons interface, contributing to a reduction in interfacial tension. Additionally, the anionic surfactant SDBS displays slightly less ordered interfacial orientation than the cationic surfactant CTAB, resulting in a stronger ability to decrease interfacial tension for CTAB at higher concentrations compared to SDBS.

Regarding the diffusion of C4 hydrocarbons, the introduction of surfactants significantly influences the diffusion of isobutane and 2-butene in the interfacial regions compared to pure H_2SO_4 systems. Surfactants can limit the diffusion of both hydrocarbons to some extent, and higher surfactant concentrations further decrease their dynamics at the interface. Moreover, the increased SDBS concentration results in a higher I/O ratio compared to the pure H_2SO_4 system. There are slight differences in trends between CTAB and DDAB cationic surfactants. Overall, surfactants can effectively increase the acid/hydrocarbon interfacial thickness of reaction regions and improve the interfacial I/O ratio while reducing the diffusion of C4 hydrocarbons.

The diffusion coefficients (D) of isobutane and 2-butene at the interface are presented in Table 6.5. For all systems, the D of isobutane is observed to be smaller than that of 2-butene, indicating that butene diffuses faster than isobutane in H_2SO_4. Overall, the introduction of surfactants tends to restrict the diffusion of both isobutane and 2-butene in the interfacial regions to some extent compared to pure H_2SO_4 systems. Additionally, higher surfactant concentrations further decrease the diffusion of isobutane and 2-butene at the interface. This is probably attributed to the slower motion of the longer alkyl chain at the interface, which hinders the motion of C4 hydrocarbons. At equivalent concentrations, SDBS slows the diffusion of C4 hydrocarbons more significantly compared to CTAB, while DDAB has a lesser influence. Furthermore, CTAB generally positively impacts the D ratio of isobutane to 2-butene at the interface, whereas SDBS exerts a negative influence. Lower DDAB concentrations enlarge the

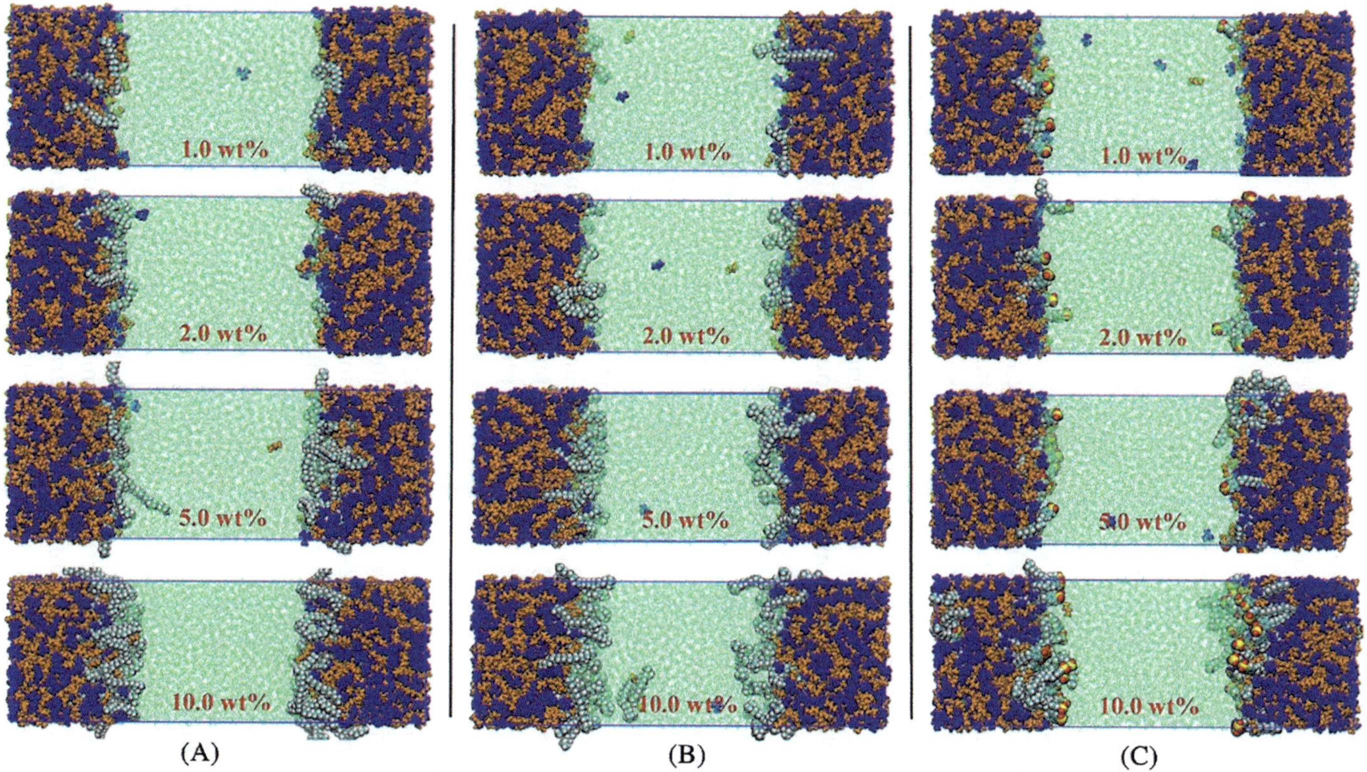

FIG. 6.18 Equilibrated snapshots of the surfactant distribution with different concentrations at the acid/hydrocarbons interface. (A) CTAB, (B) DDAB, and (C) SDBS. *Green color* represents the H_2SO_4 molecules in line style. *Blue color* represents isobutane molecules in VDW style. *Brown color* means 2-butene molecules in the VDW style. The molecules with long alkyl chains in VDW style at the interface are surfactants. CTAB, hexadecyltrimethylammonium bromide; DDAB, dihexadecyldimethylammoniumchloride; SDBS, sodium dodecyl benzene sulfonate. *From reference W. Zheng, Z. Wang, W. Sun, L. Zhao, F. Qian, H2SO4-catalyzed isobutane alkylation under low temperatures promoted by long-alkyl-chain surfactant additives, AIChE J. 67 (2021) e17349.*

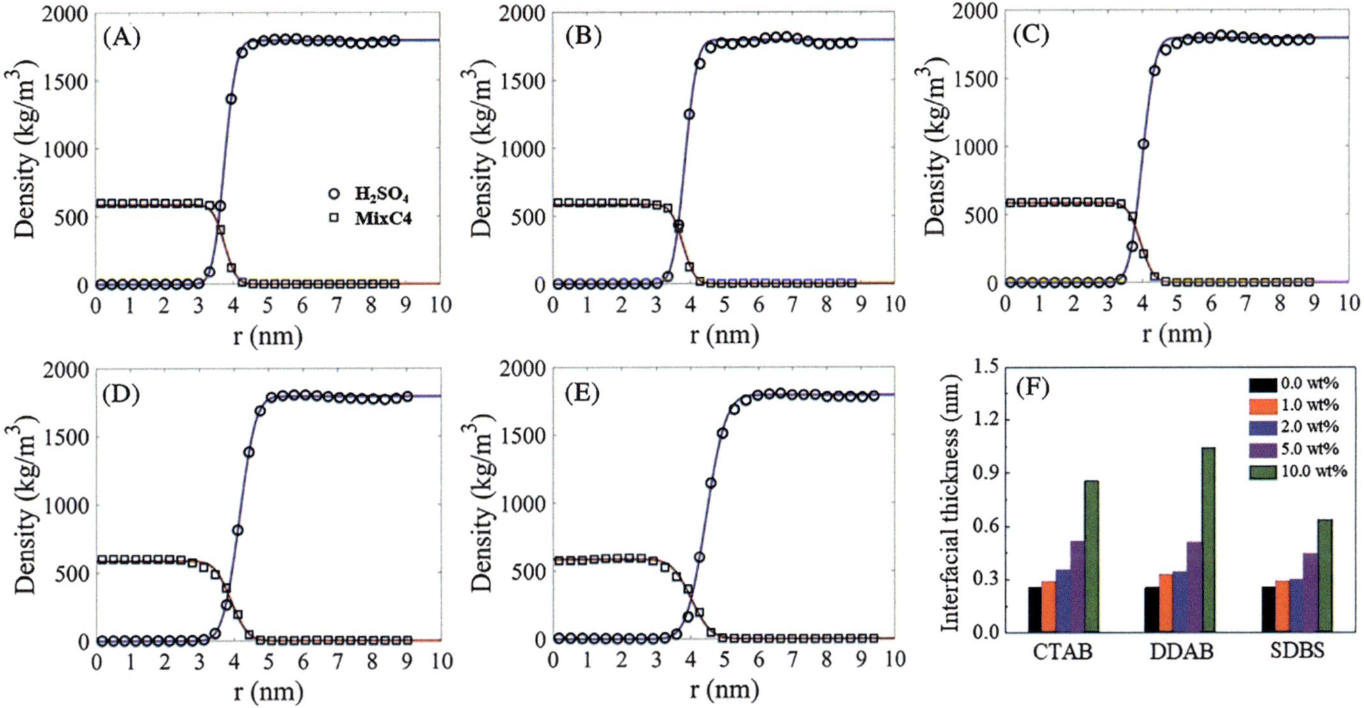

FIG. 6.19 The fitting results of H_2SO_4 and MixC4 mass density for the calculation of interfacial thickness in the SDBS-aiding systems with different concentrations (A) 0.0 wt%, (B) 1.0 wt%, (C) 2.0 wt%, (D) 5.0 wt%, (E) 10.0 wt%, and (F) the obtained interfacial thickness for different surfactants as a function of concentrations. SDBS, sodium dodecyl benzene sulfonate. *From reference W. Zheng, Z. Wang, W. Sun, L. Zhao, F. Qian, H2SO4-catalyzed iso-butane alkylation under low temperatures promoted by long-alkyl-chain surfactant additives, AIChE J. 67 (2021) e17349.*

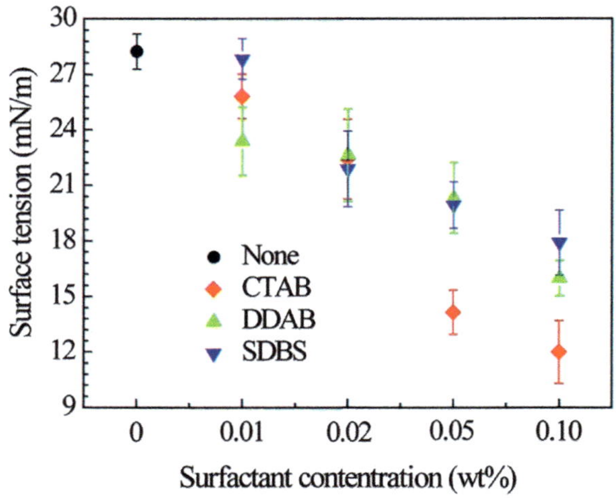

FIG. 6.20 Effect of surfactant concentrations on the interfacial tension of the acid/hydrocarbons interface. *From reference W. Zheng, Z. Wang, W. Sun, L. Zhao, F. Qian, H_2SO_4-catalyzed isobutane alkylation under low temperatures promoted by long-alkyl-chain surfactant additives, AIChE J. 67 (2021) e17349.*

TABLE 6.5 The *D* of isobutane and 2-butene, and the ratio of isobutane to 2-butente number (I/O ratio) in the interfacial regions.

Systems	Diffusion coefficients (Å/ps)			
	2-Butene	**Isobutance**	**Isobutance/2-butene**	**I/O ratio**
None	0.260	0.223	0.858	0.850
CTAB-1.0 wt%	0.212	0.181	0.854	0.832
CTAB-2.0 wt%	0.216	0.178	0.824	0.906
CTAB-5.0 wt%	0.202	0.171	0.847	0.792
CTAB-10.0 wt%	0.179	0.156	0.872	0.763
SDBS-1.0 wt%	0.214	0.179	0.836	0.796
SDBS-2.0 wt%	0.199	0.168	0.844	0.852
SDBS-5.0 wt%	0.189	0.155	0.820	0.884
SDBS-10.0 wt%	0.155	0.124	0.800	0.890
DDAB-1.0 wt%	0.225	0.196	0.871	0.772
DDAB-2.0 wt%	0.237	0.202	0.852	0.952
DDAB-5.0 wt%	0.200	0.167	0.835	0.806
DDAB-10.0 wt%	0.193	0.161	0.834	0.766

Data from reference W. Zheng, Z. Wang, W. Sun, L. Zhao, F. Qian, H_2SO_4-catalyzed isobutane alkylation under low temperatures promoted by long-alkyl-chain surfactant additives, AIChE J. 67 (2021) e17349.

D ratio of isobutane to 2-butene, whereas higher concentrations of DDAB lead to a smaller *D* ratio.

Using regular solution theory, Camper et al. studied the solubility of hydrocarbons in imidazolium-based ILs at low pressure [19]. Aschauer and colleagues also determined the solubility of isobutane, but they only conducted measurements in the neutral chloroaluminate melt [BMIm][$AlCl_4$] ($x_{AlCl3} = 0.5$) [20]. Additionally, Hu et al. measured isobutane solubility in NMA-$AlCl_3$-based ILs [21].

The physical solubility of 2-butene in the composite IL (CIL) is challenging to determine due to the rapid reaction between 2-butene and the CIL. Liu et al. successfully measured the solubility of isobutane by determining the weight increase of a given amount of CIL after it was saturated with isobutane [22]. Table 6.6 presents the data of isobutane solubility in CIL and H_2SO_4. The solubility determination was carried out in a bulb system as depicted in Fig. 6.21. The diffusivity of isobutane in the CIL (*D*) ranges from 1.0×10^{-6} to $7.0 \times 10^{-6} m^2/s$. This diffusivity increases with rising temperature, reaching about $2.4 \times 10^{-6} m^2/s$ at 5°C.

Zheng et al. conducted a comprehensive study using MD simulations to explore the solubility, diffusion, and permeability properties of isobutane and 2-butene in various 1-alkyl-3-methylimidazolium ionic liquids (ILs) with different anions and alkyl chain lengths, as well as in H_2SO_4 [23] (Figs. 6.22 and 6.23). The MD simulations revealed that ILs demonstrate higher solubility and diffusion of isobutane compared to H_2SO_4. Solvation free energy calculations show satisfactory agreement between the predicted isobutane solubility in ILs and experimental data. Specifically, the solubility trend of isobutane in ILs with different anions followed the sequence: [BMIm][Al_2Cl_7] > [BMIm][$AlCl_4$] > [BMIm][Tf_2N] > [BMIm][PF_6] > [BMIm][BF_4]. Additionally, among ILs with varying alkyl chain lengths, the trend is observed as [OMIm][$AlCl_4$] > [HMIm][$AlCl_4$] > [BMIm][$AlCl_4$]. The behavior of 2-butene exhibits a similar trend as isobutane, but with higher solubility, diffusion, and permeability. Analysis of the structure–property relationship of ILs indicates that weaker cation-anion interaction, shorter mean residence time, lower solubility parameters, and larger free volume contributes to the enhanced solubility and dissolution of isobutane. Furthermore, a consistent trend between isobutane diffusion and solubility is observed in ILs with different anions, while ILs with varying alkyl chain length exhibit an opposite trend.

Furthermore, MD simulations were utilized to investigate the interfacial properties between the hydrocarbon phase containing isobutane and 2-butene and the catalyst phase comprising H_2SO_4 or ILs with varying alkyl chain length on their imidazolium cations [24]. In addition to the solubility of isobutane, the diffusion of isobutane at the interface of acid/hydrocarbon dispersion is recognized as another crucial factor influencing the quality of alkylate. The diffusion coefficients (D_z) along the *z* direction

TABLE 6.6 Solubility of isobutane in CIL and H_2SO_4 (wt%).

Acid	5°C Saturated	1 atm	10°C Saturated	1 atm	15°C Saturated	1 atm	25°C Saturated	1 atm
CIL	0.192	0.088	0.185	0.084	0.180	0.081	0.167	0.065
H_2SO_4	0.090	0.048	0.088	0.046	0.085	0.043	0.074	0.025
H_2SO_4	–	–	0.089	0.046	0.084	0.044	0.074	0.025

Data from reference Y. Liu, G. Wu, X. Pang, R. Hu, Kinetics study on alkylation of isobutane with deuterated 2-butene in composite ionic liquids, Chem. Eng. J. 387 (2020) 123407.

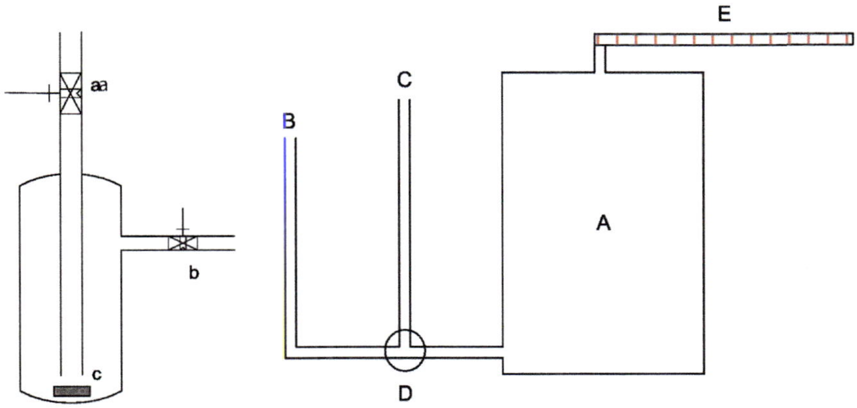

FIG. 6.21 Absorption data of isobutane in CIL at 5°C and 25°C. *From reference Y. Liu, G. Wu, X. Pang, R. Hu, Kinetics study on alkylation of isobutane with deuterated 2-butene in composite ionic liquids, Chem. Eng. J. 387 (2020) 123407.*

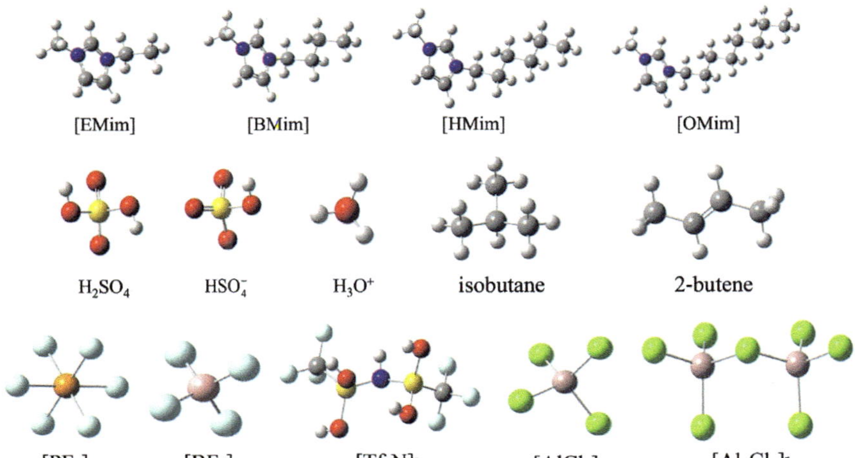

FIG. 6.22 Optimized geometries of the cation and anion as well as the H_2SO_4 for solvation free energy and diffusion calculation. *From reference W. Zheng, L. Zheng, W. Sun, L. Zhao, Screening of imidazolium ionic liquids for the isobutane alkylation based on molecular dynamic simulation, Chem. Eng. Sci. 183 (2018) 115–122.*

of 2-butene and isobutane at the interface for all systems were calculated using the modified Langevin equation, as illustrated in Fig. 6.24. While the D_z of 2-butene and isobutane remains relatively constant in the reactant phase bulk, noticeable decreases are observed at the interface for the H_2SO_4 system compared to the IL systems. For all systems, 2-butene diffuses slightly more readily than isobutane at the interface due to their different polarities. Additionally, it is evident that the diffusion rates of

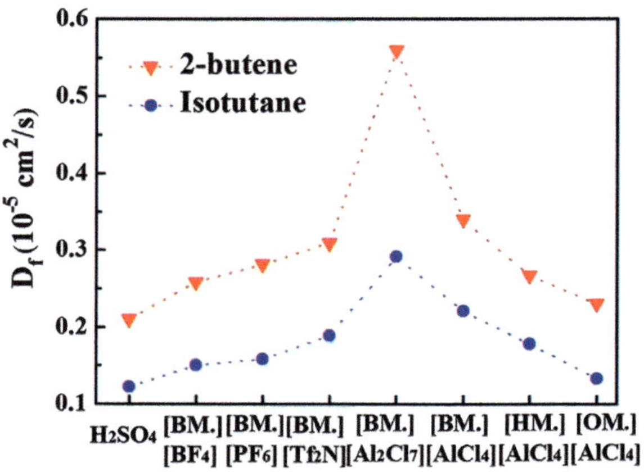

FIG. 6.23 Self-diffusion coefficients of isobutane and 2-butene in the H_2SO_4 as well as ILs at 300 K and 1 bar. *From reference W. Zheng, L. Zheng, W. Sun, L. Zhao, Screening of imidazolium ionic liquids for the isobutane alkylation based on molecular dynamic simulation, Chem. Eng. Sci. 183 (2018) 115–122.*

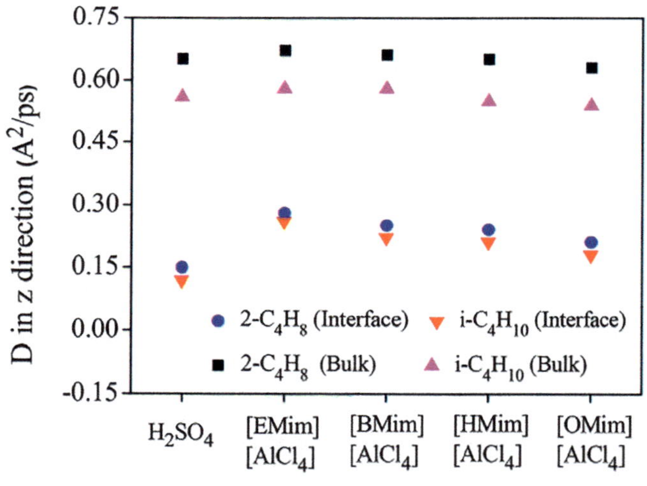

FIG. 6.24 The D_z at the interface for different simulation systems. *From reference W. Zheng, H. Wang, W. Xie, L. Zhao, W. Sun, Understanding interfacial behaviors of isobutane alkylation with C4 olefin catalyzed by sulfuric acid or ionic liquids, AIChE J. 64 (2018) 950–960.*

both 2-butene and isobutane at the interface gradually decrease with an increase in alkyl chain length, indicating that ILs with longer alkyl chains exhibit stronger interaction with the reactant mixtures and result in reduced interface tension (Fig. 6.25). This behavior is attributed to increased steric hindrance and restricted mobility of longer alkyl chains.

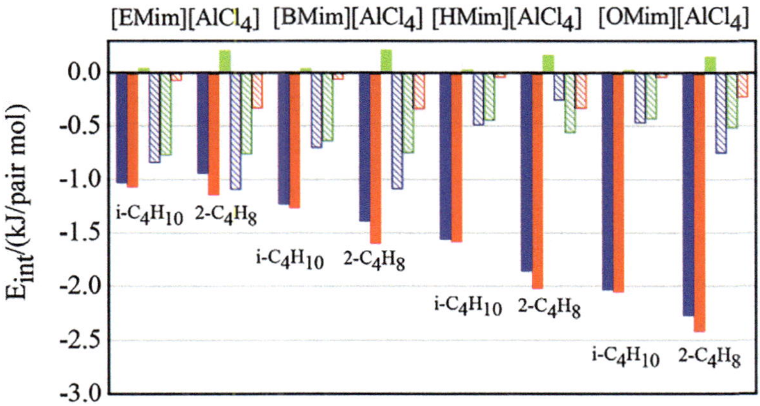

FIG. 6.25 IL-reactant mixtures interaction energy for different simulation systems. In *blue*, the total potential energy; in *red*, the Van der Waals energy; in *green*, the electrostatic contribution to the interaction energy. *From reference W. Zheng, H. Wang, W. Xie, L. Zhao, W. Sun, Understanding interfacial behaviors of isobutane alkylation with C4 olefin catalyzed by sulfuric acid or ionic liquids, AIChE J. 64 (2018) 950–960.*

6.3 Intensification mechanism of interfacial behaviors

The isobutane alkylation catalyzed by H_2SO_4 represents a typical liquid–liquid heterogeneous interfacial reaction. Effective mass transfer of isobutane to the acid phase in the interfacial regions plays a critical role in facilitating hydride transfer from isobutane to TMP^+, thereby accelerating the production of the desired C_8 fraction and suppressing cracking and oligomerization side reactions. This ultimately enhances TMP selectivity, improves alkylate quality, and reduces the formation of by-products. However, owing to polarity differences, olefins dissolve much more rapidly than alkanes. Consequently, olefins tend to be rich near the interfaces in acid/hydrocarbon dispersions during the initial stages of alkylation. The abundance of olefins accelerates fast polymerization reactions, resulting in the generation of HEs and LEs, consequently diminishing alkylate quality and catalyst lifespan. Thus, the solubility of isobutane in the acid phase is pivotal to the alkylation process. Although strong agitation is commonly employed to increase interfacial areas in acid/hydrocarbon dispersions, it does not alter the inherent dissolution characteristics of C4 hydrocarbons in the acid phase.

The impact of alkyl chain length of the cation (C_n, $n=4, 8, 12, 16$) and the anion type (SbF_6^-, NTf_2^-, PF_6^-) on alkylation performance was examined through alkylation reactions and interfacial tension measurements. The results presented in Table 6.7 revealed that additives with C_{12} alkyl chains demonstrate the most significantly enhanced quality of alkylate among all

TABLE 6.7 Ionic liquids with different alkyl chains and anions for alkylation.

	Alktlate composition (%)				
IL	C_5–C_7	C_8	C_{9+}	TMP	TMP/DMH
ADM-C_4-SbF_6	7.7	78.3	14.0	71.8	15.1
ADM-C_8-SbF_6	9.3	73.8	16.9	68.0	13.7
ADM-C_{12}-SbF_6	3.2	84.5	12.4	80.1	21.3
ADM-C_{16}-SbF_6	6.3	82.3	11.5	77.0	19.4
ADM-C_4-NTf_2	7.6	67.4	26.3	62.3	12.9
ADM-C_8-NTf_2	7.6	79.8	12.6	72.6	16.4
ADM-C_{12}-NTf_2	5.8	85.9	8.3	80.9	19.4
ADM-C_{16}-NTf_2	8.9	65.8	25.3	58.8	11.5
ADM-C_4-PF_6	9.2	75.1	15.6	69.9	16.0
ADM-C_8-PF_6	8.1	81.0	10.9	74.9	15.8
ADM-C_{12}-PF_6	5.8	86.2	8.0	80.7	16.3

Data from reference L. Wang, G. Zhao, X. Yao, B. Ren, S. Zhang, Adamantane-based cation and [MFn]− anion synergistically enhanced catalytic performance of sulfuric acid for isobutane alkylation, Ind. Eng. Chem. Res. 56 (2017) 7920–7929.

studied ADM-ILs with the same anion. This suggests that increasing the chain length within an appropriate range is favorable for the reaction, while further increases have an adverse effect. Subsequently, isopentane replaced isobutane, and 2,2,4-TMP substituted the alkylate to measure the biphasic interfacial tension between the investigated catalytic systems and isopentane or 2,2,4-TMP. Increasing the alkyl chain of ADM-IL leads to reduced interfacial tension for all investigated catalytic systems, isopentane, and 2,2,4-TMP. However, the values of the interfacial tension and the reduction rates with increasing alkyl chain length varies significantly for the ADM-ILs with different anions. The substantial enhancement in TMP production is likely attributed to the involvement of ADM-ILs as hydride donors in the hydride-transfer steps (illustrated in Fig. 6.26), which reduces the reaction barrier and accelerates alkylation kinetics, consequently influencing alkylate distribution.

The interfacial tensions for the H_2SO_4/ADM-C_8-SbF_6 system with isopentane and 2,2,4-TMP were 13.61 and 19.64 mN/m, respectively. In contrast, the corresponding interfacial tensions were 12.73 and 13.04 mN/m, respectively, for H_2SO_4 and 10.99 and 12.14 mN/m, respectively, for the H_2SO_4/ADM-C_8-PF_6 system. Additionally, ADM-C_8-SbF_6 acts as an emulsion breaker, exacerbating the interfacial properties between the hydrocarbon and acid phases, while ADM-C_8-PF_6 improves the biphasic

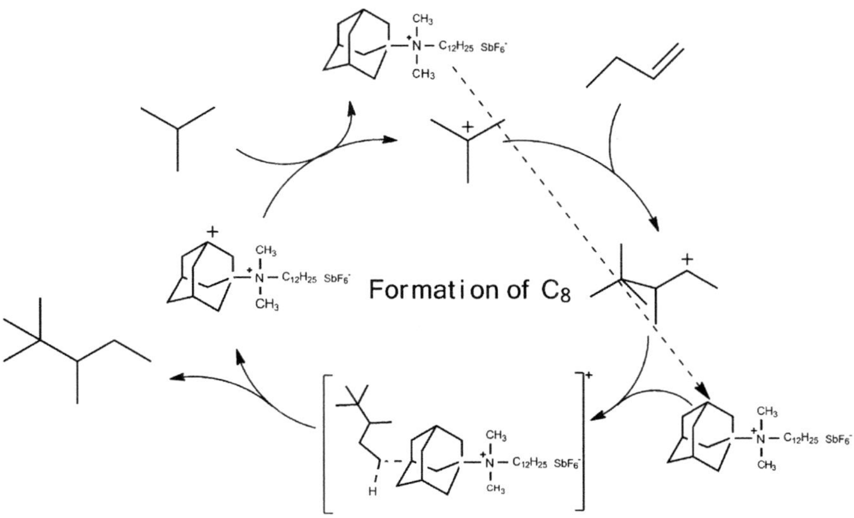

FIG. 6.26 Possible mechanism of alkylation with added ADM-ILs. *From reference L. Wang, G. Zhao, X. Yao, B. Ren, S. Zhang, Adamantane-based cation and [MFn]− anion synergistically enhanced catalytic performance of sulfuric acid for isobutane alkylation, Ind. Eng. Chem. Res. 56 (2017) 7920–7929.*

interfacial properties by functioning as a surfactant. On the other hand, ADM-C_{12}-SbF_6 exhibits desirable properties by decreasing the interfacial tension of the reactant-acid phases and increasing the one of the product-acid phases. Consequently, ADM-C_{12}-SbF_6 is found to be beneficial for forming acid-hydrocarbon emulsions and accelerating the separation of isooctane from the acid phase, thereby contributing to enhanced alkylation performance, which is consistent with the alkylation results.

Moreover, the solubility of the chosen additive in the oil phase is another essential factor influencing the quality of the alkylate and the stability of the catalytic system. ESI-MS (cation mode) was employed to determine the solubilities of ADM-C_{12}-SbF_6 in the oil phase in the presence of only ADM-IL and in the presence of H_2SO_4/0.5 wt% ADM-IL, both of which are below the detection limit. This indicates that ADM-C_{12}-SbF_6 is not easily entrained in the oil and has little effect on the oil products.

In addition, using alchemical free energy calculation, MD simulations were employed to compute the solvation free energy of isobutane and butene in the catalytic system using H_2SO_4 with and without the formyl functional $[N_{1,1,1,1}][C_{10}SO_4]$ additive [25]. This aims to elucidate the enhanced mechanism of ASO and the formyl functional $[N_{1,1,1,1}]$ $[C_{10}SO_4]$ additive.

The impact of ASO and the formyl functional $[N_{1,1,1,1}][C_{10}SO_4]$ additive on the solvation free energy of isobutane and 2-butene in the acid phase was investigated, as presented in Fig. 6.27, where a more negative value

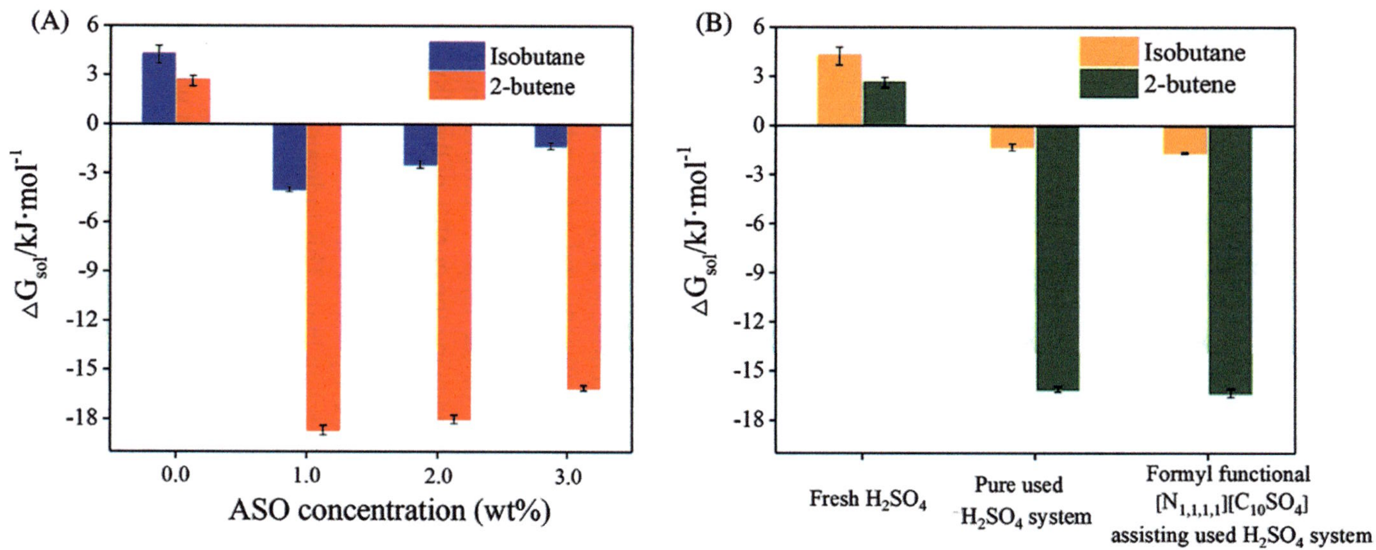

FIG. 6.27 Simulated solvation free energy of isobutane and 2-butene as a function of ASO amount (A) as well as comparison of each catalytic system (B) at 300 K and 1 bar. *From reference Z. Ma, W. Zheng, W. Sun, L. Zhao, Enhanced catalytic performance of H_2SO_4-catalyzed C4 alkylation by formyl functional $[N_{1,1,1,1}][C_{10}SO_4]$ additive, AIChE J. 69 (2023) e18179.*

of solvation free energy indicates more favorable solvation. The solvation free energy trend for isobutane differs from that of 2-butene. As the concentration of ASO increases from 0 to 1.0 wt%, the solvation free energy of isobutane decreases from 4.23 to −4.00 kJ/mol, becoming more negative with increased ASO concentration in H_2SO_4. However, at an ASO concentration of 3.0 wt%, the solvation free energy of isobutane increases to −1.33 kJ/mol. This suggests that while ASO reduces the solvation free energy of isobutane in the acid phase, higher concentrations have an adverse effect. Similarly, the solvation free energy of 2-butene follows a similar trend, indicating a preference for dissolution in the acid phase due to its larger polarity.

Furthermore, the addition of the formyl functional $[N_{1,1,1,1}][C_{10}SO_4]$ in the H_2SO_4 system with 3.0 wt% ASO promotes the solvation free energy of isobutane, reducing it to −1.68 kJ/mol compared to −1.33 kJ/mol in the pure H_2SO_4 system without the additive. Additionally, this additive significantly enhances the relative solubility of isobutane to 2-butene, having a higher value of 0.102 compared to 0.08 in the pure H_2SO_4 system, positively impacting the quality of alkylate. These findings align with experimental results, showing that the formyl functional $[N_{1,1,1,1}][C_{10}SO_4]$ can improve the selectivity of targeted C_8 components and reduce reaction time.

For C4 alkylation catalyzed by H_2SO_4, the promotion of isobutane solubility in H_2SO_4 is of major importance to the enhancement of quality of alkylate. ASO plays an important role in improving the solubility of isobutane and optimizing the interfacial characteristics of acid/hydrocarbon dispersions. The solubilities of isobutane in H_2SO_4 in the presence of different amounts of typical ASO compounds at different temperatures were calculated using the COSMO-RS method. These results were used to reveal the effects of the ASO molecular structures (in Table 6.8) on the solubility of isobutane in H_2SO_4 and to select the ASO compounds that can be utilized as cocatalysts to improve the H_2SO_4-catalyzed alkylation.

The enhanced quality of alkylate relies heavily on increasing the solubility of isobutane in H_2SO_4. This improvement facilitates hydrogen transfer reactions, boosts selectivity for primary products like TMP, and hinders the formation of undesired secondary products such as isomerization and polymerization. Using COSMO-RS, the solubilities of isobutane in H_2SO_4 with varying amounts of different ASO compounds were quantitatively calculated. Fig. 6.28 illustrates the impact of ASO15 on isobutane solubility in H_2SO_4. With the ASO15 mass percentage increasing from 0 to 3 wt%, the isobutane solubilities rise from 0.196, 0.134, 0.112, 0.095, 0.060, 0.046, 0.036 g/100 g solvent to 0.221, 0.154, 0.130, 0.111, 0.072, 0.056, and 0.044 g/100 g at temperatures of −20°C, −10°C, −5°C, 0°C, 15°C, 25°C, and 35°C, respectively. This results in an increase in isobutane solubility by 12.993%, 15.012%, 15.945%, 16.826%, 19.214%, 20.604%, and 21.849%, respectively. These findings demonstrate that the presence of

TABLE 6.8 Optimized molecular geometries and surface polarization charge densities σ of ASO molecules.

Rank	Molecule's name	Structure's formula	Molecular geometry optimization by Gaussian	Molecular surface polarization charge density σ
1	2,2,5-Trimethylhexane (ASO15)	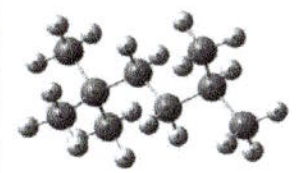	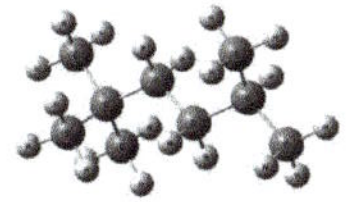	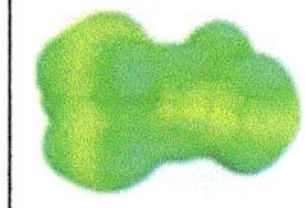
2	2,2,6-trimethyloctane (ASO17)	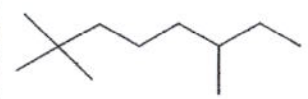	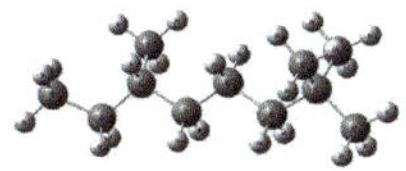	
3	2,5,5-Trimethyl-1,3-cyclopenta	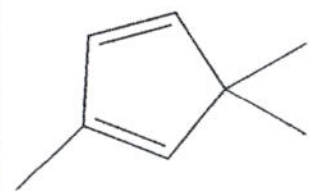	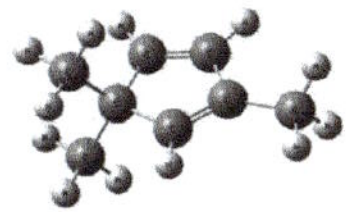	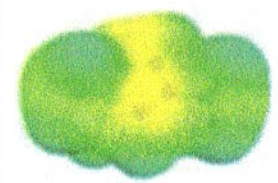
4	(*E*)-2-Ethyl-1,7,7-trimethyl-dicyclic [2.2.1]heptane (ASO5)	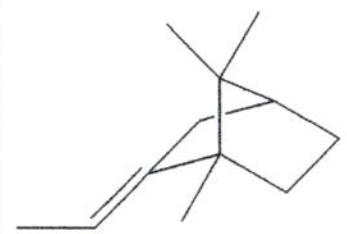	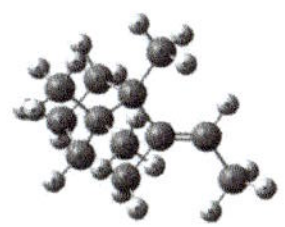	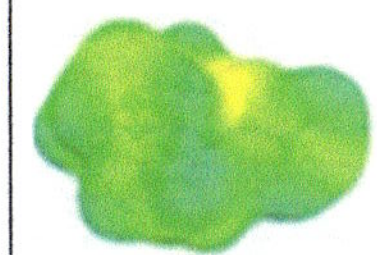
5	1,2,5,5-Tetramethyl-1,3-cyclopentadiene (ASO6)	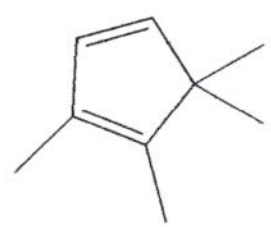	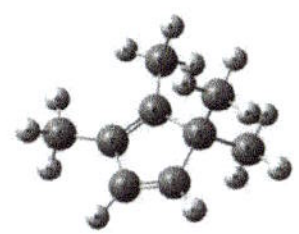	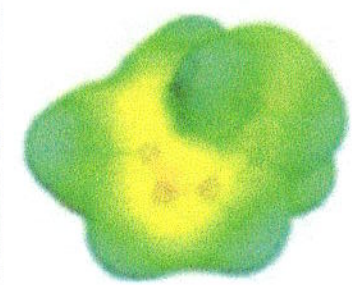

6	1,2,3,4,5-Pentamethyl-cyclopentadiene (ASO4)	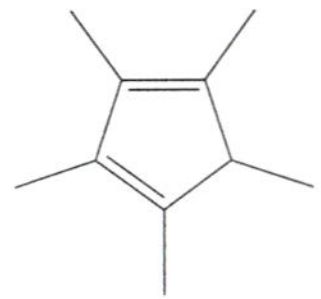	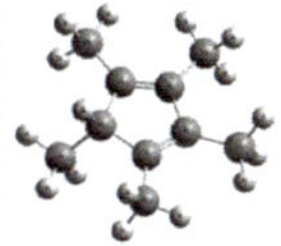	
7	1,3,5,5,6,6-Hexamethyl-1,3-cyclohexadiene (ASO2)	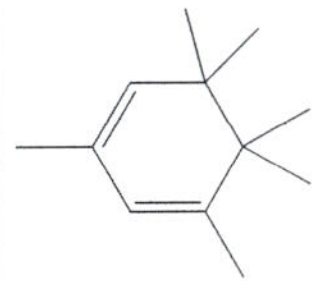	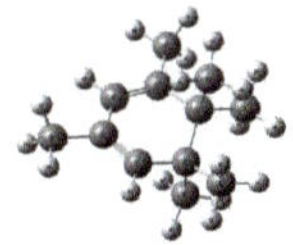	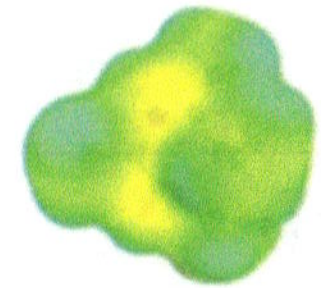
8	1,2,3,4,5,6-Hexamethyl-1,3-cyclohexadiene (ASO3)	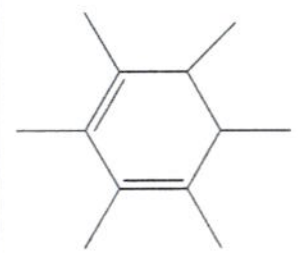	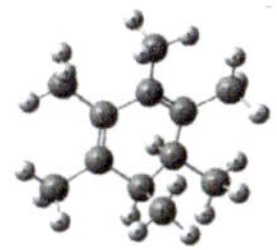	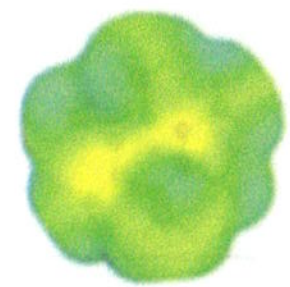

Data from reference Y. Xin, Y. Hu, Y. Wang, S. Jiang, M. Li, K. Chi, S. Zhang, F. Gao, C. Ren, S. Zhang, Alkylation catalyzed by H_2SO_4: promoting effects of ASO on the solubility of isobutane in acid phase, Fuel 326 (2022) 125034.

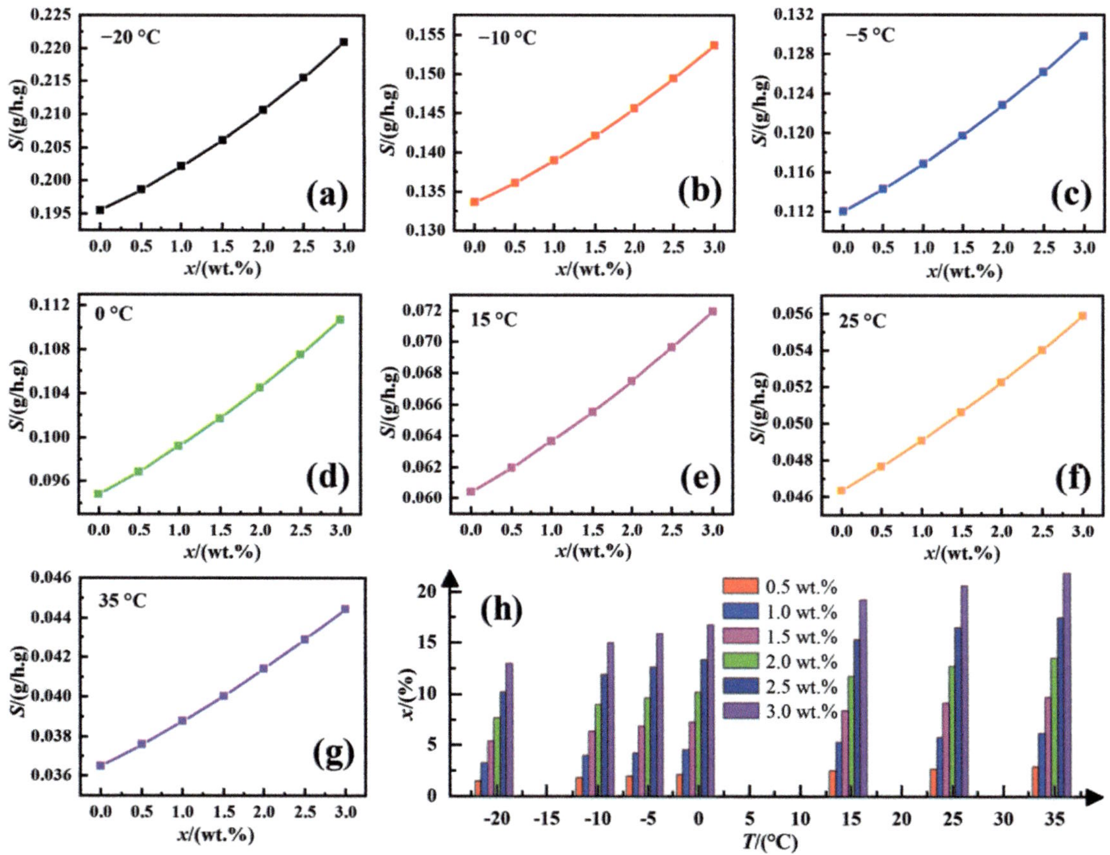

FIG. 6.28 The effects of ASO15 on the solubility of isobutane in H_2SO_4 at different temperatures: (A) –20°C, (B) –10°C, (C) –5°C, (D) 0°C, (E) 15°C, (F) 25°C, (G) 35°C, and (H) the change rate of the isobutane solubility in H_2SO_4 as a function of temperature and mass percentage of ASO15. *From reference Y. Xin, Y. Hu, Y. Wang, S. Jiang, M. Li, K. Chi, S. Zhang, F. Gao, C. Ren, S. Zhang, Alkylation catalyzed by H_2SO_4: promoting effects of ASO on the solubility of isobutane in acid phase, Fuel 326 (2022) 125034.*

ASO15 significantly boosts the solubility of isobutane in H_2SO_4, with the effect depending on the amount of ASO15 and temperature. Specifically, the solubility of isobutane in H_2SO_4 shows a linear increase with higher amounts of ASO15 for all temperatures. Similar results were observed for the effects of ASO1 ~ 7, 12, and 17 on isobutane solubility in H_2SO_4.

The effects of various ASO compounds on the solubility of isobutane in H_2SO_4 at different temperatures are shown in Fig. 6.29. This figure illustrates that the influence of ASO compounds on the solubility of isobutane in H_2SO_4 is strongly dependent on their molecular structure, quantity, and temperature. When the mass percentage of ASO increases from 0 to 3 wt% at −20°C, ASO1 ~ 7, 12, 15, and 17 consistently demonstrate a significant enhancing effect on isobutane solubility, denoted as effect I. Conversely, ASO8 and ASO13 consistently exhibit a diminishing effect, effect II, causing a decrease in isobutane solubility with increasing ASO mass percentage. ASO9, 11, and 19 generally present a negative effect (effect III), showcasing a nonlinear relationship between their amount and isobutane solubility. The effects of ASO10, 14, 16, and 18 on isobutane solubility initially appear negative and then turn positive as their mass percentage increases (effect IV).

At −10°C and −5°C, ASO1–7, 10, 12, and 14–18 exhibit effect I, while ASO8 demonstrates effect II. At −10°C, ASO13 and ASO19 display effect III, and ASO9 and ASO11 showcase effect IV. At −5°C, ASO9, 11, and 19 show effect IV, and ASO13 presents effect III. At 0°C, ASO1–7, 10–12, and 14–18 have effect I, while ASO9, 13, and 19 present effect IV, and ASO8 exhibits effect II. At temperatures of 15°C, 25°C, and 35°C, all ASO compounds except ASO8 notably increase isobutane solubility in H_2SO_4.

The results indicate that ASO15 > ASO17 > ASO12/ASO5 > ASO6 > ASO4 > ASO2/ASO3/ASO1 > ASO7 > ASO18 > ASO14/ASO10/ASO16 > ASO11 > ASO9 > ASO19/ASO13 > ASO8 in terms of promoting isobutane solubility in 98 wt% H_2SO_4. This order suggests that hexane with methyl substituents has the highest solubilizing effect, followed by octane with methyl substituents. Cyclopentadiene, cyclohexadiene, and dicyclic heptane with methyl substituents have relatively low solubilizing effects. Additionally, shorter chain paraffins and cyclodiolefins with fewer substituents exhibit more prominent solubilizing performances than other counterparts. Moreover, cyclohexanes with long side alkyl chains and hydroxyl groups may also enhance isobutane solubility in H_2SO_4.

6.4 Design of interfacial intensification additives

Prior research indicates that ASO compounds consist of uncopolymerized or copolymerized substances containing numerous double bonds and/or conjugated double bonds, along with substituents such as alkyl, carbonyl, and hydroxyl groups [26–29]. These compounds feature

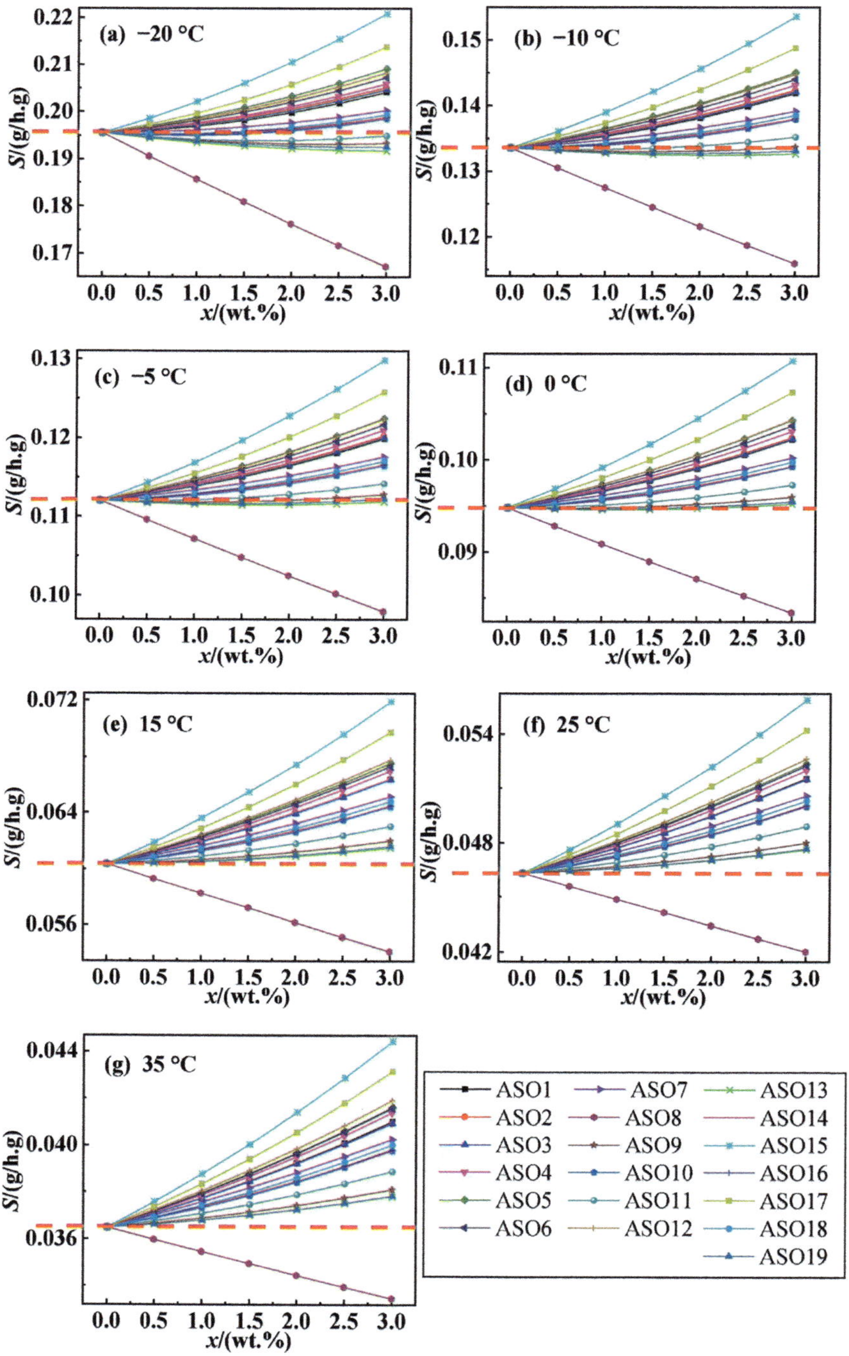

FIG. 6.29 The effects of ASO compounds on the solubility of isobutane in H_2SO_4 at different temperatures: (A) –20°C, (B) –10°C, (C) –5°C, (D) 0°C, (E) 15°C, (F) 25°C, and (G) 35°C. *From reference Y. Xin, Y. Hu, Y. Wang, S. Jiang, M. Li, K. Chi, S. Zhang, F. Gao, C. Ren, S. Zhang, Alkylation catalyzed by H_2SO_4: promoting effects of ASO on the solubility of isobutane in acid phase, Fuel 326 (2022) 125034.*

five-membered or six-membered cyclic and/or acyclic structures. Within suitable concentration ranges, ASO compounds can act as reaction intermediates and reservoirs of hydride ions. This accelerates the formation of *tert*-butyl carbocations, facilitates hydrogen-ion transfer, enhances alkylation kinetics, and increases the production of desired C8 and TMP selectivity. Additionally, ASO compounds have the potential to enhance the solubility of isobutane in H_2SO_4, as well as alter the viscosity and surface tension of H_2SO_4. Therefore, ASO compounds provide valuable insight for screening and designing cocatalysts for isobutane alkylation [26].

Using the COSMO-RS method, Xin et al. observed a significant increase in the solubility of isobutane in H_2SO_4 upon the addition of most ASO molecules [30]. Furthermore, the addition of hydrocarbons such as paraffins and cycloparaffins into H_2SO_4 can lead to improved dissolution of isobutane. Based on these findings, two cycloparaffins were selected and employed as cocatalysts (designated as cocatalysts A and B) to enhance the performance of the H_2SO_4 catalyst [30]. Alkylation experiments demonstrated that these cycloparaffins enhanced the quality of alkylates. Specifically, the addition of cocatalyst B results in a significant increase in the RON, the TMP/DMH ratio, and the contents of C_8 and TMP. Moreover, the consumption of the H_2SO_4 catalyst is reduced by 38% with the addition of cocatalyst B.

(1) Evaluation of alkylate quality for cocatalyst A

The performance of cocatalyst A was evaluated by examining the distribution, RON, and TMP/DMH ratio of the alkylate. These results are depicted in Fig. 6.30. The RON and TMP/DMH ratio are crucial metrics for assessing alkylate quality. The findings in Fig. 6.30 suggest that alkylate quality exhibits an initial increase followed by a decrease with rising quantities of cocatalyst A.

With the addition of 1.5 wt% cocatalyst A, the RON value and TMP/DMH ratio notably rise from 92.67 to 93.40 and 6.51 to 6.67, respectively. Additionally, the C_8 and TMP contents increase significantly from 68.25% and 59.10% to 72.30% and 62.85%, respectively. Optimal alkylate quality is attained under these experimental conditions. Moreover, the proportions of C_5, C_6, C_7, and C_9^+ components decrease from 3.92%, 5.24%, 6.11%, and 16.48% to 3.62%, 3.74%, 6.05%, and 14.28%, respectively. However, when the quantity of cocatalyst A in H_2SO_4 is increased to 3 wt%, the RON, TMP/DMH ratio, and the contents of C_8 and TMP decline to 91.98, 5.90, 67.33%, and 57.50%, respectively. These results indicate that the addition of an appropriate amount of cocatalyst A can significantly boost the desired C_8 content, TMP selectivity, and suppress olefin oligomerization reactions.

(2) Evaluation of alkylate quality for cocatalyst B

The findings obtained from using cocatalyst B are displayed in Fig. 6.31. It is evident that there is also a trend of increasing alkylate quality followed

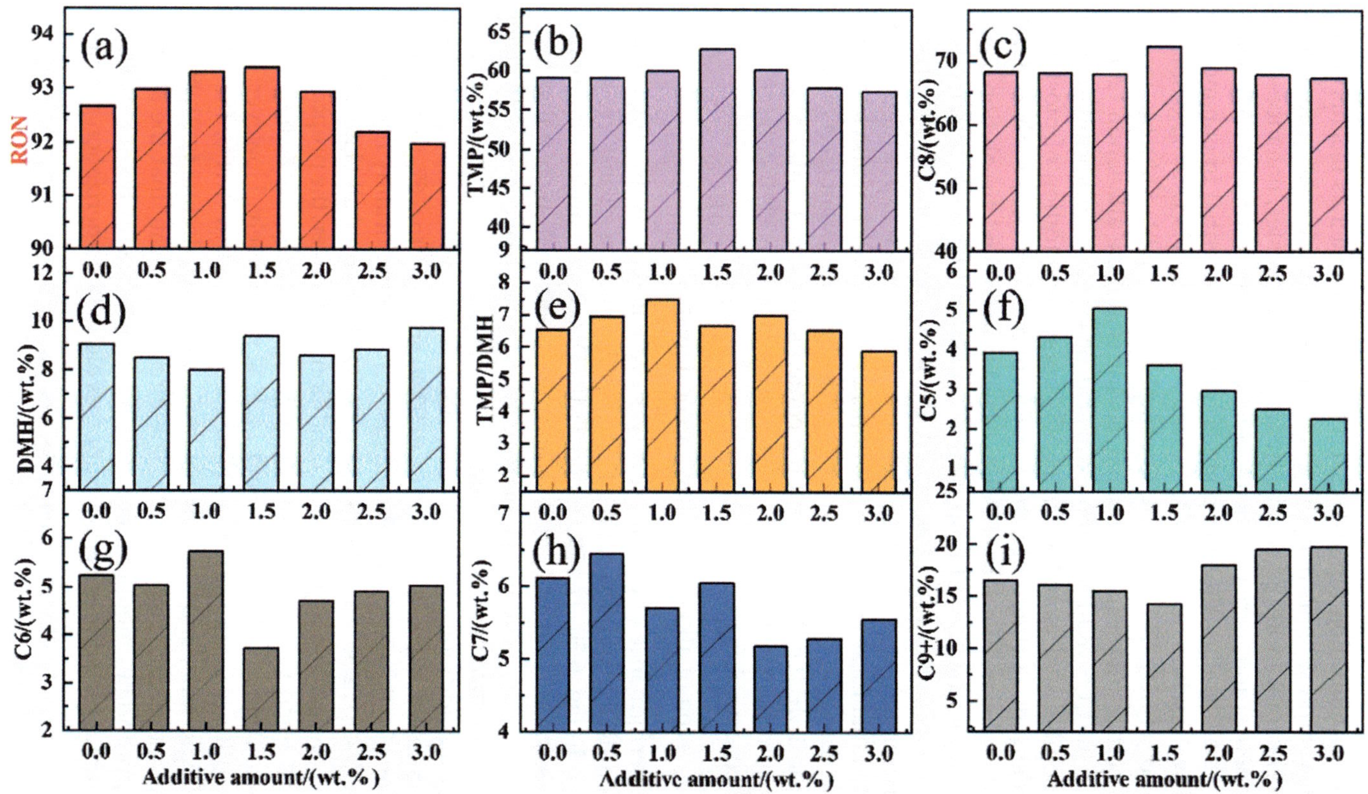

FIG. 6.30 Effects of cocatalyst A on the quality of alkylate. *From reference Y. Xin, Y. Hu, Y. Wang, S. Jiang, M. Li, K. Chi, S. Zhang, F. Gao, C. Ren, S. Zhang, Alkylation catalyzed by H_2SO_4: promoting effects of ASO on the solubility of isobutane in acid phase, Fuel 326 (2022) 125034.*

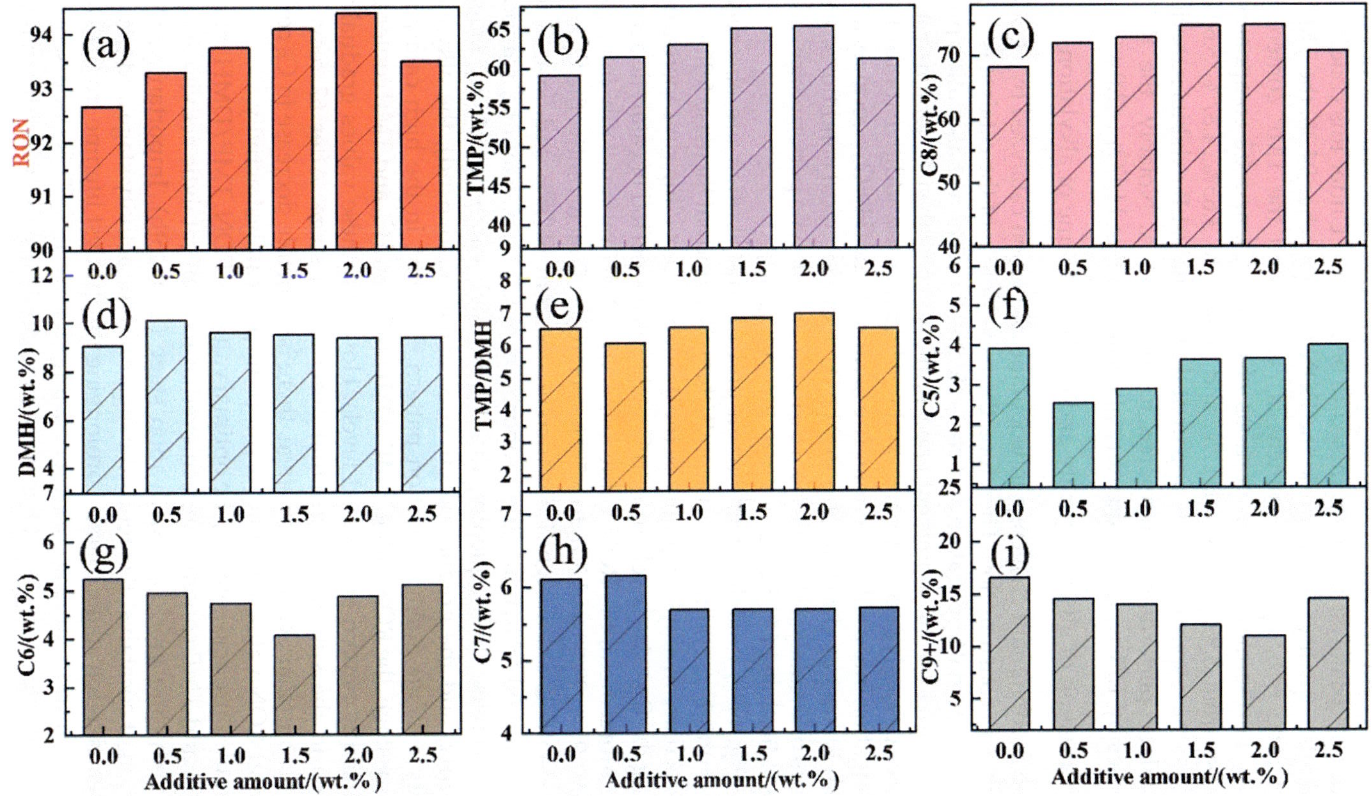

FIG. 6.31 Effects of the cocatalyst B on the quality of alkylates. *From reference Y. Xin, Y. Hu, Y. Wang, S. Jiang, M. Li, K. Chi, S. Zhang, F. Gao, C. Ren, S. Zhang, Alkylation catalyzed by H2SO4: promoting effects of ASO on the solubility of isobutane in acid phase, Fuel 326 (2022) 125034.*

by a subsequent decrease with rising quantities of cocatalyst B. The optimal amount of cocatalyst B is determined to be 2.0wt%. Under this condition, the alkylate quality is at its peak: the RON value reaches 94.38, the TMP/DMH ratio is 6.97, the C_8 content is 74.81%, and the TMP content is 65.34%. Furthermore, the proportions of C_5, C_6, C_7, and C_9^+ components decrease from 3.92%, 5.24%, 6.11%, and 16.48% to 3.65%, 4.87%, 5.69%, and 11.00%, respectively. It is apparent that cocatalyst B exhibits superior co-catalytic performance compared to cocatalyst A. Notably, the RON increases by 1.71 when the quantity of cocatalyst B in H_2SO_4 is 2.0wt%.

(3) Evaluation of alkylate quality and catalyst lifetime of alkylation

The predominant challenge in isobutane alkylation catalyzed by H_2SO_4 is high acid consumption. Therefore, the recyclability of the catalyst is crucial for industrial application and cost-effective operation of alkylation processes. Based on the aforementioned findings, cocatalyst B demonstrates superior co-catalytic performance and is selected as the additive for further investigation into catalyst lifetime. The lifetime of the (H_2SO_4+cocatalyst B) system is compared to that of pure H_2SO_4 under identical conditions. The impacts of cocatalyst B on alkylate distribution, catalytic performance, and catalyst lifetime of H_2SO_4 are illustrated in Fig. 6.32.

In Fig. 6.32, notable variations in product distribution are observed with multiple recycling runs. Overall, there is a gradual increase followed by a decrease in alkylate quality as the number of recycling runs increases. During the first six recycle runs, referred to as the period of catalytic activity reinforcement, the RON value and TMP/DMH ratio notably increase from 91.96 to 93.00 and 5.76 to 6.55, respectively. Concurrently, the C_8 and TMP contents show significant increments from 66.38% and 56.43% to 69.46% and 60.13%, while the proportions of C_5, C_6, C_7, and C_9^+ components decrease. Subsequently, during the high catalytic activity stage (7th–31st) and the gradual deactivation stage (32nd–62nd), the alkylate quality exhibits a slow variation pattern. Notably, in the high catalytic activity stage, the RON value, TMP/DMH ratio, and C_8 and TMP contents consistently remain at elevated levels. However, these values gradually decline in the subsequent gradual deactivation stage. After 62 recycle runs, in the final deactivation stage, there is a notable decrease in C_8 content and RON value, signaling a sharp drop in catalyst activity.

For the pure acid catalyst, the optimal values of RON, TMP/DMH ratio, and C_8 and TMP contents are achieved in its high catalytic activity stage (7th–20th) and gradual deactivation stage (21st–45th). Importantly, the addition of cocatalyst B extends the high catalytic activity stage and gradual deactivation stage, thereby prolonging the catalyst lifetime from 45 to 62 recycles and reducing catalyst consumption by 38%. This underscores the pivotal role of cocatalyst B in sustaining catalyst activity. Additionally, the inclusion of cocatalyst B in H_2SO_4 results in increased desired C_8 and

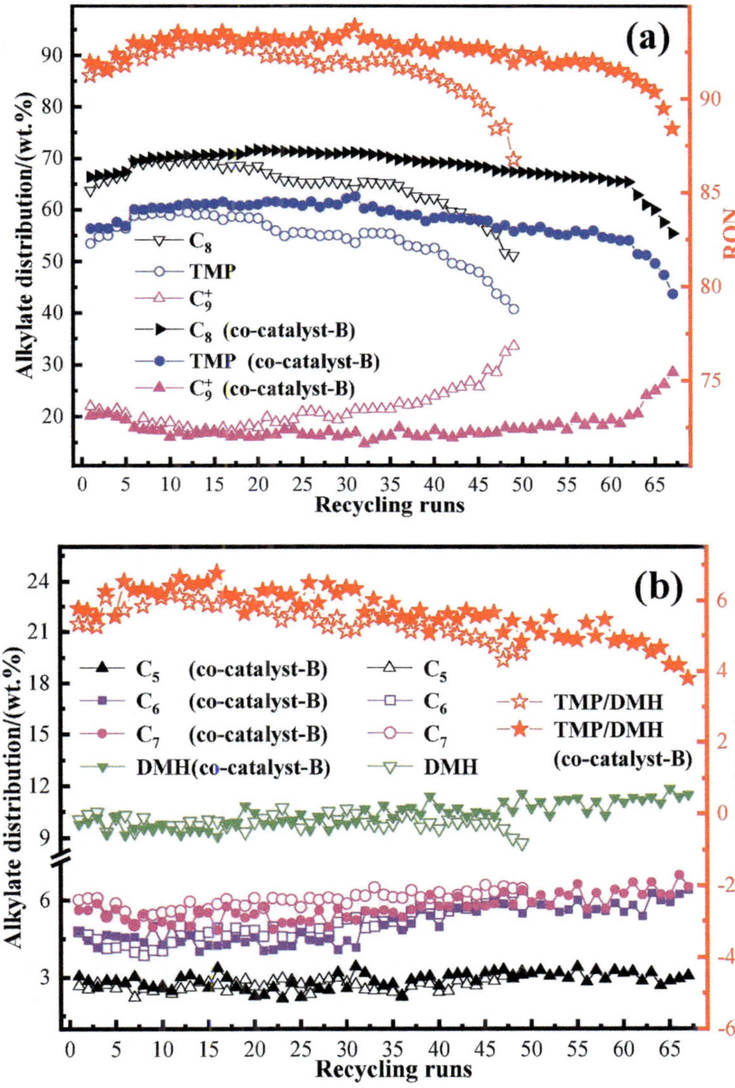

FIG. 6.32 Effects of the cocatalyst B on the alkylate distribution and the catalyst lifetime of H_2SO_4. *From reference Y. Xin, Y. Hu, Y. Wang, S. Jiang, M. Li, K. Chi, S. Zhang, F. Gao, C. Ren, S. Zhang, Alkylation catalyzed by H_2SO_4: promoting effects of ASO on the solubility of isobutane in acid phase, Fuel 326 (2022) 125034.*

TMP selectivity, enhanced RON value, and overall improvement in alkylate quality.

By investigating the effects of two cycloparaffins, cocatalysts A and B, on alkylate distribution, catalytic performance, and catalyst lifetime, it is discovered that the addition of 1.5 wt% cocatalyst A in H_2SO_4 improves the quality of alkylate. Furthermore, the inclusion of 2.0 wt% cocatalyst B in H_2SO_4 yields the best alkylate quality, prolongs the catalytic system lifetime, and contributes to the 38% reduction in catalyst consumption. Moreover, the addition of cocatalyst B in H_2SO_4 leads to the increased desired C_8 and TMP selectivity, along with an enhanced RON value, thus improving alkylate quality.

Machine learning (ML) is a powerful method for constructing quantitative structure-activity relationship (QSAR) models [31,32]. In fact, ML has demonstrated a clear benefit in creating QSAR models for a range of IL properties and gas solubility in ILs, including but not limited to melting points, viscosity, electrical conductivity, thermal decomposition temperature, critical properties, toxicity, H_2S solubility, CO_2 solubility, and water solubility [13,33–40].

Among these reports, the structure features of ILs are usually represented by group contribution, molecular fingerprint, and molecular descriptor [41]. In group contribution, ILs can be decomposed into several groups, which linearly combine with the specific property [42]. However, it is often necessary to combine molecular descriptors with the physicochemical properties of ILs in order to develop QSAR models. It is challenging to deal with novel ILs that lack available physicochemical properties. Molecular fingerprinting breaks down the structural characteristics of ILs into binary vectors (0 s and 1 s [43]), making it useful for virtual screening, similarity searching, and clustering. Physicochemical properties of ILs are calculated using semi-empirical DFTs to provide various quantum chemical and molecular orbital descriptors. These molecular descriptors, combined with ML, have been successfully used to develop QSAR models. The liquid/liquid interfacial properties of H_2SO_4 alkylation were investigated using ML in conjunction with MD simulations [44]. The intensified interfacial properties of H_2SO_4 alkylation with 11,000 ILs as additives are predicted and screened using MD simulations, semi-empirical DFTs, and ML.

Fig. 6.33A illustrates representatives of 11,000 ILs, which encompass 220 cations and 50 anions. The cations include imidazolium, pyrrolidinium, pyridinium, ammonium, phosphonium, sulfonium, guanidinium, and isoquinolinium, along with their alkyl derivatives. The anions consist of hexafluorophosphate, tetrafluoroborate, tetrachloroaluminate, heptachlorodialuminate, bistrifluoromethanesulfonylimide, alkylcarbonate, alkylsulfonate, alkylsulfate, and others. To screen ideal IL additives to intensify interfacial properties of H_2SO_4 alkylation on a large scale, MD simulations were initially conducted to obtain the interfacial thickness and tension of the H_2SO_4 alkylation containing 1496 ILs (comprising 44 cations and 34 anions). Semi-empirical DFTs were performed for the cations and anions of the total 11,000 ILs to extract various quantum chemical and molecular orbital-based descriptors, including HOMO/LUMO energies,

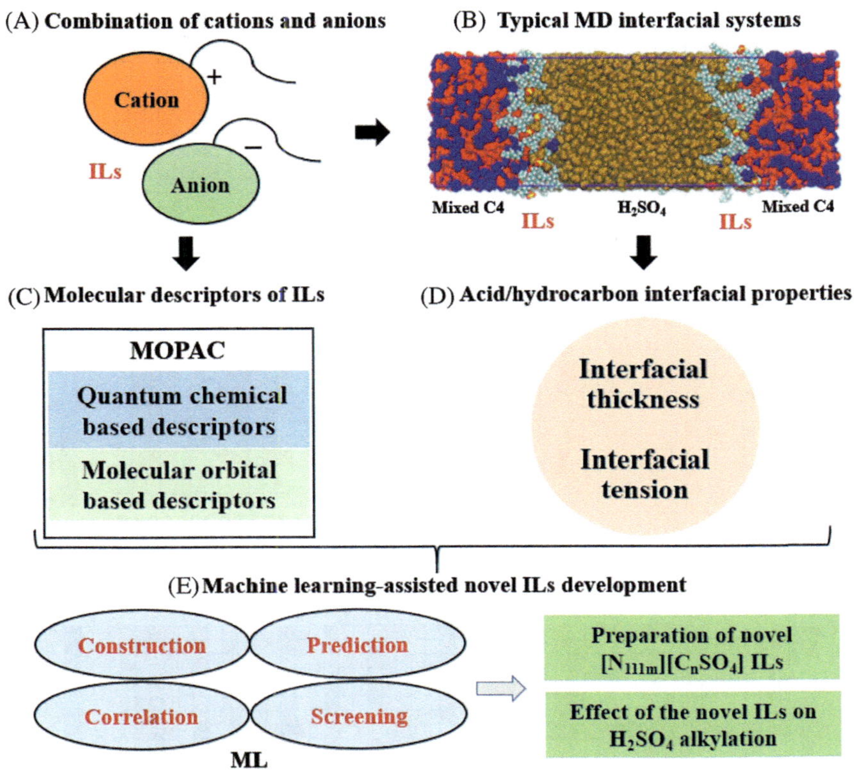

FIG. 6.33 Schematic representation of the general research strategy. *From reference W. Zheng, Z. Ma, W. Sun, L. Zhao, Target high-efficiency ionic liquids to promote H_2SO_4-catalyzed C4 alkylation by machine learning, AIChE J. 68 (2022) e17698.*

polarizabilities, charges, vibrational frequency-based eigenvalues, entropies, enthalpies, and other geometrical indices. Subsequently, ML methods such as support vector machine, random forest (RF), gradient boosting decision tree, and ANN were employed to correlate structure–function relationships with molecular descriptors of ILs as features and interfacial thickness/tension as labels to predict the intensified interfacial properties of H_2SO_4 alkylation with the total 11,000 ILs. Furthermore, the correlation between alkyl chain length of cations and anions and interfacial thickness/tension was analyzed separately.

To demonstrate the impact of ILs on the interfacial properties of H_2SO_4 alkylation, the interfacial properties of systems with ILs were normalized by dividing those in the pure H_2SO_4 system without ILs. The normalized interfacial thickness and tension of the systems with ILs are presented in Fig. 6.34. From Fig. 6.34, it is evident that the majority

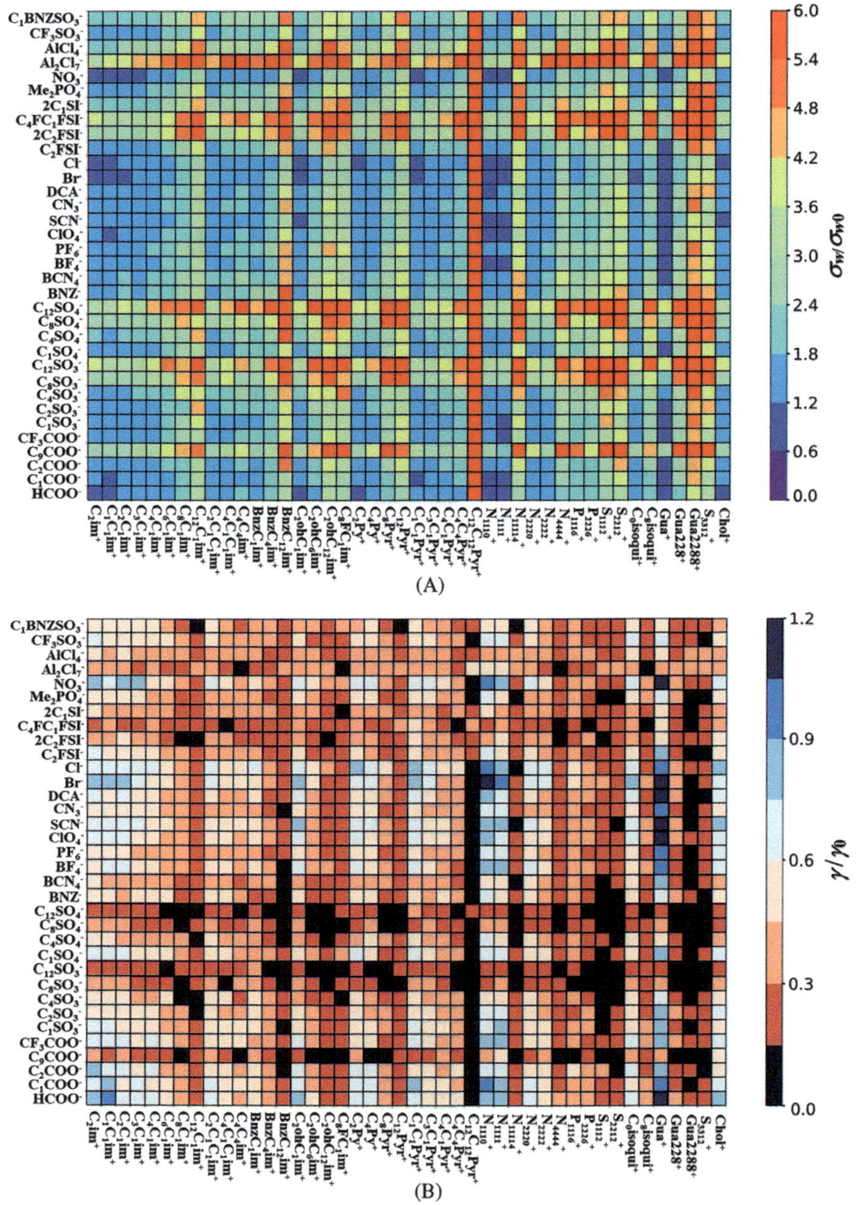

FIG. 6.34 Normalized interfacial thickness (A) and interfacial tension (B) of the systems with ionic liquids (ILs) in comparison with the pure H_2SO_4 system without IL. *From reference W. Zheng, Z. Ma, W. Sun, L. Zhao, Target high-efficiency ionic liquids to promote H_2SO_4-catalyzed C4 alkylation by machine learning, AIChE J. 68 (2022) e17698.*

of ILs can increase the interfacial thickness (σ_w) and decrease the interfacial tension (γ), indicating that IL additives can provide a larger reaction area and better biphasic dispersion for H_2SO_4 alkylation, which is beneficial for the quality of alkylate. However, the cations with shorter alkyl chains or smaller size counterparts, along with certain anions, lead to a reduction in σ_w and an increase in γ. Increasing the alkyl chain length of cations or anions contributes significantly to increasing σ_w and reducing γ. For instance, the σ_w/σ_{w0} value enlarges up to 4.0 when the carbon number of the alkyl chain increases from 1 to 12, and the γ/γ_0 value decreases to 0.3 times. $C_{12}C_{12}Pyr^+$ with two longer alkyl chains results in a significantly larger σ_w/σ_{w0} value and extremely smaller γ/γ_0 value, regardless of anion types. Certain anions, such as $Al_2Cl_7^-$, fluorinated $2C_2FSI^-$, and $C_4FC_1FSI^-$, play a crucial role in intensifying interfacial properties. The anions with longer alkyl chains also present a much larger σ_w/σ_{w0} value and smaller γ/γ_0 value. The synergistic intensification of both longer alkyl-chain cations and anions on the interfacial properties is evident, and the interfacial thickness and tension can be efficiently tuned by adjusting the alkyl chain length of either cations or anions.

Based on the MD simulation-calculated interfacial properties, IL screening was conducted using the criteria: $\sigma_w/\sigma_{w0}>6$ and $\gamma/\gamma_0<0.1$. These criteria ensure that the screened IL additives can significantly increase interfacial thickness and reduce interfacial tension.

ILs, as amphiphilic molecules, consist of polar and nonpolar regions, with the length of the alkyl chain in cations or anions exerting a decisive influence on the nonpolarity of the ILs. Fig. 6.34 illustrates the relationship between the alkyl chain length of both cations and anions and the interfacial thickness and tension based on the dataset predicted by the RF model. From Fig. 6.34A, it is evident that the introduction of ILs with shorter alkyl chain cations and anions does not significantly enhance interfacial thickness. However, increasing the alkyl chain length of either cations or anions leads to improved interfacial thickness, demonstrating a synergistic effect of both longer alkyl chain cations and anions on interfacial thickness. Similar trends are observed for interfacial tension in Fig. 6.34B. Thus, the correlation between the alkyl chain length of both cations and anions and interfacial properties provides valuable insight for designing novel ILs with different alkyl chain lengths for H_2SO_4 alkylation. Based on the screening criteria ($\sigma_w/\sigma_{w0}>6$ and $\gamma/\gamma_0<0.1$), 149 ILs were screened from the total 11,000 ILs. This large-scale screening can effectively guide the design of novel and efficient ILs to significantly improve the quality of alkylate for H_2SO_4 alkylation (Fig. 6.35).

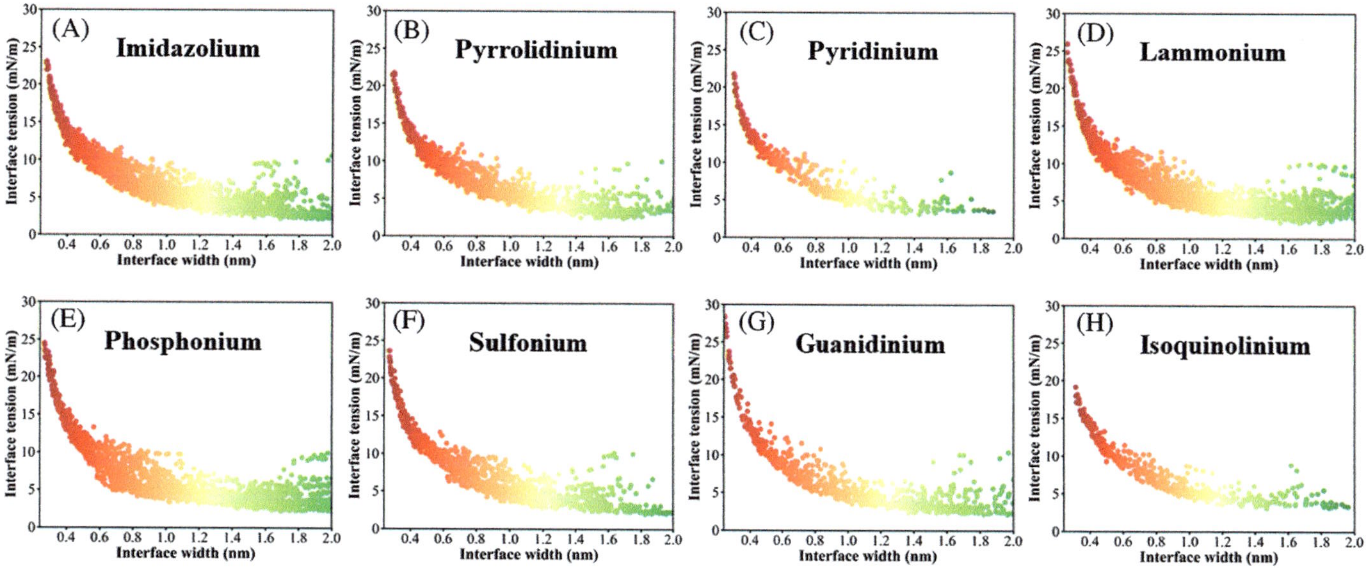

FIG. 6.35 Correlation plots for the interfacial thickness and tension based on the 11,000 datasets predicted by the random forest model. *From reference W. Zheng, Z. Ma, W. Sun, L. Zhao, Target high-efficiency ionic liquids to promote H_2SO_4-catalyzed C4 alkylation by machine learning, AIChE J. 68 (2022) e17698.*

References

[1] N. García, M.J. Fernández-Torres, J.A. Caballero, Simultaneous environmental and economic process synthesis of isobutane alkylation, J. Clean. Prod. 81 (2014) 270–280.
[2] A. Azapagic, Life cycle assessment and its application to process selection, design and optimisation, Chem. Eng J. 73 (1999) 1–21.
[3] P. Erol, J. Thöming, ECO-design of reuse and recycling networks by multi-objective optimization, J. Clean. Prod. 13 (2005) 1492–1503.
[4] N. García, J.A. Caballero, Economic and environmental assessment of alternatives to the extraction of acetic acid from water, Ind. Eng. Chem. Res. 50 (2011) 10717–10729.
[5] N. García, J.A. Caballero, How to implement environmental considerations in chemical process design: an approach to multiobjective optimization for undergraduate students, Educ. Chem. Eng. 7 (2012) e56–e67.
[6] R. Smith, Chemical Process: Design and Integration, Wiley, 2005.
[7] P. Dragnea, C.S. Bîldea, Process optimization of a butane-butene alkylation plant, UPB Sci. Bull. 76 (2014) 71–84.
[8] Bird, J., Seillier, D., Teders, M. Optimization of Refining Alkylation Process Unit Using Response Surface Methods.
[9] A. Nurmakanova, E.N. Ivashkina, E.D. Ivanchina, I. Dolganov, S. Boychenko, Predicting alkylate yield and its hydrocarbon composition for sulfuric acid catalyzed isobutane alkylation with olefins using the method of mathematical modeling, Procedia Chem. 15 (2015) 54–64.
[10] E. Ivashkina, E. Ivanchina, A. Nurmakanova, S. Boychenko, A. Ushakov, I. Dolganova, Mathematical modeling sulfuric acid catalyzed alkylation of isobutane with olefins, Procedia Eng. 152 (2016) 81–86.
[11] Z. Liu, X. Meng, R. Zhang, C. Xu, H. Dong, Y. Hu, Reaction performance of isobutane alkylation catalyzed by a composite ionic liquid at a short contact time, AIChE J 60 (2014) 2244–2253.
[12] Y. Tian, Y. Wan, L. Zhang, G. Chu, A.C. Fisher, H. Zou, Optimized artificial neural network for evaluation: C4 alkylation process catalyzed by concentrated sulfuric acid, ACS Omega 7 (2021) 372–380.
[13] W. Zheng, Z. Wang, W. Sun, L. Zhao, F. Qian, H2SO4-catalyzed isobutane alkylation under low temperatures promoted by long-alkyl-chain surfactant additives, AIChE J 67 (2021) e17349.
[14] W. Zheng, D. Li, W. Sun, L. Zhao, Multi-scale modeling of isobutane alkylation with 2-butene using composite ionic liquids as catalyst, Chem. Eng. Sci. 186 (2018) 209–218.
[15] J. Mosby, L. Albright, Alkylation of isobutane with 1-butene using sulfuric acid as catalyst at high rates of agitation, Ind. Eng. Chem. Res. 5 (1966) 183–190.
[16] D.J. Ende, R.E. Eckert, L.F. Albright, Interfacial area of dispersions of sulfuric acid and hydrocarbons, Ind. Eng. Chem. Res. 34 (1995) 4343–4350.
[17] W.-S. Chen, Solubility measurements of isobutane/alkenes in sulfuric acid: applications to alkylation, Appl. Catal. A – Gen. 255 (2003) 231–237.
[18] K. Li, R.E. Eckert, L.F. Albright, Alkylation of isobutane with light olefins using sulfuric acid. Operating variables affecting both chemical and physical phenomena, Ind. Eng. Chem. Process 9 (1970) 441–446.
[19] D. Camper, C. Becker, C. Koval, R. Noble, Low pressure hydrocarbon solubility in room temperature ionic liquids containing imidazolium rings interpreted using regular solution theory, Ind. Eng. Chem. Res. 44 (2005) 1928–1933.
[20] S.J. Aschauer, A. Jess, Effective and intrinsic kinetics of the two-phase alkylation of i-paraffins with olefins using chloroaluminate ionic liquids as catalyst, Ind. Eng. Chem. Res. 51 (2012) 16288–16298.

[21] P. Hu, Z. Wu, J. Wang, Y. Huang, Y. Deng, S. Zhou, Analysis of long term catalytic performance for isobutane alkylation catalyzed by NMA–$AlCl_3$ based ionic liquid analog, Chin. J. Chem. Eng. 27 (2019) 1857–1862.
[22] Y. Liu, G. Wu, X. Pang, R. Hu, Kinetics study on alkylation of isobutane with deuterated 2-butene in composite ionic liquids, Chem. Eng. J. 387 (2020) 123407.
[23] W. Zheng, L. Zheng, W. Sun, L. Zhao, Screening of imidazolium ionic liquids for the isobutane alkylation based on molecular dynamic simulation, Chem. Eng. Sci. 183 (2018) 115–122.
[24] W. Zheng, H. Wang, W. Xie, L. Zhao, W. Sun, Understanding interfacial behaviors of isobutane alkylation with C4 olefin catalyzed by sulfuric acid or ionic liquids, AIChE J 64 (2018) 950–960.
[25] Z. Ma, W. Zheng, W. Sun, L. Zhao, Enhanced catalytic performance of H_2SO_4-catalyzed C4 alkylation by formyl functional $[N_{1,1,1,1}][C_{10}SO_4]$ additive, AIChE J 69 (2023) e18179.
[26] L. Wang, G. Zhao, X. Yao, B. Ren, S. Zhang, Adamantane-based cation and $[MF_n]^-$ anion synergistically enhanced catalytic performance of sulfuric acid for isobutane alkylation, Ind. Eng. Chem. Res. 56 (2017) 7920–7929.
[27] Y. Xin, Y. Hu, M. Li, K. Chi, S. Zhang, F. Gao, S. Jiang, Y. Wang, C. Ren, G. Li, Isobutane alkylation catalyzed by H_2SO_4: effect of H_2SO_4 acid impurities on alkylate distribution, Energy Fuel 35 (2020) 1664–1676.
[28] X. Zhang, R. Zhang, X. Meng, H. Liu, Z. Liu, H. Ma, C. Xu, P.A. Klusener, Deactivation mechanism and activity-recovery approach of composite ionic liquids for isobutane alkylation, Appl. Catal. A – Gen. 557 (2018) 64–71.
[29] X. Meng, P. Prawang, Z. Li, H. Wang, S. Zhang, Carbon-based materials enhanced emulsification to improve product distribution in isobutane/butene alkylation catalyzed by sulfuric acid, Ind. Eng. Chem. Res. 56 (2017) 7700–7707.
[30] Y. Xin, Y. Hu, Y. Wang, S. Jiang, M. Li, K. Chi, S. Zhang, F. Gao, C. Ren, S. Zhang, Alkylation catalyzed by H2SO4: promoting effects of ASO on the solubility of isobutane in acid phase, Fuel 326 (2022) 125034.
[31] W. Beker, E.P. Gajewska, T. Badowski, B.A. Grzybowski, Prediction of major regio-, site-, and diastereoisomers in Diels–Alder reactions by using machine-learning: the importance of physically meaningful descriptors, Angew. Chem. Int. Ed. 58 (2019) 4515–4519.
[32] Q. Yang, Y. Li, J.D. Yang, Y. Liu, L. Zhang, S. Luo, J.P. Cheng, Holistic prediction of the pKa in diverse solvents based on a machine-learning approach, Angew. Chem. Int. Ed. 132 (2020) 19444–19453.
[33] V. Venkatraman, S. Evjen, H.K. Knuutila, A. Fiksdahl, B.K. Alsberg, Predicting ionic liquid melting points using machine learning, J. Mol. Liq. 264 (2018) 318–326.
[34] K. Low, R. Kobayashi, E.I. Izgorodina, The effect of descriptor choice in machine learning models for ionic liquid melting point prediction, J. Chem. Phys. 153 (2020).
[35] Y. Zhao, X. Zhang, L. Deng, S. Zhang, Prediction of viscosity of imidazolium-based ionic liquids using MLR and SVM algorithms, Comput. Chem. Eng. 92 (2016) 37–42.
[36] G.V. Carrera, M. Nunes da Ponte, Machine-learning approaches to tune descriptors and predict the viscosities of ionic liquids and their mixtures, Chem.-Meth. 1 (2021) 214–223.
[37] F. Gharagheizi, M. Sattari, P. Ilani-Kashkouli, A.H. Mohammadi, D. Ramjugernath, D. Richon, A "non-linear" quantitative structure–property relationship for the prediction of electrical conductivity of ionic liquids, Chem. Eng. Sci. 101 (2013) 478–485.
[38] F. Gharagheizi, M. Sattari, P. Ilani-Kashkouli, A.H. Mohammadi, D. Ramjugernath, D. Richon, Quantitative structure—property relationship for thermal decomposition temperature of ionic liquids, Chem. Eng. Sci. 84 (2012) 557–563.
[39] Z. Wang, Z. Song, T. Zhou, Machine learning for ionic liquid toxicity prediction, PRO 9 (2020) 65.
[40] L. Cao, P. Zhu, Y. Zhao, J. Zhao, Using machine learning and quantum chemistry descriptors to predict the toxicity of ionic liquids, J. Hazard. Mater. 352 (2018) 17–26.

[41] Y. Ding, M. Chen, C. Guo, P. Zhang, J. Wang, Molecular fingerprint-based machine learning assisted QSAR model development for prediction of ionic liquid properties, J. Mol. Liq. 326 (2021) 115212.
[42] M. Sattari, A. Kamari, H. Hashemi, A.H. Mohammadi, D. Ramjugernath, A group contribution model for prediction of the viscosity with temperature dependency for fluorine-containing ionic liquids, J. Fluor. Chem. 186 (2016) 19–27.
[43] D. Rogers, M. Hahn, Extended-connectivity fingerprints, J. Chem. Inf. Model. 50 (2010) 742–754.
[44] W. Zheng, Z. Ma, W. Sun, L. Zhao, Target high-efficiency ionic liquids to promote H_2SO_4-catalyzed C4 alkylation by machine learning, AIChE J 68 (2022) e17698.

CHAPTER

7

Alkylation process safety

For the industrial C4 alkylation process, the main processes include H_2SO_4-catalyzed and HF-catalyzed technologies. IL-catalyzed alkylation has emerged as a new technology with significant developments in the past ten years. Solid acid alkylation has been commercialized with only one industrial unit. For different catalysts, the C4 alkylation process requires significantly different conditions such as temperature, pressure, time, I/O ratio, etc. Various industrial processes are adjusted in terms of process intensification, with the emphasis on reducing catalyst consumption, minimizing unit energy usage, and ensuring safe and environmentally friendly operation of the facilities. These adjustments aim to enhance the economic efficiency, reliability, and safety of unit operations.

7.1 H_2SO_4-catalyzed process

The H_2SO_4-catalyzed alkylation process usually includes an H_2SO_4 regeneration unit. Theoretically, when the concentration of H_2SO_4 reaches 90 wt%, the corrosion of steel by either the H_2SO_4 feedstock or the reaction effluent is limited. However, in practical production, the concentrated H_2SO_4 is still the primary reason for safety concerns.

7.1.1 Low construction quality

Substandard welding techniques lead to irregularities inside the welds. Concentrated H_2SO_4 typically causes minimal pipeline corrosion because it forms a passivation film on the metal surface. However, the film can be easily destroyed by variations in temperature fluctuations, flow velocity, and concentration. If the internal surface of the pipeline is not smooth, the rate of corrosion can significantly increase. Most incidents of acid media pipeline corrosion and leakage occur near welded sections.

Aliphatic Alkylation
https://doi.org/10.1016/B978-0-443-21984-9.00004-0

7.1.2 Equipment aging

With the prolongation of equipment operating time, wear and tear are intensified by factors such as prolonged pipeline vibrations. In static equipment, concentrated H_2SO_4 accumulates in the internal packings of valves. With water absorption over prolonged operation, the concentration of H_2SO_4 decreases and results in corrosion.

In addition, seal failure may occur and pose a safety risk during valve operation. Therefore, before operating valves, it is imperative to ensure the proper use of personal protective equipment. It is advisable to augment splash guards on the sealing surfaces. For critical equipment such as hydraulic heads, seal life is affected by factors including installation quality and operational environment. Any leakage can inevitably require handling procedures for acid-containing equipment, introducing a series of risks.

7.1.3 Unsafe daily operations

Changes in feedstock composition, occurrences of accidents in upstream devices, and failures in internal dehydration units and hydrogenation units may directly or indirectly lead to changes in acid catalysts. Without adequate communication, operators could not identify risks promptly and take effective measures. During daily operations, safety precautions that seem uncritical are based on thorough summaries and must be strictly adhered to, with daily inspections and supervision strengthened. For example, improper acid sampling inspections may result in acid overflowing with foam into sampling bottles due to sudden pressure drops, posing dangers.

7.1.4 Inadequate monitoring

Equipment wear and acid accumulation are inevitable during processes such as material flow, reaction, separation, neutralization, etc. These problems, from occurrence to resulting abnormal operation, may manifest as changes in various measurable parameters. Unfortunately, these changes are not always detected immediately. Therefore, technical staff must regularly conduct data comparison and analysis to detect problems as early as possible, assess risks, devise countermeasures, and minimize the harm of emergencies.

H_2SO_4-catalyzed alkylation technology has been industrialized for a long time, accumulating rich data on susceptible areas prone to issues, and providing a foundation for safe management of equipment operation. However, it is essential for technical management personnel across different devices to maintain good communication and share information to ensure that these operational experiences are fully utilized.

7.2 HF-catalyzed process

HF-catalyzed alkylation process uses HF as the catalyst. To minimize the hazards of HF, novel processes add HF regeneration units. Due to its low boiling point, strong acidity, and the particular effect of fluoride ions on human health, there is a push for operational health protection and environmental protection, which decreases the number of HF alkylation units. The prevention of HF release is a priority, and managing the disposal of fluorinated waste generated from neutralizing small amounts of HF is also a concern. Key safety features of the HF alkylation process include:

7.2.1 Utility system protection

The HF alkylation operates at a relatively high pressure and reactor temperature is controlled by circulating cooling water. A lower shell-side circulating water pressure and reactor tube bundle leaks will lead to HF and hydrocarbons entering the circulating water system, causing corrosion and additional safety risks from explosive gases. Typically, HF alkylation units have additional circulating water systems to avoid impacts on refinery systems, and internal corrosion and safety risks by leaks are still unavoidable. Current online pH monitoring of circulating water systems can detect leaks timely, reducing safety risks, but the issues caused by leaks in circulating water systems cannot be eliminated, necessitating enhanced corrosion monitoring and maintenance of utility systems.

7.2.2 Decreased equipment reliability

Over time, containers and other equipment in HF alkylation units accumulate significant rust. Commonly, the strong descaling ability of HF is expected to clean internal rust and scale from equipment and pipes, which then gets entrained with the material and deposits in areas of low flow velocity, leading to measurement failures and difficulties in valve operation. Besides, a unique aspect of HF alkylation units is the reduction in strength of the equipment and pipe materials, leading to brittleness and frequent fractures of small-diameter pipes (DN20 and below) under external forces.

7.2.3 Risks from high-pressure acidic media

The low boiling point of HF means the equipment and pipes will carry acidic media operating under high pressure. Any issues with pipes or mechanical seal failures can easily lead to leaks of highly corrosive, flammable, and explosive media. Additionally, the probability of leaks at static seals in equipment, pipes, and valves is higher due to operating pressure and corrosion from HF. Although color-changing indicators are set up on-site, even minor leaks of media, especially in poor weather conditions like rain, will cause low-concentration hydrofluoric acid injuries.

7.2.4 Environmental risks from waste acid disposal

Trace amounts of HF carried with downstream materials eventually enter the wastewater system, where fluoride ions must be precipitated into solids before the wastewater is discharged. In case of abnormal operations, the amount of acid requiring neutralization and precipitation can spike suddenly, easily leading to secondary problems and issues with the wastewater treatment system.

7.2.5 Personal safety risks from nonstandard operations

HF alkylation units have comprehensive safety regulations, from graded protective clothing to tool use: for instance, the positioning of acid-resistant clothing and adjustments to gloves and cuffs depending on the relative position of the hands. Neglecting these requirements will easily result in injuries. Equipment maintenance is a heavy-disaster area for issues; all tools used with acid media must be neutralized before reuse to ensure that the acid effects are confined within the HF alkylation unit area.

7.3 Ionic liquids-catalyzed process

The main active component of the ILs applied in the alkylation process is aluminum chloride, with the anionic part containing metal compounds. Compared to H_2SO_4, ILs have a lower density, higher viscosity, greater solubility with hydrocarbons, and high separation difficulty in operation. The intermediate active component added during the process is tert-butyl chloride, necessitating the direct addition of tert-butyl chloride or indirect addition via gas supplementation. Key safety features of the IL-catalyzed alkylation facilities include:

7.3.1 ILs corrosion

While ILs have limited corrosiveness, hydrolysis can produce hydrochloric acid. The corrosion of equipment and pipes is significantly related to the water content, especially in alkali-washing processes. The presence of tert-butyl chloride in alkylate and the hydro-dehalogenation refining process for alkylate will cause hydrochloric acid corrosion in the upper parts of the column and associated pipes, requiring monitoring to identify potential issues early and mitigate safety risks.

7.3.2 The operation of chloride ion supplement

Alkali washing and water washing result in the loss of ILs and timely replenishment is needed to ensure catalytic activity. The replenishment of chloride ions is usually done by adding hydrogen chloride through a gas cylinder under reduced pressure, and it is necessary to strictly follow operational standards to control the risks involved.

7.3.3 Handling of alkali waste

In the IL-catalyzed alkylation process, the amount of alkali used is relatively larger compared to other alkylation processes, necessitating the drying transfer of the generated alkali residue waste, which involves mechanical operations such as air drying and bagging. This increases the risk of mechanical injury to workers and is accompanied by an adverse operating environment with unpleasant odors, requiring more refined designs to overcome these challenges.

7.4 Challenges in alkylation

7.4.1 H_2SO_4-catalyzed process

In the H_2SO_4-catalyzed alkylation, the main challenges include:

7.4.1.1 Control of feed I/O ratio

The composition of raw materials varies with changes in upstream unit operating conditions, and it cannot be guaranteed that the feed I/O ratio always be higher than 1:1. Operators need to pay attention to the trend of isobutane tank liquid level in the fractionation system and intervene promptly. In the design and construction of the unit, a separate large-capacity isobutylene tank is also necessary. The capacity of the tank should be determined by comprehensively evaluating the time required for feedstock changes and upstream unit adjustments to ensure stable production of the unit during periods of insufficient isobutane component in the raw material composition. If emergencies occur, the isobutane in the isobutane tank also needs to ensure sufficient replacement hydrocarbons. In some special cases, even the supplementation of isobutane cannot maintain the balance of the unit, and it is necessary to reduce the operating load to ensure safety.

7.4.1.2 *Control of acid/hydrocarbon ratio*

Although different H_2SO_4-catalyzed alkylation technologies have different schemes for controlling the acid/hydrocarbon ratio, the core objective is to ensure sufficient contact between olefins and isobutane in the acid catalyst. The appropriate acid/hydrocarbon ratio is the basis for good mass transfer in the reaction system. In the actual production process, without timely detection of various influencing factors and effective countermeasures, a decrease in the acid/hydrocarbon ratio causes serious consequences of acid runaway. Moreover, the acid-holding capacity of the reaction system is usually large. It often takes a long time to respond to the error after the consequences of insufficient reaction of olefins, during which the olefins feed to the reaction system must be cut out, and the unit cannot operate normally. Therefore, for the control of the acid/hydrocarbon ratio in the reaction system, it is necessary to maintain a margin, monitor relevant parameters during unit operation, intervene early when the decreasing trend is observed, and avoid serious problems.

7.4.1.3 *Stable operation of neutralization and discharge system*

The neutralization system in the unit neutralizes the waste alkali solution to the pH of about 12 with high-concentration sulfuric acid. The neutralization system also has a large capacity, and it takes a long time for uniform mixing to achieve a stable pH value. The pH value of the sewage discharge fluctuates normally between 6 and 9. In the event of a leak, if it occurs at a critical location, the unit usually needs to be shut down to vent the leak to ensure controllable safety risks. When the leaking medium contains a large amount of concentrated acid, it is difficult to treat. There are two ways for the leaked medium: one is the collection tank, where after evaporating volatile hydrocarbons, the acid is sent to the waste acid system for recovery; the other outlet is the sewage discharge pool, where after neutralization, it is discharged from the unit. The remaining small amount of acid-containing medium is directly neutralized and processed. However, even a trace of remaining acid-containing medium will have a significant impact on the neutralization system due to its large amount of high-concentration acid. In such cases, relying on the normal waste alkali of the unit is far from meeting the neutralization requirements. High-concentration alkali needs to be directly introduced into the neutralization system, but achieving uniform mixing is difficult and time-consuming. It may even force the interruption of production due to insufficient capacity of the neutralization system. During normal operation of the unit, the neutralization system should be operated with a low liquid level as much as possible. In addition, solid alkali storage is necessary for effective response to various emergencies to prevent minor problems from escalating.

7.4.1.4 Corrosion prevention in process operation

Alkylation units have good corrosion control strategies in design, but in actual operation, corresponding operating parameters must be strictly controlled to ensure that corrosion is within acceptable limits. Corrosion before the neutralization unit is controlled by maintaining an acid concentration of over 90%. The corrosion rate during production is controllable. However, before the unit is shut down for maintenance, the feed must be cut out and processed to ensure safety during maintenance, which involves the normal operation of acid-containing equipment and pipelines being neutralized with alkali solution, and the corrosion of equipment and pipelines caused by residual acid during neutralization is severe. Therefore, a rigorous and comprehensive feed return-back solution is needed to minimize the sulfuric acid remaining in the pipelines and equipment.

The focus of the return-back solutions is to ensure high flow rates during displacement, allowing most of the sulfuric acid to be carried away by the flow, and intermittent operations for the acid in the dead zones to settle before being carried downstream for collection and treatment.

Before alkali neutralization, low-flow areas in the process should be checked, and detailed confirmation and treatment plans should be refined for relevant areas to minimize the residual acid in the system and reduce the difficulty of neutralization.

The water separated from the top reflux tank of the de-isobutane tower needs to undergo iron ion content testing and analysis. This is an important indicator of the severity of corrosion. When the analysis results increase, attention should be paid and effective measures should be taken to reduce the iron ion content. If the iron ion content is found to be high, the acid concentration in the reaction part is normal, and the reaction products after alkali washing are also normal, it can be determined that the corrosion is caused by the fractionation part. In units using alkali washing and water washing processes, this part of the corrosion cannot be avoided. On the one hand, corrosion in the fractionation section can be reduced by adding corrosion inhibitors, but the pH value should be monitored.

Efforts are made to maintain the pH value of the effluent water below 7 to ensure that the alkaline corrosion inhibitor injected will not be carried from the circulating isobutane into the reaction system. On the other hand, adsorption is used to treat reaction products, controlling the water content in the reactants to prevent dew point corrosion in the fractionation section due to temperature drop.

7.4.1.5 Management of stable unit operation

The medium required for alkylation unit production is mainly C4 and concentrated H_2SO_4. From operational experience, a stable unit operation is a crucial condition for reducing various risks. It is essential to minimize significant adjustments to the unit and avoid frequent start-stop cycles.

During normal operation, the primary focus is on the C4 feedstock. Mixed C4 primarily comes from catalytic cracking units and usually requires treatment such as alkali washing and water washing to remove sulfides, thus potentially carrying water or even amine alkaline substances. If not detected in time, impurities in the feedstock can cause a sharp decrease in sulfuric acid concentration, leading to a series of side reactions.

Some units are set to optimize reaction feed composition. During abnormal conditions in the feedstock, the catalyst in the addition unit is the first to be affected, indicating a warning signal of changes in the feedstock composition. Controlling the I/O ratio in alkylation reactions is also crucial. It is necessary to maintain both the reaction conversion rate and the economic operation of the unit. Efforts should be made to ensure that olefins react completely in the reactor while enhancing the research octane number (RON) of the alkylate. This can be achieved by optimizing control through reducing isobutane circulation to avoid excessive I/O ratio and energy wastage. When there is a significant difference between the RON of the alkylate and the expected value, the influence of the reaction I/O ratio should be promptly eliminated by increasing isobutane circulation.

7.4.2 HF-catalyzed process

The reaction temperature of the HF-catalyzed alkylation is close to room temperature, but the pressure is much higher than for other alkylation processes. The catalyst regeneration process is simple, but the wastewater discharge treatment process is more complex.

7.4.2.1 Strict attention to raw material changes

Impurities in the raw materials directly affect the consumption of HF catalysts. An increase in HF consumption significantly increases the operational risks of the unit. Decreasing acid concentration may intensify the equipment corrosion, the operation of the acid regeneration unit will increase stage-specific risks, while the neutralization and sedimentation processes will escalate the workload. Therefore, technical personnel in HF alkylation units focus on the stability of upstream unit operations to ensure the safety and stability of the alkylation unit.

7.4.2.2 Control of the I/O ratio

Control of the I/O ratio is crucial for an alkylation reaction. It is essential to ensure that the unit has sufficient isobutane resources available at all times to ensure both complete reaction and safe production of the unit.

7.4.2.3 Control of the acid/hydrocarbon ratio

The acid-to-hydrocarbon ratio in HF-catalyzed alkylation is mainly controlled by gravity difference, although some units use forced circulation by acid pumps. However, due to the relatively high failure rate of acid pumps, controlling the acid/hydrocarbon ratio by gravity difference is more common. It is necessary to pay attention to the interface level of the acid-hydrocarbon separation tank, conduct regular manual confirmations, and not rely solely on instrument measurements to ensure the detection of problems.

7.4.2.4 Control of reaction temperature

The reaction temperature is controlled below 40°C and typically does not require special refrigeration equipment. However, since alkylation reactions are exothermic, it is necessary to strictly control the temperature to reduce undesired side reactions and corrosion. Therefore, even in summer, the temperature of circulating cooling water must exceed 30°C, posing a challenge to reactor temperature control. It is necessary to strictly control the flow rate of reaction materials and even appropriately reduce the unit load to reduce heat release from the reaction, thereby maintaining the reaction temperature within the appropriate range and ensuring stable unit operation. Additionally, the reactor should be operated at the lower limit of the reaction temperature when the load of unit is relatively low. Although this may increase energy consumption, it is beneficial for long-term operation and can make better overall benefits.

7.4.2.5 Stable operation of neutralization and precipitation units

The neutralization and precipitation units in HF alkylation are complex. After a solid alkali is transformed into a liquid, changes in environmental temperature or flow rate can easily cause drain blockages, making unclogging a daily task for operators. This comes with significant safety risks. Routine inspection of the unit and regular unclogging work are crucial for protecting and restoring accessories such as pipeline insulation and heat tracing, despite the increased cost of inspection and maintenance. This can reduce the occurrence of blockages, ultimately lowering the safety risks of unit operation.

7.4.3 IL-catalyzed process

IL alkylation is currently implemented only in China, with few units deployed for commercial operation. The unit consumption of the IL catalyst is higher than that of HF but much lower than that of H_2SO_4. However, the absolute loss of IL catalyst is the highest. Therefore, reducing the loss of IL catalysts is an important aspect of unit operation.

7.4.3.1 Close attention to raw material changes

Impurities such as water, alcohols, ethers, etc., in the raw materials will increase the solubility of the ILs in the hydrocarbon phase, leading to increased losses of catalyst and a decrease of RON. This can result in a series of issues such as activation, neutralization, alkali residue treatment, etc. Therefore, during unit operation, attention must be paid to the operating conditions of the raw material pretreatment section to ensure that impurities are effectively controlled through prehydrogenation treatment, and to promptly control the changes in the dehydration of reaction feed. Comprehensive judgment of raw material changes is necessary to maintain stable unit operation.

7.4.3.2 Evaluation of ionic liquid activity

Since a small number of ILs will be carried out in the effluent, especially when the activity of the ionic liquid decreases and reactions are incomplete, the amount of ILs carried out in the effluent will increase significantly.

Index

Note: Page numbers followed by *f* indicate figures, *t* indicate tables, and *s* indicate schemes.

T

U

9780443219849